Volume 5

GENETIC RESOURCES, CHROMOSOME ENGINEERING, AND CROP IMPROVEMENT

Forage Crops

Genetic Resources, Chromosome Engineering, and Crop Improvement
Volume 1: Grain Legumes
edited by Ram J. Singh and Prem P. Jauhar

Genetic Resources, Chromosome Engineering, and Crop Improvement
Volume 2: Cereals
edited by Ram J. Singh and Prem P. Jauhar

Genetic Resources, Chromosome Engineering, and Crop Improvement
Volume 3: Vegetable Crops
edited by Ram J. Singh

Genetic Resources, Chromosome Engineering, and Crop Improvement
Volume 4: Oilseed Crops
edited by Ram J. Singh

Genetic Resources, Chromosome Engineering, and Crop Improvement
Volume 5: Forage Crops
edited by Ram J. Singh

GENETIC RESOURCES, CHROMOSOME ENGINEERING, AND CROP IMPROVEMENT SERIES

Series Editor, Ram J. Singh

Volume 5

GENETIC RESOURCES, CHROMOSOME ENGINEERING, AND CROP IMPROVEMENT

Forage Crops

EDITED BY

RAM J. SINGH

CRC Press
Taylor & Francis Group
Boca Raton London New York

CRC Press is an imprint of the
Taylor & Francis Group, an **informa** business

CRC Press
Taylor & Francis Group
6000 Broken Sound Parkway NW, Suite 300
Boca Raton, FL 33487-2742

First issued in paperback 2019

© 2009 by Taylor & Francis Group, LLC
CRC Press is an imprint of Taylor & Francis Group, an Informa business

No claim to original U.S. Government works

ISBN-13: 978-1-4200-4739-4 (hbk)
ISBN-13: 978-0-367-38602-3 (pbk)

Library of Congress Cataloging-in-Publication Data

Forage crops / editor, Ram J. Singh.
    p. cm. -- (Genetic resources, chromosome engineering, and crop improvement series ; v. 5)
    Includes bibliographical references and index.
    ISBN 978-1-4200-4739-4 (alk. paper)
    1. Forage plants. 2. Field crops. I. Singh, Ram J. II. Title. III. Series.

SB193.5.F67 2009
633.2--dc22
                                      2008043506

Visit the Taylor & Francis Web site at
http://www.taylorandfrancis.com

and the CRC Press Web site at
http://www.crcpress.com

Dedication

This book is dedicated to Emeritus Professor William F. Grant for his research in cytogenetics and mutagenesis at the international level. His major research in birdsfoot trefoil (*Lotus corniculatus*) utilizing interspecific hybridization with wild diploid species has shown a way to transfer genes from wild species via diploid hybridization and amphidiploidy to produce an indehiscent cultivar. He has carried out studies in *Betula, Lotus,* and *Manihot* using cytological, cytophotometric, chromatographic, and molecular techniques. He showed that mosaic virus-resistant plants of *Manihot esculentum* were distinguishable from susceptible plants by chromatographic techniques. He pioneered the use of higher plant species (*Allium cepa, Hordeum vulgare, Tradescantia, Vicia faba*) for monitoring and testing for mutagenic effects of environmental pollutants. He was elected a fellow of the AAAS and the Royal Society of Canada and has received an honorary DSc degree from McMaster University. He was presented with the first lifetime achievement award for his scientific contributions and service to the Genetics Society of Canada. He is a past editor of the *Canadian Journal of Genetics and Cytology* (now *Genome*) and the *Lotus Newsletter.*

Preface

What is a weed? A plant whose virtues have not been discovered.

Ralph Waldo Emerson (1803–1882)

Domestication of crops from weeds and the taming of wild animals have been an integral part of human civilization since time immemorial. Herbivorous animals, including prehistoric animals, grazed on leaves and succulent stems of weeds as well as leaves of shrubs and trees. Humans from the Old World tamed herbivorous animals such as horses, cattle, sheep, goats, water buffaloes, elephants, swine, domestic fowl, and various others, depending upon their needs. These animals consume the vegetative parts of cereals and legumes after grains are harvested. However, many crops of grass and of the legume family are used exclusively for livestock feed. These crops are used for grazing or are harvested for green-chop feeding, silage, or hay. Grasses such as bermudagrass and ryegrass are used for turf and forage.

Most of the forages belong to the grass family Poaceae (Gramineae) or the legume family Fabaceae (Leguminosae). Six crops from the grass family (wheatgrass, wildrye grasses, *Brachiaria*, bahiagrass, bermudagrass, and ryegrass) and three crops from the legume family (alfalfa, birdsfoot trefoil, and clover) have been included in *Forage Crops*, volume 5 of the series *Genetic Resources, Chromosome Engineering, and Crop Improvement*. Forage does not include plants cut or chopped and fed to animals such as hay, silage, or freshly cut grass. These are known as fodders. However, in developing countries, green pods are hand-picked from common bean, pea, cowpea, faba bean, chickpea, lentil, mungbean, and azuki bean; green leaves and stems are used as fodder for livestock. Goats love green leaves of pigeonpea—my personal observation while living in my hometown (the village of Sirihara in the state of Uttar Pradesh) in India. Lupin is used for forage in Europe and for grain in Australia. These crops have been covered in *Grain Legumes*, volume 1 of this series. Volume 2 (*Cereals*) included chapters on wheat, rice, barley, oat, maize, sorghum, pearl millet, rye, and man-made triticale. These are cultivated for grain. Straw is fed to livestock worldwide. Soybean, groundnut, and sunflower—covered in *Oilseed Crops*, volume 4 of this series—were initially used as forage crops but are now grown for oil and meal.

The majority of forage crops (alfalfa, clover, birdsfoot trefoil, bermudagrass) were domesticated in the Old World because they were fed to or allowed to be grazed on by tamed animals. By contrast, the New World inhabitants, the American Indians, did not tame herbivore animals. More than two thirds of wheatgrasses and wildrye grasses are native to Eurasia; from 22 to 30 grasses are considered native to North America and these are distributed throughout the vast prairies of the northern Great Plains of the United States and Canada. Bahiagrass originated in tropical South America.

Several books on forage crops are in print:

Forage Crops (G. H. Ahlgren), published in 1956;
Theory and Dynamics of Grassland Agriculture (J. R. Harlan), published in 1956;
Grass and People (Charles Morrow Wilson), published in 1961;
Genetic Resources of Forage Plants (J. G. McIvor and R. A. Bray, eds.), published in 1983;
Forage Crops (D. A. Miller), published in 1984;
Pastures (R. H. M. Langer, ed.), published in 1990;
Genetic Resources of Mediterranean Pasture and Forage Legumes (S. J. Bennett and P. S. Cocks, eds.), published in 1999;
Forages—An Introduction to Grassland Agriculture, volume I (R. F. Barnes, C. J. Nelson, M. Collins, and K. J. Moore, eds.), published in 2003; and
Forages—The Science of Grassland Agriculture, volume II (R. F. Barnes, C. J. Nelson, K. J. Moore, and M. Collins, eds.), published in 2007.

The latter two books are in their sixth editions. However, none of the books mentioned here or elsewhere has attempted to assemble the comprehensive information on genetic resources, gene pools, cytogenetics, and varietal improvement that has been compiled in *Forage Crops*.

The intensive varietal improvement of forage crops for high yield and improved nutritional quality (or elimination of antinutritional quality) and high palatability are the primary breeding objectives of various national programs (public institutions and private industries). Centro Internacional de Agricultura Tropical (CIAT) has a mandate for collection, maintenance, and varietal improvement of *Brachiaria*. Most genetic improvement of forage crops has been accomplished by conventional breeding assisted by germplasm resources, cytogenetics, plant pathology, entomology, agronomy, cell and tissue cultures, and molecular biology. Three forage legumes (*Medicago truncatula*, *Lotus japonicus*, and *Trifolium pretense*) are considered model crops for molecular genetics and genome sequencing.

The introductory chapter of this book (Chapter 1) summarizes landmark research done in the nine forage crops discussed in this book. Successive chapters provide a comprehensive account of the origin of each crop, its genetic resources in various gene pools, basic and molecular cytogenetics, conventional breeding, and the modern tools of molecular genetics and biotechnology. Appropriate germplasm collections can be an excellent resource for genetic enhancement of various traits in forage crops and for broadening their genetic bases. The genetic bases of forage crops are extremely narrow. In view of the narrow genetic base of oilseed crops, three gene pools (GPs) have now been identified by scientists for each crop: primary (GP-1), secondary (GP-2), and tertiary (GP-3). The recommendation is to use GP-2 and GP-3 resources in producing widely adapted varieties. The utilization of these resources (wide hybridization) in producing high-yielding cultivars that are resistant to abiotic and biotic stresses and have improved nutritional qualities is discussed in this book.

Eight major forage crops—alfalfa (Chapter 2), wheatgrass and wildrye grasses (Chapter 3), bahiagrass (Chapter 4), *Brachiaria* (Chapter 5), birdsfoot trefoil (Chapter 6), clover (Chapter 7), bermudagrass (Chapter 8), and ryegrass (Chapter 9)—are included in this book. Other minor grasses and legumes are not included in this volume.

Each chapter has been written by experts in the field. I am extremely grateful to all the authors for their outstanding contributions and to the reviewers of all the chapters. I have been fortunate to know them both professionally and personally, and our communication has been very cordial and friendly. I am thankful to Byron Byrson for identifying several forage crop researchers and encouraging them to contribute their expertise to this book. I am particularly indebted to Govindjee, William Grant, and Joseph Nicholas for their comments and suggestions.

This book is intended for scientists, professionals, and graduate students whose interests center upon genetic improvement of crops in general and major forage crops in particular. This book is also intended as a reference for plant breeders, taxonomists, cytogeneticists, germplasm explorers, pathologists, entomologists, physiologists, agronomists, molecular biologists, food technologists, and biotechnologists. Graduate students in these disciplines who have an adequate background in genetics, as well as other researchers interested in biology and agriculture, will also find this volume a worthwhile source of reference. I sincerely hope that the information assembled here will help in the much needed genetic amelioration of forage crops to feed livestock because an ever expanding global population depends on livestock. I anticipate that this book will enhance awareness of raising livestock in open pastures and feeding them grasses grown in nature rather than meals enriched with hormones and other animal by-products.

I end this preface with a quotation from Jack R. Harlan (1917–1998):

> The wider crosses not only take patience but determination, commitment, and, frequently, a good deal of skill and ingenuity. (*Crop Science* 16:329–333, 1976)

Ram J. Singh
Urbana-Champaign, Illinois

The Editor

Ram J. Singh, MSc, PhD, is an agronomist–plant cytogeneticist in the Department of Crop Sciences, at the University of Illinois at Urbana-Champaign. He received his PhD degree in plant cytogenetics under the guidance of the late Professor Takumi Tsuchiya from Colorado State University, Fort Collins, Colorado. He has benefited greatly from the expertise of and working association with Drs. T. Tsuchiya, G. Röbbelen, and G. S. Khush.

Dr. Singh conceived, planned, and conducted pioneering research related to cytogenetic problems in barley, rice, rye, oat, wheat, and soybean. Thus, he isolated monotelotrisomics and acrotrisomics in barley, identified them by Giemsa C- and N-banding techniques, and determined chromosome arm-linkage group relationships. By using pachytene chromosome analysis, Dr. Singh identified 12 possible primary trisomics in rice and determined chromosome-linkage group relationships. In soybean [*Glycine max* (L.) Merr.], he established genomic relationships among species of the genus *Glycine* and assigned genome symbols to all species. Dr. Singh constructed, for the first time, a soybean chromosome map based on pachytene chromosome analysis that laid the foundation for creating a global soybean map. By using fluorescent *in situ* hybridization, he confirmed the tetraploid origin of the soybean.

Dr. Singh developed a methodology to produce fertile plants with $2n = 40$ chromosomes from an intersubgeneric cross between soybean and a perennial wild species, *Glycine tomentella* ($2n = 78$); this invaluable invention has been awarded a patent (US-2007-0261139-A1). He has published 70 research papers in reputable international journals, including the *American Journal of Botany, Chromosoma, Critical Review in Plant Sciences, Crop Science, Euphytica, Genetics, Genome, Journal of Heredity, Plant Breeding, The Nucleus,* and *Theoretical and Applied Genetics.* In addition, he has summarized his research results by writing 13 book chapters. His book *Plant Cytogenetics* (first edition in 1993 and second edition in 2003) is widely used for teaching graduate students. He has presented research findings as an invited speaker at national and international meetings.

Dr. Singh is a member of the Crop Science Society of America and the American Society of Agronomy, the chief editor of the *International Journal of Applied Agricultural Research* (IJAAR), and editor of *Plant Breeding.* In 2000 and 2007, he received the Academic Professional Award for Excellence: Innovative & Creativity from the Department of Crop Sciences, the University of Illinois at Urbana-Champaign. He was invited to become a visiting professor by Dr. Kiichi Fukui (Osaka University, Osaka, Japan; October 12, 2004–January 12, 2005) and Dr. Gyuhwa Chung (Chonnam National University, Yeosu, Chonnam, South Korea; October, 2006). Dr. Singh is an editor for the series entitled *Genetic Resources, Chromosome Engineering, and Crop Improvement* and has published *Grain Legumes* (volume 1), *Cereals* (volume 2), *Vegetable Crops* (volume 3), and *Oilseed Crops* (volume 4). *Forage Crops* (volume 5) and *Medicinal Crops* (volume 6) are expected in print by 2009 and 2010, respectively.

Contributors

Carlos A. Acuña
Agronomy Department
University of Florida
Gainesville, Florida

Reed E. Barker
Grass Genomic Testing
Oregon State University
Corvallis, Oregon

Gary R. Bauchan
USDA-ARS
Electron and Confocal Microscopy Unit
Soybean Genomics and Improvement Lab
Beltsville, Maryland

Ann R. Blount
North Florida Research and Education
    Center—Marianna
University of Florida
Marianna, Florida

Cacilda Borges do Valle
EMBRAPA/CNP Gado de Corte
Campo Grande, Brazil

William F. Grant
Department of Plant Science
McGill University, Macdonald Campus
Montreal, Canada

Kevin B. Jensen
USDA-ARS Forage and Range Research Lab
Utah State University
Logan, Utah

Geunhwa Jung
Department of Plant, Soil, and Insect Sciences
University of Massachusetts
Amherst, Massachusetts

J. B. Morris
USDA-ARS Plant Genetic Resources
    Conservation Unit
University of Georgia
Griffin, Georgia

Minoru Niizeki
Faculty of Agriculture and Life Science
Laboratory of Plant Breeding and Genetics
Hirosaki University
Hirosaki, Aomori-ken, Japan

Maria Suely Pagliarini
Departamento de Biologia Celular e Genética
Universidade Estadual de Maringá
Maringá, Brazil

G. Pederson
USDA-ARS Plant Genetic Resources
    Conservation Unit
University of Georgia
Griffin, Georgia

K. Quesenberry
Department of Agronomy
University of Florida
Gainesville, Florida

Ram J. Singh
Department of Crop Sciences, National
    Soybean Research Laboratory
University of Illinois
Urbana, Illinois

Charles M. Taliaferro
Plant and Soil Science Department
Oklahoma State University
Stillwater, Oklahoma

Scott Warnke
USDA/ARS
Floral and Nursery Plants Research Lab
Beltsville, Maryland

M. L. Wang
USDA-ARS Plant Genetic Resources
    Conservation Unit
University of Georgia
Griffin, Georgia

Richard R.-C. Wang
USDA-ARS Forage and Range Research Lab
Utah State University
Logan, Utah

Yanqi Wu
Plant and Soil Sciences Department
Oklahoma State University
Stillwater, Oklahoma

Contents

CHAPTER 1

Landmark Research in Forage Crops

Ram J. Singh and William F. Grant

CONTENTS

1.1 INTRODUCTION

Herbivore animals have preferred to graze certain grasses, shrubs, and leaves of trees since prehistoric times. Humans selected, domesticated, cultivated, and harvested grains from the cereals, legumes, oilseed, and other crops for their use and simultaneously tamed the animals, including ruminant livestock, by feeding them vegetative parts also known as straw (roughages). Furthermore, domesticated animals graze on wastelands, on fallows, under trees and groves, and in pastures. Pasture crops include alfalfa, clover, birdsfoot trefoil, wheatgrass, wildrye grasses, bermudagrass, brachiaria, and ryegrass, as well as other minor crops such as vetches (*Vicia* spp.) and brassicas. The classic example of a cultivated crop is alfalfa, a forage legume that has been cultivated since 1,300 B.C. Its importance to feeding horses is described in Greek literature between 440 and 322 B.C. (Chapter 2). With the establishment of human civilization, animals started to be confined by the erection of barriers (stones, hedges) followed by fencing. Humans started to raise crops, tend to animals, and cultivate pastures—a triangular symbiotic relationship since the beginning of human civilization. Freshly cut or uncut grasses, legumes, and other crops were converted into hay or silage

Figure 1.1 **(See color insert following page 274.)** A herd of sheep and goats returning home after a day of grazing in a forest in Udaipur, Rajasthan, India.

(known as fodder) and were brought to the settlement to feed livestock. Feeding fodder became a common practice in raising livestock.

Forage crops belong mostly to the grass (Poaceae, formerly Gramineae) and legume (Fabaceae, formerly Leguminosae) families. Legumes and grasses grown in rotation enrich the soil because root nodules of legumes fix atmospheric nitrogen. Forage crops cover the soil, preventing soil erosion and degradation. In developing countries, deforestation is usually common in order to develop pasture, and livestock frequently overgraze the forest and pasture, disturbing agro-ecosystems. Figure 1.1 shows a herd of sheep and goats being taken home after a day of grazing in a forest at Udaipur, Rajasthan, India. Overgrazing of the forest is a common practice in India, and herds include cows, water buffaloes, sheep, and goats. It is necessary that regional as well as global genetic resources be appropriately collected, maintained, and preserved before extinction occurs (Singh and Lebeda 2007).

Nine primary forage crops are included in *Forage Crops,* volume 5 of the series *Genetic Resources, Chromosome Engineering, and Crop Improvement.* They are alfalfa (Chapter 2), wheatgrass and wildrye grasses (Chapter 3), bahiagrass (Chapter 4), *Brachiaria* (Chapter 5), birdsfoot trefoil (Chapter 6), clover (Chapter 7), bermudagrass (Chapter 8), and ryegrass (Chapter 9). Forage crops have been characterized into warm-season and cool-season crops. Warm-season crops (bahiagrass, brachiaria, and bermudagrass) are grown in tropical areas of the world; cool-season crops are alfalfa, wheatgrass, wildrye grass, birdsfoot trefoil, clover, and rye grass. Alfalfa is the fourth most widely grown crop in the United States behind maize, wheat, and soybean (Chapter 2). The main focus of this chapter is to summarize recent knowledge and achievement on genetic resources and their taxonomy, diversity, collection, conservation, evaluation, and utilization in breeding of these nine major forage crops of economic importance.

1.2 IMPORTANCE OF FORAGE CROPS

Forage crops are exclusively used for livestock feed. However, alfalfa, birdsfoot trefoil, and clover have many traits of economic importance. Alfalfa contains 15–22% crude protein and is an excellent source of vitamins and minerals. It is consumed directly by humans in the form of alfalfa sprouts and alfalfa juice and is beginning to be used as a biofuel for the production of electricity, ethanol, and bioremediation of soils with high levels of nitrogen, and as a factory for the production of industrial enzymes (Chapter 2). Alfalfa (Chapter 2), birdsfoot trefoil (Chapter 6), and clover (Chapter 7) flowers attract honey-producing bees. Birdsfoot trefoil has high nutritive value and palatability, does not cause bloat, and is used extensively for soil improvement (Chapter 6). Clovers are an important crop grown worldwide for forage and as a sustainable crop to manage agricultural pests. Red clover reduces the incidence of a range of cancers and *Escherichia coli* in cattle (Chapter 7).

Three forage legume species, *Medicago truncatula* (Chapter 2), *Lotus japonicus* (Chapter 6), and *Trifolium pratense* (Chapter 7), are model crops for genome research and genome sequencing and the genome of *M. truncatula* has been completely sequenced, a closely related species to alfalfa (http://www.medicago.org/, http://www.kazusa.or.jp/lotus). The genome size of *L. japonicus* is 472.1 Mb; *T. pratense* is 468 Mb (Chapter 6), and *M. truncatula* is ~454–526 Mb (Chapter 2).

Grasses are used for forage and turf. Bermudagrass is widely used in tropical and warmer temperate regions of the world for livestock grazing and fodder. It has an excellent sod-forming habit and is used as turf on home and institutional lawns; parks; athletic fields; golf course fairways, tees, putting greens, and roughs; cemetery grounds; and roadside rights-of-way. Bermudagrass provides excellent protection against soil erosion from wind and water (Chapter 8).

Wheatgrass and rye grasses are valued throughout the temperate regions of the world as forage for livestock, habitat for wildlife, and other qualities relating to aesthetics, soil stabilization, weed control, watershed management, and low-maintenance turfgrass. The biofuel industry is exploring the use of these grasses as alternative energy sources (Chapter 3).

Brachiaria is the single most important genus of forage grasses for pastures in the tropics for low-fertile and acid soil requirements. It is adapted to shade, drought, water logging, and low fertility and has high tolerance to aluminum and heavy defoliation, good seed production, and apomixis rendering uniform pastures. *Brachiaria* grasses have drastically modified the landscape in Central Brazil and in most tropical Latin American pastures (Chapter 5).

1.3 ESTABLISHMENT OF INTERNATIONAL AND NATIONAL PROGRAMS

The following international and national institutes have been established for major forage crops research:

1. Centro Internacional de Agricultura Tropical (CIAT) (International Center for Tropical Agriculture), Cali, Columbia (http://www.ciat.cgiar.org/): In addition to rice, bean, cassava, and tropical fruits, CIAT collects, maintains, and conducts varietal improvement of all grasses of the genus *Brachiaria*.
2. International Livestock Research Institute, Nairobi, Kenya, Africa (http://www.ilri.org/): The mandate of the institute is livestock improvement, primarily for Africa. Maize-based crop–livestock systems provide the livelihoods of the majority of smallholder families in east and southern Africa (maize is described in *Cereals*, volume 2 of this series).
3. National Programs: National (public institutes) and private industries worldwide have forage crop improvement programs. The Indian Grassland and Fodder Research Institute (IFRI), Jhansi, Uttar Pradesh, India (http://www.igfri.ernet.in/aicrp_fc.htm), was established in 1962 and focuses on all aspects of forage research. EMBRAPA at Campo Grande, Mato Grosso do Sul (MS), Brazil (http://www.cnpgc.embrapa.br), has extensive varietal improvement programs for *Brachiaria*, bahiagrass, and *Stylosanthes macrocephala* M.B. Ferreira & Sousa Costa (forage legume). In the United States, federal research stations (USDA/ARS; http://www.ars.usda.gov) have affiliations with public and independent universities and with private industries and are actively engaged in collecting, maintaining, and improving forages for their region.

1.4 GENE POOLS OF FORAGE CROPS

Based on a review of the literature on hybridization, Harlan and de Wet (1971) proposed a three-gene-pool (GP) concept—primary (GP-1), secondary (GP-2), and tertiary (GP-3)—for

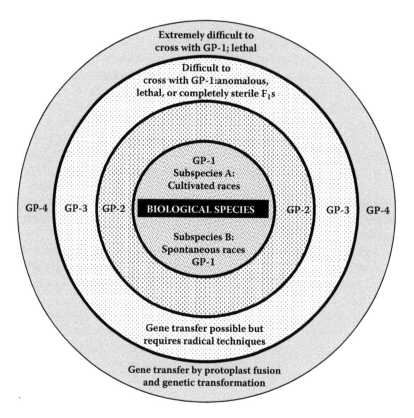

Figure 1.2   Four gene pools (GP-1 to GP-4) have been identified for forage crops.

utilization of germplasm resources for crop improvement. However, four gene pools (GP-1 to GP-4) have been identified for forage crops, particularly for clover, with integrated, multidisciplinary approaches through plant exploration, plant introduction, taxonomy, genetics, cytogenetics, plant breeding, microbiology, plant pathology, entomology, agronomy, physiology, distant hybridization, and molecular biology, including cell and tissue culture and genetic transformation (Figure 1.2). These efforts have produced superior forage crop cultivars with resistance to abiotic and biotic stresses and improved forage quality and quantity. The concept of primary, secondary, tertiary, and quaternary gene pools and genetic transformation has played a key role in improving forage crops (Chapters 2–9).

1.4.1 Primary Gene Pool

The primary gene pool (GP-1) consists of landraces and biological species and has been identified for forage crops described in this volume (Figure 1.2). Wild progenitors of cultivated forage crops are identified, postulated, and proposed based on geographical distribution, classical taxonomy, cytogenetics, and molecular methods (Chapters 2, 3–5, and 7). Each chapter describes primary gene pools in detail. For example, GP-1 for alfalfa ($2n = 2x = 16$) consists of cultivated, naturalized, and wild forms of *Medicago sativa* ssp. *sativa,* including modern and obsolete cultivars, landraces, and ecotypes. In addition, *M. sativa* ssp. *falcata, M. sativa* nssp. *varia, M. sativa* ssp. *glutinosa, M. sativa* nssp. *tunetana, M. sativa* ssp. *coerulea* ($2n = 2x = 16$), and *M. sativa* ssp. *glomerata* can be easily crossed with *M. sativa* ssp. *sativa* with normal bivalent pairing, but with an occasional quadrivalent configuration, typical of an autotetrapoid (Chapter 2). The primary gene pool of clover has been divided into five subgroups based on their value to humans (Chapter 7).

1.4.2 Secondary Gene Pool

The secondary gene pool (GP-2) includes all species that can be hybridized with GP-1 with the aid of embryo culture rescue and possessing at least some fertility in the F_{1s} that results in gene transfer (Figure 1.2). GP-2 species have been clearly identified for alfalfa (Chapter 2), wheatgrass and wild rye grasses (Chapter 3, Table 3.5), and clover (Chapter 7, Table 7.2). Identification of GP-2 species for *Brachiaria* (Chapter 5), birdsfoot trefoil (Chapter 6), bermudagrass (Chapter 8), and ryegrass (Chapter 9) has not yet been clearly identified.

1.4.3 Tertiary Gene Pool

The tertiary gene pool (GP-3) is the outer limit of potential genetic resources. Prezygotic barriers can cause partial or complete hybridization failure, inhibiting introgression between GP-1 and GP-3 (Singh 2003; Figure 1.2). Tertiary gene pools are present for some forage crops (Chapters 2–9). The tertiary gene pool in alfalfa includes 36 annual *Medicago* species that are closely related to alfalfa. However, no hybrids have been produced that are fertile and produce progeny (Chapter 2). Genomes of six genera belonging to GP-3 for wheatgrass and wild rye have been designated (Chapter 3, Table 3.5). Ploidy levels, hybrid incompatibility, and apomixis play a major role in identifying GP-3 for *Brachiaria* (Chapter 5). The tertiary gene pool in *Lotus* includes nine diploid species closely related to *L. corniculatus* (Grant and Small 1996). Some fertile hybrids and progeny have been obtained (Grant 1999). Chapter 7 identifies 13 GP-3 species for various clovers.

1.4.4 Quaternary Gene Pool

The quaternary gene pool (GP-4) is the extreme outer limit of genetic resources and requires radical methods to produce F_1 hybrids such as protoplast fusion and genetic transformation (Figure 1.2). The quaternary gene pool would include all other nonrelated and noncultivated species in the genus *Trifolium* (Chapter 7). Chromosome number in the genus *Trifolium* ranges from $2n = 10$ to 180. In Chapter 6, protoplast fusion attempts in birdsfoot trefoil are given; fusants have been obtained between birdsfoot trefoil and soybean, rice, and alfalfa.

1.5 GERMPLASM RESOURCES FOR FORAGE CROPS

National and international institutes and private industries for the nine forage crops presented in this volume collect, maintain, disseminate, and develop breeding lines and varieties. Cultivars with traits such as resistance to abiotic and biotic stresses, high silage and hay yield, and improved nutritional quality and quantity are developed and maintained. Plant exploration of landraces and wild relatives of the nine forage crops described in this book is extensive, and they have been characterized by a multidisciplinary approach such as classical taxonomy, cytogenetics, and molecular methods (Chapters 2–9).

Of the nine forage crops, alfalfa (Chapter 2), *Brachiaria* (Chapter 5), birdsfoot trefoil (Chapter 6), clover (Chapter 7), and bermudagrass (Chapter 8) originated in the Old World. Bahiagrass originated in the New World (Chapter 4). The majority of Triticeae species originated in the Old World (Eurasia and Asia), and 22–30 species are native to North America (Chapter 3, Table 3.2).

The current U.S. collection of alfalfa (Chapter 2) includes 4,151 *M. sativa* accessions and 4,009 accessions representing other perennial and annual species (http://sun.ars-grin.gov/npgs/). Montpellier, France, has a large collection, also: 2,633 *M. sativa* accessions (http://www.ecpgr.cgiar.org/databases/Crops/Medicago.htm). Plant Gene Resources of Canada holds 1,620 accessions of *M.*

sativa (http://pgrc3.agr.ca/). *Ex situ* collection of wheatgrasses and wildryes (Chapter 3) includes 5,279 accessions being maintained at Pullman, Washington (http://www.ars-grin.gov/). The germplasm information site of the CGIAR (Consultative Group on International Agricultural Research; http://singer.grinfo.net) reports the existence of 1,805 accessions for the *Brachiaria* (Chapter 5) world collection, of which 63% are in CIAT (Colombia: 1,133 accessions) and 37% in ILRI (Kenya/Ethiopia: 672 accessions). The U.S. collection of *Lotus* taxa (Chapter 6) includes 883 accessions, of which 57% are those of birdsfoot trefoil (http://www.ars-grin.gov/npgs/acc/acc_queries.html). Plant Gene Resources of Canada (http://pgrc3.agr.gc.ca/cgi-bin/npgs/html/acc_search.pl?accid=Lotus) lists 171 accessions of *Lotus* and their origin. The *Trifolium* collection contains more than 5,000 accessions, representing about 200 species collected or donated from more than 90 countries (http://www.ars-grin.gov/npgs/searchgrin.html; Chapter 7). The genus *Cynodon* includes nine species and 10 varieties and ploidy ranging from diploid ($2n = 2x = 18$) to triploid ($2n = 3x = 27$), tetraploid ($2n = 4x = 36$), pentaploid ($2n = 5x = 45$), and hexaploid ($2n = 6x = 54$) (Chapter 8).

Plant exploration, collection, introduction, and germplasm preservation have helped adaptation and distribution of forage crops worldwide. Use of cytogenetics, classical taxonomy, and molecular cytogenetics has helped establish the genomic relationships among several forage crops species. Three forage crops, *Medicago truncatula, Lotus japonicus,* and *Trifolium pratense,* are considered model crops for genome sequencing because of their small genome size (Chapters 2, 6, and 7).

1.6 GERMPLASM ENHANCEMENT FOR FORAGE CROPS

Varietal improvement programs to develop elite breeding lines with high yield are dependent upon breeding objectives, type of forage crops (cross- and self-pollinated, long- and short-term reproductive cycles, perennial or annual, legumes or grasses, warm or cold season, forage and fodder, turf, and sod), end-use products (honey, ornamentals) and by-products (biodiesel), breeding methods (conventional vs. molecular-aided breeding), and loss versus benefit of raising a particular crop (supply and demand market value). Other important qualities include inheritance patterns and heritability of the selected characters, such as antitoxin (nutritional) factors (condensed tannins, hydrogen cyanide, and proanthocyanidins—all in birdsfoot trefoil). Good-quality forage crops may be improved from available germplasm, selection from introduced germplasm, breeding methodologies, mutation breeding, wide hybridization, and genetic transformation (Chapters 2–9). GMO alfalfa with the Roundup Ready gene has been produced (Chapter 2).

The core collection concept has been utilized in germplasm collection, maintenance, and breeding programs for several forage crops. The core collection approach to germplasm evaluation is a two-stage approach. The first stage involves examining all accessions in the core collection for a desired trait. This information is then used to determine which clusters of accessions in the entire germplasm collection should be examined during the second stage of screening. Theoretically, the probability of finding additional accessions with a desired characteristic should be the highest priority in these clusters (Chapters 2 and 7).

For example, there are two core subsets related to *Medicago:* one each for the perennial and annual *Medicago* species (Chapter 2). Of the 1,105 accessions representing 47 countries and four perennial species, the core subset has 200 accessions with representatives from each species and geographical area. The annual *Medicago* core subset of 211 accessions was selected from 1,220 accessions representing 34 *Medicago* annual species. The South Australian *Medicago* Genetics Resource Center in Adelaide, South Australia, holds more than 22,000 annual *Medicago* species accessions and the core collection consists of 1,705 accessions (Chapter 2). There are currently three clover core collections for *Trifolium alexandrinum* (24 accessions), *T. resupinatum* (32 accessions), and *T. subterraneum* (40 accessions) (Brad Morris, personal communication to RJS, January 18, 2008). Each of these cores was developed by stratification from country of origin. Accessions were selected at random from each group using a 10% proportional selection allocation. A minimum of one accession was selected from each country.

1.6.1 Breeding Forage Crops for High Yield

Forage (pasture, hay, and silage) is measured by the yield of dry matter and biotic resistance (pest and pathogens), abiotic tolerance (cold hardiness, winter hardiness, adverse soil adaptation, drought and flood tolerance), quality (dry matter digestibility and crude protein, lignin content, palatability), and, finally, animal performance (Chapters 2–9).

The present genetic base of forage crops is extremely narrow because breeders have largely confined their varietal improvement programs to GP-1. Legume forage breeders still have to exploit wild progenitors and wild perennial species to broaden the genetic base of modern forage cultivars. Self-pollinated clover cultivars have been produced by phenotypic selection. Cultivars of cross-pollinated clover have been produced by mass selection, phenotypic recurrent selection, and back-crossing. Wide hybridization among clover species has produced lines with improved yield, forage quality, and seed production. Cell and tissue cultures including genetic transformation have not produced improved forage crop cultivars, with the exception of Roundup Ready alfalfa (http://monsanto.mediaroom.com/index.php?s=43&item=469). The first rhizomatous birdsfoot trefoil cultivar has been produced. This trait is beneficial for plant survival and plant growth but does not ensure performance or survival (Chapter 6; http://www.inia.org.uy/sitios/lnl/vol34/grant.pdf).

Of the six forage grasses included in this volume, a varietal improvement program is in progress for *Brachiaria*, a warm-season grass, but the program is restricted to Brazil and CIAT. Initial cultivar development has depended on direct selection from introduced germplasm from Africa and Australia. Diploid *Brachiaria* are sexual, whereas tetraploid cytotype is apomictic, which inhibits genetic recombination in interspecific hybridization between $2x$ and $4x$ species (Chapter 5). Bermudagrass, also a warm-season grass, is cultivated for forage, hay, and turf in the United States, China, and Australia. Early, high-yielding bermudagrass was selected for specific environments—ecotype selection—from introduced germplasm. Clonally and seed-propagated high-yielding superior forage and turf hybrid bermudagrass cultivars possessing cold hardiness have been produced after intervarietal and interspecific crosses. Genetic transformation has not as yet produced a high-yielding bermudagrass (Chapter 8). High-yielding varieties of wheatgrasses and wildrye grasses have been obtained primarily by selection from introduced germplasm and interspecific crosses (Chapter 3).

1.6.2 Breeding Forage Crops for Antinutritional Elements

The ultimate aim of forage crop breeders is to produce forages for pastures and silage with high nutritive qualities, digestibility, palatability, and excellent turf quality. Forage legumes and grasses harbor antinutritional chemicals harmful to animals. For example, cyanogenic glycosides, hydrogen cyanide (HCN), proanthocyanidins, and condensed tannins found in birdsfoot trefoil and clover are harmful to animals. *Lotus uliginosus* does not contain HCN, and this trait can be transferred to *L. corniculatus* through hybridization and selection (Chapter 6). Bermudagrass is high yielding but animal performance (daily weight gain) is low because of low digestibility; however, several digestibility-improved bermudagrass varieties have been released (Chapter 8). Phytoestrogens found in legume forages cause marked changes to structure and function of the reproductive organs, especially in the female. The amount of phytoestrogens can be reduced through breeding (Familton 1990).

Forages deficient in certain chemicals such as cobalt, iodine, zinc, magnesium, and selenium and the interaction among copper, molybdenum, and sulfur are detrimental to animals. Nitrate (NO_3^-) and nitrite (NO_2^-) poisoning is usually common by consuming sudangrass [*Sorghum bicolor*

(L.) Moench], sorghums (*Sorghum* spp.), and maize (*Zea mays* L.). Perennial grasses cause a lesser problem because nitrogen fertilizer is applied at a lower level, and legumes do not cause nitrate poisoning because they fix their own nitrogen. Chemicals–mycotoxins synthesized by fungi that attack grasses, also cause many disorders of grazing animals (Mayland and Cheeke 1995).

1.6.3 Breeding Apomictic Forage Crops

Apomixis is usually defined as asexual reproduction through seeds and is prevalent in the tropical forage grasses. Several tetraploid ($2n = 4x = 40$) apomictic bahiagrass cultivars have been released (Chapter 4); they are generated by crossing tetraploid sexual and apomictic genotypes (Figure 4.3). EMBRAPA is engaged in transferring apomictic genes to diploid ($2n = 2x = 18$), natural, and sexual *Brachiaria ruziziensis* from tetraploid ($2n = 4x = 36$) apomictic *B. decumbens* (Chapter 5). Clonally propagated bermudagrass has been developed and used for pasture and turf (Chapter 8).

1.6.4 Development of Breeding Methods

Breeding methodologies have been developed depending upon the nature of pollination. Forage legumes (alfalfa, birdsfoot trefoil, and clover) are often cross-pollinated and honeybees play a major role in pollination. Conventional breeding methods have produced forage crops with high biomass yield, excellent turf quality, resistance to biotic stresses, and tolerance to abiotic stresses. Clonal hybrid bermudagrass cultivars have been developed by hybridization and selection (Chapter 8). Cell and tissue culture and genetic transformation methods have been initiated and are in progress in forage legumes (Chapters 2, 6, and 7) and grasses (Chapters 3–5 and 8). Stable glyphosate-tolerant alfalfa known as Roundup Ready® has been produced by genetic transformation (Chapter 2).

1.7 CONCLUSIONS

1. The credit for selecting forage crops should go to herbivore animals prior to the time they were domesticated by humans. Depending on the agro-ecosystem, cattle, water buffaloes, sheep, goats, horses, and elephants may graze in the forest and in the pasture. In developing countries, animals routinely graze in the forest. At the same time, the forest may become converted (deforestation) into pasture by planting forage crops. Nine forage crops (alfalfa, Chapter 2; wheatgrass and wildrye grass, Chapter 3; bahiagrass, Chapter 4; *Brachiaria*, Chapter 5; birdsfoot trefoil, Chapter 6; clover, Chapter 7; bermudagrass, Chapter 8; and ryegrass, Chapter 9) have been included in *Forage Crops*, volume 5 of the series *Genetic Resources, Chromosome Engineering, and Crop Improvement*.
2. Three forage legume species, *Medicago truncatula* (Chapter 2), *Lotus japonicus* (Chapter 6), and *Trifolium pratense* (Chapter 7), are model crops for genome research and genome sequencing (http://www.medicago.org/, http://www.kazusa.or.jp/lotus).
3. Forage crops are used for turf and for producing honey and pharmaceutical products. Legume forages enrich the soil by fixing atmospheric nitrogen through root nodule bacteria. Forages may also be used under arid, drought, and saline conditions and to prevent soil erosion.
4. Forage crop breeders have confined their efforts to the primary gene pool (GP-1). Exploitation of the secondary (GP-2), tertiary (GP-3), and quaternary (GP-4) gene pools is hampered because of pre- and postzygotic barriers. Researchers must continue to conduct extensive plant exploration to collect primitive cultivars, landraces, and wild relatives before the spread of high-yielding varieties, over-grazing of forest and deforestation, and environmental factors make them extinct. As acquired and after evaluation, these invaluable genetic materials are being deposited in gene banks for medium- and long-term storage.
5. Forage crop breeders from national and international institutes and private industries are utilizing conventional breeding methods and genetic transformation to produce high-quality forages for pasture, turf, hay, and silage and also to remove antinutritional factors from forages.

REFERENCES

Familton, A. S. 1990. Animal disorders arising from consumption of pasture. In *Pastures: Their ecology and management,* ed. R. H. M. Langer, 284–298. Oxford: Oxford University Press.

Grant, W. F. 1999. Interspecific hybridization and amphidiploidy of *Lotus* as it relates to phylogeny and evolution. In *Trefoil: The science and technology of Lotus*, ed. P. R. Beuselinck, 43–60. Madison, WI: American Society of Agronomy and Crop Science Society of America, CSSA Special Publ. No. 28.

Grant, W. F., and Small, E. 1996. The origin of the *Lotus corniculatus* (Fabaceae) complex: A synthesis of diverse evidence. *Canadian Journal of Botany* 74:975–989.

Harlan, R. R., and de Wet, J. M. J. 1971. Toward a rational classification of cultivated plants. *Taxon* 20:509–517.

Mayland, H. F., and Cheeke P. R. 1995. Forage-induced animal disorders. In *Forages. Volume II: The science of grassland agriculture,* 5th ed., ed. R. F. Barnes, D. A. Miller, and C. J. Nelson, 121–135. Ames: Iowa State University Press.

Singh, R. J. 2003. *Plant cytogenetics,* 2nd ed. Boca Raton, FL: CRC Press Inc.

Singh, R. J., and Lebeda, A. 2007. Landmark research in vegetable crops. In *Genetic resources, chromosome engineering, and crop improvement, vegetable crops,* Vol. 3, ed. R. J. Singh, 1–15. Boca Raton, FL: Taylor & Francis Group.

CHAPTER 2

Alfalfa (*Medicago sativa* ssp. *sativa* (L.) L. & L.)

Gary R. Bauchan

CONTENTS

2.1 INTRODUCTION

Alfalfa (*Medicago sativa* L.) is an autotetraploid ($2n = 4x = 32$), allogamous, cool-season perennial legume. Sometimes referred to as Lucerne, alfalfa is called the "queen of the forages." Alfalfa hay is used primarily as animal feed for dairy cows, but also for horses, beef cattle, sheep, chickens, turkeys, and other farm animals. It is the fourth most widely grown crop in the United States, behind corn, wheat, and soybean, and occupies double the cotton acreage. There are 8.7 million ha of alfalfa cut for hay with an average yield of 8.4 Mg/ha (USDA National Agricultural Statistics Service 2007). Alfalfa can be compressed into cubes and/or pellets, which are used in feed for smaller pet animals, including rabbits, hamsters, guinea pigs, etc. Alfalfa is sometimes grown in mixtures with forage grasses and other legumes. The acreage of all hay harvested per year, including alfalfa, is 25 Mg/ha (USDA National Agricultural Statistics Service 2007).

Alfalfa seed is primarily grown in the northwestern areas of the United States, primarily in the states of California, Idaho, Nevada, Oregon, Wyoming, and Washington. The approximate yield of alfalfa seed for the United States is 5.2 Mg. A fringe benefit to the production of alfalfa seed is the production of honey from bees, because alfalfa requires bees for pollination. Alfalfa seeding rates depend on the climate and soil conditions, but, on average, 11–23 kg/ha are used (Tesar and Marble 1988).

Alfalfa is also important due to its high biomass production. Dormant-type alfalfa can be cut for hay three to four times within a growing season where cold seasonal climates occur; in warmer climates where cold seasons do not occur, the nondormant-type alfalfa can be cut five to seven times a season. The record yield of one acre of alfalfa is 10–22 Mg/ha without irrigation and 54 Mg/ha with irrigation (Hill, Shenk, and Barnes 1988). Alfalfa is a widely adapted crop, energy efficient and an important source of biological nitrogen fixation. Alfalfa forms a symbiotic relationship with *Sinorhizobia meliloti*. The average hectare of alfalfa will fix about 80 kg of nitrogen per year, thus reducing the need to apply expensive nitrogen fertilizers (Heichel and Barnes 1984).

One of the most important characteristics of alfalfa is its high nutritional quality as animal feed. Alfalfa contains between 15 and 22% crude protein and serves as well as an excellent source of vitamins and minerals. Specifically, alfalfa contains vitamins A, D, E, K, U, C, B1, B2, B6, B12, niacin, panthothanic acid, inocitole, biotin, and folic acid. Alfalfa also contains the minerals phosphorus, calcium, potassium, sodium, chlorine, sulfur, magnesium, copper, manganese, iron, cobalt, boron, and molybdenum, as well as trace elements such as nickel, lead, strontium, and palladium (Bickoff, Kohler, and Smith 1972). Alfalfa is directly consumed by humans in the form of alfalfa sprouts. Alfalfa juice is also used in some health food products.

In addition to its traditional use as an animal feed, alfalfa is beginning to be used as a biofuel for the production of electricity and ethanol, for bioremediation of soils with high levels of nitrogen, and as a factory for the production of industrial enzymes such as lignin peroxidase, alpha-amylase, cellulase, and phytase.

Medicago truncatula Gaertner, an annual species of *Medicago* (referred to as annual medics) that is closely related to perennial alfalfa, has become the model plant for genomic studies. It is a forage crop, as are other medics used in sheep and cattle production in Australia and in countries surrounding the Mediterranean Sea (Table 2.1). The key attributes are that it has a small diploid genome (-5×10^8 bp), is self-fertile and thus autogamous, is a prolific seed producer with a short regeneration time, and fixes its own nitrogen (Cook 1999). It has been proposed to sequence the entire genome of *M. truncatula* with an initial sequence published online in 2004 (http://www.medicago.org). Sequencing data and cytogenetic evidence have shown that the genomes of *M. truncatula* and *M. sativa* are very similar and thus the genomic information obtained from *M. truncatula* can have direct impact on the improvement of alfalfa.

The objective of this chapter is to present information on the origin and domestication of alfalfa, genetic resources of alfalfa and related species, and the available cytogenetic, breeding, and molecular genetic tools available for the improvement of alfalfa.

Table 2.1   Genus *Medicago*[a] with Geographic Distribution, Gene Pool Designation, and Chromosome Number

Species	Subspecies	Geographic Distribution	Gene Pool	Chrom. No.
		Perennials		
M. arborea L. S.		Europe, Turkey	GP-2	32, 48
M. cancellata M. Bieb.		European Russia	GP-2	48
M. carstiensis Jacq.		E Europe	GP-2	16
M. daghestanica Rupr. ex Boiss.		Dagestan	GP-2	16
M. edeworthii Sirjaev		Himalayas	GP-2	16
M. hybrida (Pourret) Trautv.		France, Spain	GP-2	16
M. marina L.		S Europe, N Africa, Middle East, Crimea	GP-2	16
M. papillosa	*ssp. macrocarpa (Boiss) Urban*	Turkey	GP-2	16
M. papillosa	*ssp. papillosa Bioss.*	Turkey, S Russia	GP-2	16
M. papillosa Boiss.		Turkey, S Russia	GP-2	16
M. pironae Vis.		Italy	GP-2	16
M. platycarpa L.		Trautv. China, Mongolia, C Asia	GP-2	16
M. prostrata Jacq.		Albania, E Europe, Italy	GP-1	16
M. rhodopea Velen.		Bulgaria	GP-2	16
M. rupestris M. Bieb.		Crimea, W Caucasus	GP-2	16
M. ruthenica L. Trautv.		China, Korea, Mongolia, Russia	GP-2	16
M. sativa	*ssp. coerulea* Schmalh.	E Turkey, Iran, Former USSR	GP-1	16
M. sativa	*ssp. sativa* (L.) L & L.	Possible native range— Middle East, C Asia, now widely distributed	GP-1	32
M. sativa	*ssp. glomerata*	Balbis. S Europe, N Africa, Caucasus	GP-1	16, 32
M. sativa	*ssp. falcata* (L.) Arcangeli	S Europe, Former USSR	GP-1	16, 32
M. sativa	*var. falcata* Archang.	N Eurasia	GP-1	16, 32
M. sativa	*var. viscosa* Posp.	S Europe, N Africa, Caucasus	GP-1	16
M. sativa	*ssp. × varia* Martin	Europe, Iran, Syria, Turkey, Caucasus	GP-1	16, 32
M. sativa	*ssp. × tunetana* Murbeck	S Europe, N Africa, Caucasus	GP-1	16
M. saxatilis M. Bieb.		Crimea	GP-2	48
M. secundiflora Durieu		S Europe, N Africa	GP-2	16
M. suffruticosa Raymond ex DC.		France, Spain, Morocco	GP-2	16
		Annuals		
M. arabica (L.) Hudson		Europe, N Africa, Middle East, Crimea, Caucasus	GP-3	16

Continued

Table 2.1   Genus *Medicago*[a] with Geographic Distribution, Gene Pool Designation, and Chromosome Number (*Continued*)

Species	Subspecies	Geographic Distribution	Gene Pool	Chrom. No.
M. archiducis-nicolai Sirjaev		C China, NE Tibet	GP-3	16
M. arenicola (Huber-Mor)		Turkey	GP-3	16
M. astroites (Fisch. & Mey.)		E Mediterranean	GP-3	16
M. biflora (Griseb.) E Small		Turkey, Iran, S Transcaucasus	GP-3	16
M. blancheana Boiss.		E Mediterranean	GP-3	16
M. brachycarpa		Turkey, Lebanon, Iraq, Transcaucasia	GP-3	16
M. carica (Huber-Mor. & Sirjaev) E. Small		SW Anatolia	GP-3	16
M. ciliaris (L.) Krocker		Mediterranean basin, Iraq	GP-3	16
M. constricta Durieu		E Mediterranean basin, Iran, Iraq	GP-3	14
M. coronata (L.) Bartal.		E Mediterranean basin, Iran, Iraq	GP-3	16
M. crassipes (Boiss.) E. Small		Anatolia, Lebanon, N Iraq, W Iran	GP-3	16
M. cretacea M. Bieb.		Russia, Former USSR	GP-3	16
M. disciformis DC.		Mediterranean basin	GP-3	16
M. doliata Carmign.		Italy, Spain, Algeria, Morocco	GP-3	16
M. fischeriana (Ser.) Trautv.		Turkey, Iraq, Iran, European USSR	GP-3	16
M. granadensis Willd.		Egypt, Israel, Syria, Turkey	GP-3	16
M. halophila (Boiss.) E Small		S Anatolia	GP-3	16
M. heldreichii (Boiss.) E. Small		Turkey	GP-3	16
M. heyniana Greuter		Greece	GP-3	16
M. huberi E. Small		SW Anatolia	GP-3	16
M. hypogaea E. Small		SE Mediterranean	GP-3	16
M. intertexta (L.) Miller		W. Mediterranean basin	GP-3	16
M. isthmocarpa (Boiss. & Bal.) E. Small		C Anatolia	GP-3	16
M. italica (Miller) Fiori		Mediterranean basin	GP-3	16
M. laciniata (L.) Miller		N Africa, Arabian peninsula, India, Pakistan, Afghanistan	GP-3	16
M. lanigera Winkler & B. Fedtsch.		Afghanistan, Turkmenistan, Tajikistan	GP-3	16
M. laxispira		Heyn. Iraq	GP-3	16
M. lesinsii E. Small		Mediterranean basin	GP-3	16
M. littoralis Rohde ex Lois		Mediterranean basin, E Europe, Caucasus	GP-3	16
M. lupulina L.		Europe, N Africa, Middle East, Asia	GP-3	16, 32

Table 2.1   Genus *Medicago*[a] with Geographic Distribution, Gene Pool Designation, and Chromosome Number (*Continued*)

Species	Subspecies	Geographic Distribution	Gene Pool	Chrom. No.
M. medicaginoides (Retz.) E. Small		SE Europe, SW USSR, Turkey, Iran, S and C Asia	GP-3	16
M. minima (L.) Bartal.		Europe, N Africa, India, Russia	GP-3	16
M. monantha (C. V. Meyer) Trautv.		Middle East, S and C Asia	GP-3	16
M. monspeliaca (L.) Trautz.		W Europe, N Africa, E Mediterranean, Jordan, Iraq, Iran, USSR, C Asia	GP-3	16
M. murex Willd.		Mediterranean basin	GP-3	14
M. muricoleptis Tineo		France, Italy	GP-3	16
M. nöana Boiss.		Iraq, Turkey	GP-3	16
M. orbicularis (L.) Bartal.		Mediterranean basin, Middle East, C Asia	GP-3	16
M. orthoceras (Kar. & Kir.) Trautv.		Middle East, S and C Asia	GP-3	16
M. ovalis (Boiss.) Sirjaev		S Spain, Morocco	GP-3	16
M. pamphylica (Huber-Mor. & Sirjaev) E. Small		S Anatolia	GP-3	16
M. persica (Boiss.) E. Small		Iran, Iraq	GP-3	16
M. phrygia (Boiss. & Bal.)		Turkey, Syria, Iraq, Iran	GP-3	16
M. plicata (Boiss.) Sirj.		Turkey	GP-3	16
M. polyceratia (L.) Trautv.		W Mediterranean	GP-3	16
M. polymorpha L.		Europe, N Africa, Middle East, Crimea, Caucasus, C Asia	GP-3	14
M. popovii Sirj.		C Asia	GP-3	16
M. praecox DC.		S and E Europe, Cyprus, Turkey, Crimea	GP-3	14
M. radiata L.		Middle East, Russia, C Asia	GP-3	16
M. retrorsa (Boiss.) E. Small		Afghanistan	GP-3	16
M. rhytidiocarpa (Boiss. & Bal.) E. Small		S Anatolia	GP-3	16
M. rigida (Boiss. & Bal.)		S Anatolia	GP-3	16
M. rigidula (L.) All.		Mediterranean basin, E Europe, Caucasus, C Asia	GP-3	14
M. riguloides Small.		E Mediterranean, Middle East	GP-3	14
M. rostrata (Boiss. & Bal.)		Turkey	GP-3	16
M. rotata Boiss.		Cyprus, Iraq, Israel, Jordan, Lebanon, Syria, Turkey	GP-3	16

Continued

Table 2.1   Genus *Medicago*[a] with Geographic Distribution, Gene Pool Designation, and Chromosome Number (*Continued*)

Species	Subspecies	Geographic Distribution	Gene Pool	Chrom. No.
M. rugosa Desr.		Mediterranean basin	GP-3	30
M. sauvagei Negre		Morocco	GP-3	16
M. scutellata (L.) Miller		Mediterranean basin, S Ukraine, Crimea	GP-3	30
M. shepardii Post.		Ex Boiss. Turkey	GP-3	16
M. soleirolii Duby		France, Italy, Algeria, Tunisia	GP-3	16
M. strasseri Greuter, Matthas & Risse		Crete	GP-3	16
M. tenoreana Ser.		France, Italy, Yugoslavia	GP-3	16
M. truncatula Gaertner		Mediterranean basin, E Europe, Caucasus	GP-3	16
M. turbinata (L.) All.		Mediterranean basin	GP-3	16

[a] According to Small, E. and M. Jomphe. 1989. *Canadian Journal of Botany* 67:3260–3294.

2.2 DOMESTICATION AND DISSEMINATION OF ALFALFA

Alfalfa is recognized as the oldest plant grown solely for forage (Bolton 1962). Alfalfa (purple flowered) originated from the Near East and central Asia (Asia Minor, Transcaucasia, Iran, and Turkistan), with its geographical center in northern Iran (Ivanov 1977). These areas have cold winters and hot, dry summers with soils typically well drained and near neutral in pH with subsoils with high lime content (Sinskaya 1950). A second center of origin as proposed by Sinskaya (1950) is near Turkistan and Central Asia, which practiced irrigation with locations that have moderately cold winters, low humidity, and hot, dry summers. Yellow-flowered *M. sativa* ssp. *falcata,* a subspecies that naturally hybridizes with purple-flowered alfalfa, has significantly contributed to the evolution and eventual domestication of alfalfa. Subspecies *falcata* has a wider distribution than purple flowered alfalfa and possesses winter hardiness, drought tolerance, creeping root habit (rhizomes), and important disease resistance (Barnes et al. 1977; Barnes, Goplen, and Baylor 1988).

The oldest known reference to alfalfa is from Turkey as early as 1300 B.C. The value of alfalfa as a feed for horses is described in Greek literature between 440 and 322 B.C., when it was introduced into Greece by armies returning from battle in the Middle East (Michaud et al. 1988). The Romans acquired alfalfa from the Greeks and spread the crop throughout the Roman Empire. The Romans are credited with being the fathers of forage culture because of their advanced knowledge of seeding rates, harvest schedules, fertility, soil requirements, and other aspects of growth and culture. About the time alfalfa arrived in Italy, the Chinese sent out an expedition along the Silk Highway to obtain the highly prized Iranian horses, which came with alfalfa forage to support their newly obtained horses. Thus, by the time of the Christian era, alfalfa was spread throughout Europe and into China.

Natural hybrids of purple and yellow forms of *M. sativa* spread into more northern latitudes and eventually spread from its center of origin into Europe, North Africa, and Arabia (Michaud, Lehman, and Rumbaugh 1988). Although alfalfa was introduced into Spain by the Romans, the Moorish invasions into Spain from Africa reintroduced alfalfa into Western Europe. After the fall of the Roman Empire, during the Dark Ages the use of alfalfa virtually disappeared in Europe. Alfalfa was reintroduced into Europe during the sixteenth century. Alfalfa was distributed to the New World into South America in the eighteenth century by invading armies, explorers, and missionaries as

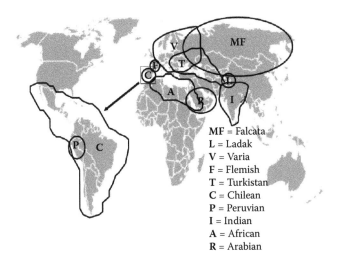

MF = Falcata
L = Ladak
V = Varia
F = Flemish
T = Turkistan
C = Chilean
P = Peruvian
I = Indian
A = African
R = Arabian

Figure 2.1 World map with the locations of the historical alfalfa germplasm sources used in North American alfalfa cultivars.

feed for horses and other livestock. In 1736, European colonists brought alfalfa to the eastern United States as they immigrated to the New World. These introductions generally were not successful, except for those planted on well-drained limestone soils. Alfalfa was well suited to the dry climates and irrigated soils of the western United States, where it was introduced from Mexico by Spanish missionaries as early as the 1830s. Alfalfa eventually spread eastward to the intermountain region and the southern Great Plains. Movement into areas with severe winters was limited by the lack of winter hardiness in the primarily Spanish-derived (nondormant) germplasm. The introduction of four winter-hardy (dormant) types (cv. Grimm, cv. Ontario Variegated, cv. Baltic, and cv. Cossack) from northern Europe to the north central states between 1858 and 1910 allowed successful alfalfa culture in the colder and more humid areas of the midwestern and northeastern United States (Barnes et al. 1977, 1988). An excellent review of the history of alfalfa has been written by Russelle (2001).

Barnes et al. (1977) identified nine distinct germplasm sources introduced into the United States from different regions of the world in the early 1900s (Figure 2.1). These germplasm sources have been incorporated into the development of modern alfalfa varieties. They are described in descending order from the most winter hardy (most fall dormant) to the least winter hardy (least fall dormant): 'Falcata,' 'Varia,' 'Ladak,' 'Turkistan,' 'Flemish,' 'Chilean,' 'Peruvian,' 'Indian,' and 'African.' Since this time, additional germplasm sources have been recognized for the very nondormant characteristic found in the Arabian peninsula (Smith, Al-Doss, and Warburton 1991) (Figure 2.1).

Before 1925, most alfalfa breeding efforts in North America were directed toward selecting strains that were more winter hardy. During the next 30 years, emphasis was placed on developing cultivars that combined winter hardiness and resistance to bacterial wilt. During the late 1950s, the emphasis was placed on developing cultivars resistant to other diseases and several insect pests (Barnes et al. 1988). Beginning in the 1950s, emphasis was put on breeding alfalfa with multiple pest resistance through recurrent-selection breeding schemes. Most modern varieties of commercial alfalfa have moderate to high levels of resistance to major diseases and insect and nematode pests. Scientists of the North American Alfalfa Improvement Conference have developed standardized test procedures for screening alfalfa for its major diseases, insects, and nematodes (http://naaic.org/stdtests/index.html).

Recent improvements of alfalfa in the 1990s and 2000s have focused on improved quality traits; new uses, such as grazing tolerance, resistance to a major insect pest (the potato leafhopper), and, through the use of genetic engineering technologies, herbicide resistance (Roundup Ready);

and improved forage quality through the introduction of genes for reduced lignin content, which improves fiber digestibility and the efficiency of protein utilization.

2.3 BOTANY

The taxonomy of alfalfa is as follows:

order: Fabales
family: Fabaceae (Leguminosae)
subfamily: Faboideae
tribe: Trifolieae
section: Medicago
genus: *Medicago* L.
species: *sativa*
subspecies: *sativa*

2.3.1 Taxomomy

Alfalfa is a member of the genus *Medicago,* which contains 33 perennial and 66 annual species (Table 2.2).

2.3.2 General Description of Alfalfa

According to Lesins and Lesins (1979), alfalfa plants are perennial, with stems 30–120 cm long; stems are procumbent, ascending to erect and arising from the crown. The root system is primarily at tap root, which can reach 6 m, with effective fibrous secondary branching in the top 2 m. Roots form a symbiotic relationship with *Sinorhizobium meliloti* for nitrogen fixation. Vegetative parts are covered to varying degrees with simple appressed hairs. New cultivars of alfalfa with potato leaf-hopper resistance have glandular hairs. Stipules are entire or toothed in their basal part. The first true leaf arising from the seed epicotyl is usually unifoliolate with an orbicular blade and stipules. The second and subsequent leaves are normally pinnately trifoliolate with slender stipules adnate to the petiole. Multifoliolate forms exist with four to seven leaflets. Leaflets are 8–28 mm in length by 3–15 mm in width, obovate (at lower nodes), cuneate or linear-oblanceolate (at upper nodes), and serrate in their apical part; the midrib ends in a terminal tooth.

Alfalfa plants have 7- to 35-flowered peduncles that are several times longer than the corresponding petiole, with a terminal cusp. Florets are 6–12 mm long, usually in an elongate raceme. The pedicel is equal to or longer than the calyx tube; the bract is plus or minus the length of the tube. The corolla is violet or lavender and rarely pink or white; the standard is twice or more long as wide, with parallel sides in its middle part and wings longer than the keel. Young pods rise from the calyx and then bend sideways; there are appressed simple (rarely glandular) hairs. It is spineless, with one to five coils and turning clockwise, 3–9 mm in rotation. Many veins start from the ventral suture and run obliquely, branching somewhat and then anastomosing in the outer part of the coil face. Seeds are yellow or brownish or greenish-yellow, 1.2–2.5 mm in length by 1.0–1.5 mm in width. Seed weight is 1.0–2.5 g/1,000 seeds. The radicle is slightly more than half the seed length.

2.4 GERMPLASM RESOURCES

The largest collection of alfalfa germplasm exists in the United States in the National Plant Germplasm System Collection of *Medicago* germplasm. This germplasm collection is located at the Western Regional PI Station in Pullman, Washington. The current collection contains a total of

Table 2.2　List of Cultivated *Medicago* Species, Common Names, Usage, and Growth Cycles

Cultivated Species of *Medicago*	Common Name	Use	Growth Cycle
Medicago arabica (L.) Huds.	Spotted bur clover, spotted medic	Animal forage	Annual
Medicago arborea L.	Tree alfalfa, tree medic	Animal forage, ornamental	Perennial
Medicago italica (Mill.) Fiori	Disc medic	Animal forage, soil improvement, companion crop	Annual
Medicago littoralis Rohde ex Loisel.	Strand medic	Animal forage, soil improvement, companion crop	Annual
Medicago lesinsii E. Small		Animal forage, soil improvement, companion crop	Animal
Medicago lupulina L.	Black medic	Animal forage, soil improvement, companion crop	Annual
Medicago minima (L.) Bartal.	Little bur-clover	Animal forage	Annual
Medicago murex Willd. Annual		Animal forage, soil improvement, companion crop	Annual
Medicago orbicularis (L.) Bartal.	Button clover	Animal forage, soil improvement, companion crop	Annual
Medicago polymorpha L.	Toothed bur-clover, toothed medic, California bur-clover	Animal forage, soil improvement, companion crop	Annual, biennial, perennial
Medicago rigidula (L.) All.	Tifton bur-clover, Tifton medic	Animal forage, soil improvement, companion crop	Annual
Medicago rugosa Desr.	Gama medic	Animal forage, soil improvement, companion crop	Annual
Medicago sativa nothosubsp. *varia* (Martyn) Arcang.	Sand lucerne, variegated lucerne	Animal forage	Perennial
Medicago sativa subsp. *falcata* (L.) Arcang.	Yellow-flower alfalfa, sickle alfalfa, yellow Lucerne	Animal forage	Perennial
Medicago sativa subsp. *sativa*	Alfalfa, Lucerne	Animal fodder and forage, soil improvement, human food-sprouts, medicinal tea	Perennial
Medicago scutellata (L.) Mill.	Snail medic	Animal forage, soil improvement, companion crop	Annual
Medicago truncatula Gaertn.	Barrel medic	Animal forage, soil improvement, companion crop	Annual

8,160 accessions, with 4,151 *M. sativa* accessions and 4,009 accessions representing other perennial and annual species (http://sun.ars-grin.gov/npgs/). Montpellier, France, has a large collection also, with 2,633 *M. sativa* accessions (http://www.ecpgr.cgiar.org/databases/Crops/Medicago.htm). Plant Genetics Resources of Canada holds 1,620 accessions of *M. sativa* (http://pgrc3.agr.ca/).

2.4.1 Collection Preservation

The U.S. *Medicago* collection contains original seed packets and regeneration seed lots kept at 4 or 5°C with 28% relative humidity. Seed lots that are distributed are kept in the same conditions. Seed counts have been obtained on all packets except for original seed. Approximately 84% of the accessions in the collection have a safety backup sample stored at the National Seed Storage Lab (NSSL), Fort Collins, Colorado. Accessions not backed up are those with very few seeds or that are currently queued for regeneration.

From 1979 to 1997, the majority of perennial accessions were regenerated using the original or the oldest existing seed lot, under isolation in cages using up to 200 plants per population (this number varies with amount of seed available). From 1987 to 1998, most of the annual medic species were increased at Riverside, California (Bauchan and Greene 2002). A small amount of seed can be obtained through the U.S. National Germplasm System (http://www.ars-grin.gov/npgs).

2.4.2 Evaluation of the Collection

Since 1981, approximately one third of the perennial *Medicago* collection has been evaluated for (Bauchan and Greene, 2002)

diseases (anthracnose, aphanomyces, bacterial wilt, downey mildew, fusarium, common leaf spot, lepto leaf spot, phytophthera root rot, rhizoctonia, spring black stem, stemphyllium leaf spot, and verticillium wilt);

insects (blue aphid, lygus bug, pea aphid, potato leafhopper, root curcullio, snout beetle, spotted alfalfa aphid, and white fly);

chromosome number;

agronomic traits (crown stem diameter, crown depth, crown origin, crown budding, days to maturity, determinate tap root percentage, determinate tap root position, fall growth, root number, seedling vigor, seedling year productivity, seed production seed weight, stem–leaf ratio, tap root, unifoliate internode length, grazing tolerance, fall dormancy, winter survival, multifoliolate leaf expression);

feed quality traits (acid detergent fiber of leaves, acid detergent fiber of stems, crude protein of leaves, crude protein of stems, neutral detergent fiber of leaves, neutral detergent fiber of stems, and bypass protein); and

abiotic stress tolerant traits (acid-soil tolerance, frost damage, germination of seedlings in salt, tolerance to salt tolerance, and winter injury).

All of the disease resistance, insect resistance, feed quality, and stress tolerance evaluations were conducted using standardized tests developed by research scientists in the federal government, universities, and private industry and approved by the members of the North American Alfalfa Improvement Conference (1999). These standardized tests are used by the alfalfa seed industry to evaluate and characterize new alfalfa varieties. Scientists in the federal government, universities, and private industry conducted the tests and provided the data, which were imported into the Germplasm Resources Information Network (GRIN) (http://www.ars-grin.gov/npgs).

2.4.3 Core Subsets

A core subset is a small subset of accessions from the entire collection that (1) contains most of the genetic variability that exists in the overall collection; (2) identifies duplications in the collection; (3) recognizes where there is a lack of germplasm representing a species or geographic location; (4) simplifies evaluation, especially for difficult and/or expensive traits; and (5) increases the utilization of the collection (Brown 1989; Diwan, McIntosh, and Bauchan 1995). Thus, if researchers are looking for a trait of interest or are testing a new trait and cannot screen the entire collection, the core subsets are a good place to begin the search. There are currently two core subsets related to *Medicago*: one for the perennial and one for the annual *Medicago* species.

2.4.3.1 *Perennial* Medicago *Core Subset*

Basigalup, Barnes, and Stucker (1995) developed a perennial *Medicago* core subset in Minnesota from 1989 through 1991. The core subset was selected from 1,105 accessions representing 47 different countries and four species (*M. sativa* including the six subspecies [ssp. *sativa*, ssp. *varia*, ssp. *falcata*, ssp. *coerulea*, ssp. *falcata*, and ssp. *ambigua*]), *M. cancellata*, *M. platycarpa*, and *M. ruthenica*. The 1,105 accessions were assembled into 18 different groups based on the original collection site and their geographic proximity (i.e., one group was accessions collected in Hungary, Greece, Italy, and Romania). These accessions were evaluated in Rosemount, Minnesota, in 1989 for 21 morphological traits and eight quality traits. Additional evaluation data on these accessions were obtained from the GRIN system (http://www.ars-grin.gov/npgs), including information on the resistance to 10 diseases and five insects and salt and acid tolerance. Selection of the core was made by using cluster analysis to designate the most diverse accessions within each geographic proximity group. The core subset has 200 accessions (~18% of the evaluated accessions) with representatives from each species and geographic area.

2.4.3.2 *Annual* Medicago *Core Subset*

Between 1990 and 1992 the annual *Medicago* core subset was developed in Maryland (Diwan, Bauchan, and McIntosh 1994). The medic core subset was selected from 1,220 accessions representing 34 annual *Medicago* species. These accessions were evaluated using 15 morphological and agronomic traits (days to flower, days to full pod, biomass production within a species, biomass production among species, plant height, plant width [spread], growth habit, middle leaflet length, middle leaflet width, third internode length, pod production, pod spinyness, number of flowers/raceme, number of pods/raceme, and seed size). All of these data are available in the GRIN system (http://www.ars-grin.gov/npgs). The selection for the core was made using cluster analysis; the final selection of core accessions within each cluster was based on geographic location. The annual medic core subset has 211 accessions (~17% of the evaluated accessions), and 34 annual *Medicago* species are represented (Diwan et al. 1995). The core collection was evaluated in six locations across the United States (Athens, GA; Beltsville, MD; Ithaca, NY; Logan, UT; St. Paul, MN; and Tucson, AZ) for 11 traits (days to flower, days to full pod stage, full pod stage, growth habit, biomass within species, biomass among species, pod production, pod spinyness, plant height, plant width, and winter hardiness). All of these data are available in the GRIN system (http://www.ars-grin.gov/npgs).

The South Australian *Medicago* Genetics Resource Center in Adelaide, South Australia, holds a very large (more than 22,000 accessions) collection of annual *Medicago* species. The collection has been evaluated for up to 27 different agronomic traits, which have been assembled into a database. Skinner et al. (1999) developed a core subset of this large collection using the evaluation data available and Euclidean distances to identify the low, high, and middle ranges of each trait, which, when combined with geographic data, yielded a sampling of the variation expressed for each character. The core collection consists of 1,705 accessions for all the species represented.

2.5 GENE POOLS OF ALFALFA

Medicago species gene pools can be grouped into three classes according to Harlan and de Wet (1971). The primary gene pool (GP-1) consists of species that can be crossed to produce vigorous hybrids that exhibit normal meiotic chromosome pairing and possess total seed fertility. The secondary gene pool (GP-2) consists of species that, when crossed to alfalfa or species within GP-1, exhibit at least some seed fertility but may require tissue culture methods to rescue hybrid plants. The hybrid plants typically show some infertility problems. The tertiary gene pool (GP-3) contains species that, when crossed with alfalfa using extreme techniques, will produce plants that are completely sterile and lethal genes are expressed. These species are at the outer limits of hybridization with alfalfa.

2.5.1 Primary Gene Pool

The primary gene pool of alfalfa consists of cultivated, naturalized, and wild forms of *Medicago sativa* ssp. *sativa,* including modern and obsolete cultivars, landraces, and ecotypes. In addition, *M. sativa* ssp. *falcata, M. sativa* nssp. *varia, M. sativa* ssp. *glutinosa, M. sativa* nssp. *tunetana, M. sativa* ssp. *coerulea* ($2n = 2x = 16$), and *M. sativa* ssp. *glomerata* can be easily crossed with *M. sativa* ssp. *sativa* with normal bivalent pairing with an occasional quadrivalent configuration typical of an autotetrapoid. These species constitute the *Medicago sativa* complex (Quiros and Bauchan 1988). Utilization of hybridization between various ploidy levels—especially diploid ($2n = 16$) and tetraploid ($2n = 4x = 32$) plants—has been a useful mechanism for hybridization between these species (Cleveland and Stanford 1959; Bingham and Saunders 1974).

2.5.2 Secondary Gene Pool

Several perennial *Medicago* species cannot be hybridized with *M. sativa* without the aid of ovule-embryo culture rescue (Quiros and Bauchan 1988). McCoy (1983) and McCoy and Smith (1986) developed a method that allowed for the recovery of hybrids between alfalfa and *M. cancellata, M. daghestanica, M. dzhawakhetica, M. hybrida, M. marina, M. papillosa, M. rupestris, M. rhodopea,* and *M. saxatilis.* Some hybridizations required $2x$–$4x$ crosses in order to recover viable offspring (McCoy and Smith 1986).

2.5.3 Tertiary Gene Pool

Thirty-six species of annual *Medicago* species are closely related to alfalfa; however, there are only a few reports of hybrids between alfalfa and these annual species. All of the reported successful hybrids have utilized the tetraploid species *M. scutellata* (Sangduen, Sorensen, and Liang 1982) and *M. rugosa* (Arcioni et al. 1993). Chromosome imbalance is the suspected reason for infertility because both have 30 chromosomes rather than the 32 found in *M. sativa* (Bauchan and Elgin 1984). No hybrids have been found to be fertile and produce progeny. Recently, protoplast fusion has been used to introduce portions of the annual *Medicago* species genome with limited success (Mizukami et al. 2006; Tian et al. 2002)

2.6 CYTOGENETICS

Cultivated alfalfa is an autotetrapoid (Stanford 1951) ($2n = 4x = 32$) (Figure 2.2); however, a majority of the species in the genus *Medicago* are diploid ($2n = 2x = 16$) (Table 2.2). *Medicago sativa* ssp. *coerulea* is proposed to be the diploid ($2n = 2x = 16$) (Figure 2.3) progenitor of *M. sativa* (Lesins

Figure 2.2 Tetraploid (*2n* = *4x* = 32) somatic chromosomes of alfalfa (*M. sativa* ssp. *sativa*).

and Lesins 1979). Brummer, Cazarro, and Luth (1999) utilized flow cytometry to determine that *M. sativa* ssp. *falcata* can be found as either diploid or tetraploid (Table 2.2) in the U.S. *Medicago* collection. Bingham and Saunders (1974) were able to scale *M. sativa* from diploid all the way up to octoploid (*2n* = *8x* = 64). The basic genome number for the genus *Medicago* is *x* = 8, although six species (*M. constricta, M. lesinsii, M. polymorpha, M. praecox, M. rigidula,* and *M. riguloides*) are diploids with *2n* = *2x* = 14. Utilizing pachytene analysis, Lesins et al. (1974) determined that these 14 chromosome species originated from chromosome translocation in which two chromosomes joined and the centromere of one of the chromosomes was lost. The appearance of an exceptionally long chromosome in each of these species is noticeable in karyotypes (Lesins and Lesins 1979).

In addition to *M. sativa* subspecies, perennial species of *Medicago* that are tetraploid with *2n* = *4x* = 32 include *M. arborea, M. dzhawakhetica, M. papillosa,* and *M. prostrata* (Table 2.2). In addition, there are two hexaploid perennial species: *M. cancellata* and *M. saxatilis. Medicago arborea* accessions have been found to have both tetraploid and hexaploid accessions. *Medicago cancellata* and *M. saxatilis* are thought to be alloautohexaploids with two genomes from sativa and four from *M. rhodopea* (*M. saxatilis*) or *M. rupestris* (*M. cancellata*) (Lesins and Gillies 1972; Lesins and Lesins, 1979).

Figure 2.3 Diploid (*2n* = *2x* = 16) somatic chromosome of alfalfa (*M. sativa* ssp. *coerulea*).

Two species of annual *Medicago* species are tetraploid: *M. rugosa* and *M. scutellata* with $2n =$ $4x = 30$ (Table 2.2). The nucleolus organizer region (NOR) chromosomes of *M. scutellata* have two pairs of NOR chromosomes and *M. rugosa* has only one pair. The two species could have arisen by hybridization of a $2n = 14$ and a $2n = 16$ species, followed by polyploidization or, in the case of *M. rugosa,* the loss of two NOR chromosomes (Bauchan and Elgin 1984). Phenolic compounds and restriction fragment length polymorphism (RFLP) molecular markers have revealed close relationships between these tetraploid annual *Medicago* species and diploid species with both $2n = 16$ and $2n = 14$ (Classen et al. 1982; Simon 1976; Mariani, Pupilli, and Calderini 1996).

Comprehensive chapters have been written on the cytogenetics of alfalfa (see Lesins and Gillies 1972; Stanford, Clement, and Bingham 1972; Lesins and Lesins 1979; McCoy and Bingham 1988, 1991; McCoy and Echt 1992). The most recent progress for *M. sativa* can be characterized by the advances in the use of computerized image analysis, chromosome banding, and *in situ* hybridization.

Bauchan and Campbell (1994) developed and utilized a computerized image analysis system to critically measure, analyze, and construct the karyotype of diploid alfalfa. Since that time, the development of faster computers (20–600 Mhz) with larger memory (512 kB–128 MB RAM) has allowed for increased resolution (1,048 × 960 pixels to 2,096 × 1,920 pixels) and direct connections with Windows-based software such as spreadsheets for data analysis and presentation programs for production of pictures for analysis and publication. These advances have made it possible for even more precise karyotypic investigations, and several cells per hour can be analyzed rather than one or two per day (Bauchan and Hossain 2001a) (Figure 2.4).

Johnson et al. (1984) and McCoy and Bingham (1988) stated that their attempts to band the chromosomes of *Medicago* were either unsuccessful or produced bands at the centromeres only and thus this technique did not aid in the advancement of *Medicago* cytogenetics. Masoud, Gill, and Johnson (1991) showed that the chromosomes of diploid *M. sativa* ssp. *sativa* (L.) L. & L. cv.

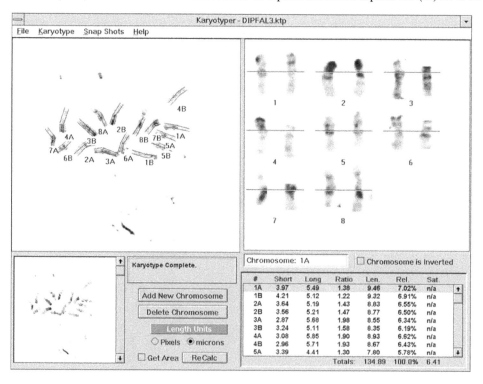

Figure 2.4   Image analysis using Karyotyper software program to karyotype C-banded diploid alfalfa.

CADL (cultivated alfalfa at the diploid level) (Bingham and McCoy 1979) had additional bands. Bauchan and Hossain (1997) perfected the C-banding technique and extensively applied the technique to prove that there are additional bands; they developed a standardized karyotype for diploid ssp. *falcata* and ssp. *coerulea*. Bauchan and Hossain (1998a) were the first to develop N-banding for the genus *Medicago*, and this was the first successful utilization of N-banding to identify individual chromosomes of a dicot plant. Falistocco, Falcinelli, and Veronesi (1995) published a karyotype of tetraploid *M. sativa* ssp. *sativa* showing that the individual chromosomes could be identified; they added additional proof that alfalfa is an autotetraploid.

These approaches have advanced the science of alfalfa cytogenetics and have led to a number of new discoveries about the molecular cytogenetics of the genus *Medicago*, particularly the species in the *Medicago sativa* complex.

2.6.1 *Medicago sativa* ssp. *coerulea*

Bolton and Greenshields (1950) were the first to publish a photomicrograph of the somatic chromosomes of diploid *M. sativa* ssp. *coerulea*. At that time they were unable to distinguish the chromosomes; however, they did recognize two chromosomes possessing satellites (or the Nuclear Organizer Region; (NOR). The first karyotype was published by Buss and Cleveland (1968), and additional karyotypes have been developed by Agarwal and Gupta (1983), Bauchan and Campbell (1994), and Bauchan and Hossain (2001a). A majority of the karyotypes of diploid *M. sativa* ssp. *coerulea* have been based on meiotic pachytene chromosomes (Buss and Cleveland 1968; Clement and Stanford 1963; Gillies 1968; Gillies and Bingham 1971; Ho and Kasha 1972). Kasha et al. (1970) developed a standardized pachytene karyotype of alfalfa that reconciled the differences found in the previous pachytene studies.

The somatic chromosome karyotype of this subspecies consists of one pair of chromosomes (chromosome 8), four pairs of submetacentric chromosomes (chromosomes 1–4), and three pairs of short metacentric chromosomes (chromosomes 5–7). Chromosome 8 possesses the NOR of the alfalfa genome; the NOR chromosomes are a diagnostic feature of the karyotype. The large submetacentric chromosome pair 1 is easy to distinguish if the chromosomes have not spiralized too much during pretreatment. Occasionally, a tertiary constriction can be found on chromosome 4. Using a computerized image analysis system, it is possible to determine the homologous chromosomes and develop a karyotype based on chromosome arm lengths, arm ratios, and total chromosome length (Bauchan and Campbell 1994; Bauchan and Hossain 2001a).

The C- (Bauchan and Hossain 1997) and N- (Bauchan and Hossain 1998a) banding patterns of *M. sativa* ssp. *coerulea* displayed several more bands than diploid ssp. *falcata;* a majority of the bands were located on their short arms. The location and intensity of the bands in each chromosome are unique, which allows for the precise identification of individual chromosomes for karyotypic studies. Generally, in addition to centromeric bands, which are the most prominent of the bands, the standardized karyotype of *M. sativa* ssp. *coerulea* exhibits telomeric bands in all the chromosomes' short arms. All of the chromosomes except chromosome 7 have interstitial bands in their short arms, and chromosomes 1, 2, and 3 each have bands on their long arms. N-banding reveals an additional interstitial band on the long arm of chromosome 5. The NOR chromosome is characterized as having a centromeric band, a prominent NOR band, and a telomeric and interstitial band on the long arm (Bauchan and Hossain 1997, 1998a). Masoud et al. (1991) reported the first C-banded karyotype of diploid *M. sativa* ssp. *sativa* cv. CADL (cultivated alfalfa at the diploid level). However, they observed mostly centromeric and telomeric bands and only a few interstitial bands.

Comparison of the C- and N-banding patterns reveals somewhat corresponding banding patterns. All of the short arms have terminal bands in both C- and N-banding; however, the intensity of the bands varies from very prominent to faint. The interstitial bands on the short arms also vary in their intensity as well as in their locations (e.g., the C-band on chromosome 1 is located near the middle of the short arm, whereas the N-band is located closer to the terminal end rather than the centromere). A characteristic feature of the N-banding pattern is a very intense, darkly stained inter-stitial band on the long arm of chromosome 3 (Bauchan and Hossain 1998a). N-banded chromo-somes exhibit very poor contrast between the darkly stained heterochromatin and the more lightly stained euchromatin; therefore, C-banding has been much more extensively used than N-banding.

Studies on the polymorphisms in banding patterns in 14 accessions, which represent the wide geographic distribution of the accessions in the U.S. germplasm collection, showed a majority of the plants containing cells identical to the standardized karyotype. However, in half of the acces-sions that exhibited polymorphisms, the heterochromatic DNA bands differed in their intensity, location, and total number of bands among and between accessions. In some accessions, telomeric bands were missing, and in others, bands were missing on only one of the chromosomes in a pair. Some plants contain chromosomes with two bands on their long arms. A very large "mega" NOR chromosome was detected (possibly due to a translocation); isochromosomes and primary trisomic and tetrasomic plants have also been observed (Hossain and Bauchan 1999).

Based on the pachytene karyotype of diploid ssp. *falcata,* Gillies (1970a) and Ho and Kasha (1972) concluded that it did not differ significantly from the pachytene karyotype of diploid ssp. *coerulea* (Gillies 1968). However, upon close observation of the pachytene karyotypes, all of the *M. sativa* ssp. *coerulea* chromosomes have heterochromatic bands on the telomere of their short arms, whereas there were no heterochromatic telomeres on the short arms of diploid ssp. *falcata* except when the entire short arm of the chromosome was heterochromatinized. Thus, it appears that dip-loid ssp. *falcata* contains a lower amount of heterochromatic DNA than ssp. *coerulea.* However, this will depend on the accession that was studied because the amount of heterochromatic DNA varies from accession to accession (Bauchan and Hossain 1999a). C- and N-banding studies of mitotic chromosomes by Bauchan and Hossain (1997, 1998a) have shown conclusively that diploid ssp. *coerulea* possesses more heterochromatic DNA than diploid ssp. *falcata.* Upon a reexamination of the heterochromatic distribution in both mitotic and meiotic chromosomes, these studies agree that diploid ssp. *falcata* contains a lower amount of heterochromatin than ssp. *coerulea.*

2.6.2 Diploid *Medicago sativa* ssp. *falcata*

Utilizing image analysis techniques, it is possible to make critical measurements of chromo-somes. Diploid ssp. *falcata* chromosomes are slightly shorter and average 1.35–2.3 μm in length (Bauchan unpublished) than ssp. *coerulea* chromosomes, which are an average of 1.50–2.4 μm in length (Bauchan and Campbell 1994). The standard C- (Bauchan and Hossain 1997) and N- (Bauchan and Hossain 1998a) banding patterns can be characterized as having bands only at the centromeric regions and a large band at the NORs of the satellited chromosome 8. Occasionally, an interstitial band occurs in the long arm of the NOR chromosome.

A comparison of the mitotic banding pattern to meiotic pachytene chromosomes (Gillies 1970a) shows a relatively good correlation between mitotic heterochromatic bands and the chromatic knobs (chromomeres) found at pachytene. Gillies (1970a) and Ho and Kasha (1972) found that a majority of the chromomeres at pachytene in *M. sativa* ssp. *falcata* were located on either side of the cen-tromeres and there were no prominent telomeric knobs except when an entire arm was heterochro-matic. In a majority of the accessions studied using C-banding of somatic chromosomes, no bands were found on the short arms of chromosomes.

However, in a survey of 17 accessions of diploid ssp. *falcata* representing the wide geographic distribution of the subspecies, Bauchan and Hossain (1999a) discovered that 10 accessions possessed 1–14 additional C-bands when compared to the standard karyotype for ssp. *falcata.* Polymorphisms for the number, locations, and intensities of C-bands were detected among and between acces-sions. Only 9% of the plants studied contained displayed polymorphisms. C-bands may exist at the terminal ends of all the chromosomes and interstitial bands were detected on both the short arms and the long arms of chromosomes 1, 2, 3, and 6. These additional bands could have preexisted, or they could be the result of crossing over from hybridization with other subspecies in the *Medicago sativa* complex. The absence of diagnostic bands on the chromosomes makes it difficult to analyze the karyotype of this subspecies critically; however, with the aid of a computerized image analysis system (Bauchan and Campbell 1994; Bauchan and Hossain 2001a), it is possible to develop karyotypes based on chromosome morphology measurements.

Chromosome modifications such as isochromsomes for the short arms of chromosomes 2 and 6 were also observed (Bauchan and Hossain 1999a). Isochromosomes are monocentric chromosomes with homologous arms that are mirror images of one another (Reiger, Michaelis, and Green 1991). Isochromosomes have the potential of determining the dosage effect of genes. No isochromosomes or telosomes have been observed for the long arm of any of the species studied thus far in the *Medicago sativa* complex. Three individual seedlings from two different accessions were found to exhibit B-chromosomes (Hossain and Bauchan 1999). B-chromosomes are usually smaller than the normal A-chromosomes and are generally heterochromatic. Normally, they do not influence the viability and phenotype of the organism, do not pair with the A-chromosomes, and affect mitotic behavior by lagging and elimination, polymitosis, or preferential distribution (Reiger et al. 1991). Because *Medicago sativa* spp. are allogamous, it is presumed that polymorphisms might exist within its genome, which could be detected by banding and molecular techniques.

2.6.3 Tetraploid *Medicago sativa* ssp. *falcata*

Preliminary studies of six accessions of tetraploid ssp. *falcata* offer a surprising result. Most of the plants possess chromosomes that have C-bands in addition to the normal centromeric bands. There is a range from an accession that contains the fewest number of additional bands; four pairs of chromosomes had an extra telomeric band on their short arms, whereas the rest of the chromosomes had only centromeric bands. Two accessions had multiple bands on each chromosome and a banding pattern similar to doubled diploid ssp. *coerulea*. In general, a majority of the heterochromatic bands appear in the short arms of the chromosomes, with only two chromosomes possessing interstitial bands on their long arms (Bauchan and Hossain unpublished). All of the accessions studied had yellow flowers with sickle-shaped pods. However, additional studies need to be conducted to prove that these accessions are not the product of hybridization with *M. sativa* ssp. *sativa*.

2.6.4 Tetraploid *Medicago sativa* ssp. *sativa*

Analysis of somatic chromosomes of tetraploid alfalfa has been previously reported (Agarwal and Gupta 1983; Falistocco 1987; Schlarbaum, Johnson, and Stuteville 1988, Falistocco et al. 1995). Agarwal and Gupta (1983) karyotyped several *Medicago* species with chromosome measurements made using an ocular micrometer; thus, the accuracy of the measurements is questionable. Falistocco (1987) measured chromosomes from photomicrographs, resulting in much larger chromosome measurements (e.g., the total length ranged between 9 and 12 μm) than what has been reported for the genus *Medicago*. Schlarbaum et al. (1988) karyotyped tetraploid alfalfa from plants that had been regenerated from tissue culture. Generally it is known that plants regenerated from a tissue culture system potentially can have chromosomes that have been altered (Nagarajan and Walton 1987).

In studies of the nine germplasm sources of alfalfa in the United States (Bauchan and Hossain 1998b, 2001b; Bauchan, Campbell, and Hossain 2002, 2003), C-banding polymorphisms were detected in the number, position, and intensity of terminal and interstitial bands within a germplasm source. Bauchan and Hossain (2001b) proposed the use of their karyotype of the 'African'-type alfalfa as a standard karyotype (Figure 2.5) due to their observation that these accessions possessed the largest number of hetrochromatic bands; thus, it was easier to compare this karyotype to all others. Morphometric measurements and C-banding studies of the nine germplasm sources reveal that alfalfa has four nearly identical sets of chromosomes; this allows for the development of an accurate

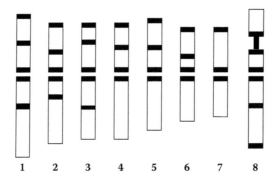

Figure 2.5   Idiogram of the C-banding standard karyotype of alfalfa.

karyotype (Falistocco et al. 1995; Bauchan and Hossain 1998b, 2001b; Bauchin et al. 2002, 2003). The similarity of the chromosome morphology and the C-banding pattern among homologous is not always perfect; however, enough resemblance between the chromosomes makes it possible to group alfalfa chromosomes into eight sets of four chromosomes.

The mitotic banding patterns of tetraploid *M. sativa* ssp. *sativa* (Figure 2.6) resemble the distribution of the heterochromatic knobs found in meiotic pachytene chromosomes (Gillies 1970b). However, the mitotic chromosomes in the tetraploid did not appear to be shorter and did not appear to possess a larger amount of heterochromatin than the diploid subspecies *M. sativa* ssp. *coerulea* observed in pachytene chromosomes (Gillies 1970b).

A wide range of differences was observed among the nine germplasm sources. The *falcata* germplasm source is strikingly different from the other germplasm sources because it has fewer terminal and interstitial bands. *Falcata* chromosomes have primarily C-bands at their centromeres. Nondormant alfalfa germplasm sources, which include 'African,' 'Chilean,' 'Peruvian,' and 'Indian' germplasm, exhibit the largest number of heterochromatic bands (Bauchan et al. 2003). The 'African' chromosomes possess the highest amount of heterochromatic DNA and resemble the banding pattern of a presumably doubled diploid *M. sativa* ssp. *coerulea* (Bauchan and Hossain 2001b). The banding patterns of 'Chilean' (Bauchan et al. 2002) and 'Peruvian' chromosomes are similar. Indian chromosomes have the fewest number of bands when compared to the other nondormant sources, especially on the long arms of their chromosomes (Bauchan et al. 2003). The ranking of germplasm sources from highest amount of heterochromatic DNA C-bands to lowest number of heterochromatic DNA C-bands studied thus far is as follows: 'African' > 'Peruvian' > 'Chilean' > 'Flemish' > 'Indian' > 'Varia' > 'Ladak' > 'Turkistan' > 'Falcata.' It is interesting to note that the

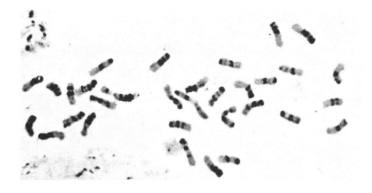

Figure 2.6   C-banded chromosomes of tetraploid alfalfa.

more winter-hardy germplasm sources have fewer C-bands, indicating the incorporation of Falcata germplasm in alfalfa (Bauchan and Hossain 1998b).

The tetraploid C-banded karyotype by Falistocco et al. (1995) differs from the karyotype of 'African' primarily in the reduced number of bands reported. A number of reasons could explain the differences. First, they used different pretreatments of the chromosomes. Our experience has shown that an ice-water pretreatment for 18 h gives the maximum number of bands and thus the ability to distinguish the individual chromosomes from each other more critically, based on the larger number of landmark bands. Two bands located next to each other may appear as a single band if the chromosomes are too contracted, thus reducing the number of detectable bands. Second, they used the Italian variety Turrena, which may have some *M. sativa* ssp. *falcata* germplasm in its background. Thus, this variety may have a reduced amount of constitutive heterochromatic DNA compared to the 'African' germplasm source.

2.6.5 Annual *Medicago* Species

The annual *Medicago* species, sometimes referred to as medics, are close relatives to alfalfa and contain a number of agronomic traits, such as insect resistance due to glandular hairs on the stems, leaves, and pods, which could be beneficial to alfalfa. The medics have not been well studied cytogenetically. The first published photomicrographs of annual *Medicago* chromosomes were presented by Lesins and Lesins (1961). Both Heyn (1963) and Lesins and Lesins (1979) provide an overview of the cytogenetics of the annual *Medicago* species. Since that time, Schlarbaum, Small, and Johnson (1984), Mariani and Falistocco (1991), and Mariani et al. (1996) have studied the $2n = 16$ species *M. arabica, M. blancheana, M. ciliaris, M. doliata, M. granadensis, M. intertexta, M. muricoleptis, M. nöana,* and *M. rotata* by developing karyotypes based on physical measurements of the chromosomes from photomicrographs.

Bauchan and Elgin (1984) discovered that the two tetraploid species *M. scutellata* and *M. rugosa* have $2n = 30$ chromosomes rather than the recorded 32. Because *M. scutellata* possesses two pairs of NOR chromosomes, it was suggested that this species was an allotetraploid derived from the hybridization of a $2n = 16$ and a $2n = 14$ species and then spontaneously chromosome doubled. This began the search for the putative progenitors of these species. Falistocco and Falcinelli (1991) and Mariani et al. (1996) developed karyotypes of the $2n = 14$ species *M. murex, M. polymorpha, M. praecox,* and *M. rigidula.* Based on karyotypic similarity and some RFLP molecular data, Mariani et al. (1996) suggested that four species—*M. intertexta, M. muricoleptis, M. polymorpha,* and *M. murex*—were the most likely candidates for progenitors to *M. scutellata* and *M. rugosa.* Studies on the banding of the annual species have yielded only bands at the centromeres (Mariani and Falistocco 1990, 1991; Falistocco and Falcinelli 1993); thus, this technique appears not to add valuable cytogenetic markers for karyotypic analysis or identification of individual chromosomes.

Medicago truncatula has become a model plant species for molecular genetic studies (Cook et al. 1997; Cook 1999). It has a relatively small genome (1.15 pg) (Blondon et al. 1994) and a diploid autogamous genetic system (Cook et al. 1997), is readily regenerated and transformed *in vitro* (Rose, Nolan, and Bicego 1999), and is an excellent model system for studying N_2 fixation (Barker et al. 1990). A project to sequence the *M. truncatula* genome is under way in which the *M. truncatula* genome will become the second plant to be completely sequenced (http://www.medicago.org). *Medicago truncatula* is closely related taxonomically to alfalfa (Lesins and Lesins 1979). Extensive interspecific homology of known genes or expressed sequence tags (ESTs) (Choi et al., 2004) has been reported for *M. truncatula* and cultivated alfalfa. In addition, Thoquet et al. (2002) have shown that molecular markers mapped in *M. truncatula* have identical or similar map locations in diploid alfalfa. *Medicago truncatula* has become a model plant for molecular studies not only for alfalfa but also for many leguminous plants (Zhu et al. 2005). Choi et al. (2004) concluded from their studies of molecular mapping and cytogenetic mapping (Kulikova et al. 2001) that the genomes of

M. truncatuala to *M. sativa* are extremely similar and proposed a JointMap for *Medicago*. Thus, it appears that many contributions made by the *M. truncatula* genomics community will be applicable to studies of the alfalfa genome.

2.6.6 Cytogenetics and Breeding

Utilization of chromosome number counting and chromosome morphology differences has been used as evidence for the production of interspecific hybrids between perennial and annual *Medicago* species (Sangduen et al. 1982; Piccirilli and Arcioni 1992), hybrids between perennial wild *Medicago* species and alfalfa (summarized by McCoy and Smith 1986), and somatic hybrids produced via cell fusion (Pupilli et al. 1995; Nenz et al. 1996; Mizukami et al. 2006; Tian et al. 2002). The cytogenetic evidence used to determine the production of the hybrids was primarily somatic chromosome numbers, with the number and morphology of NOR chromosomes as key features and chromosome pairing at metaphase I (MI) in microsporocytes. Giemsa C- (Bauchan and Hossain 1997) and N- (Bauchan and Hossain 1998a) banding techniques have been used to determine the hybridity of crosses between diploid *M. sativa* ssp. *falcata* and *M. sativa* ssp. *coerulea* because of the contrasting differences between banding patterns; ssp. *falcata* have bands only at their centromeres and ssp. *coerulea* have multiple bands, so it was easy to identify hybrids between the two species.

It was also easy to identify individual chromosomes of ssp. *coerulea* (Bauchan and Hossain 1997, 1998a). Utilizing banding techniques, it should be possible to discern the exchange of bands in the F_2 and subsequent generations as well as backcross progenies. This same technique could be used to identify interspecific hybrids between ssp. *coerulea*/ssp. *sativa* and annual species because, so far, all the annual species studied have only centromeric bands. Identification of aneuploids for specific chromosomes can be accomplished by using banding techniques. Critical studies of the aneuploids can locate genes of agronomic importance on individual chromosomes.

Alfalfa breeders have been attempting to incorporate the genes of wild species into cultivated alfalfa through the use of interspecific hybridization via sexual crosses (see Quiros and Bauchan 1988 and McCoy and Bingham 1988 for summary), embryo rescue (Bauchan 1987), ovule embryo rescue (McCoy and Smith 1986), and somatic hybridization (Pupilli et al. 1995; Nenz et al. 1996; Mizukami et al. 2006; Tian et al. 2002) with some success. A majority of the crosses made have used either ssp. *coerulea* or ssp. *sativa* as one of the parents due to their economic importance. Some of the attempted crosses could have been unsuccessful in incorporating wild genes due to the lack of pairing and thus lack of crossing over caused by large amounts of heterochromatin in the short arms of the chromosomes.

Heterochromatin is characterized by the presence of middle and highly repetitive sequences—often in tandem arrays of satellites—tightly bound to protein (Bickmore and Craig 1997). Heterochromatin interferes with crossing over and thus there is a reduction in the formation of chiasmata (Dyer 1964; John and Lewis 1965; Sybenga 1975). A better candidate for crossing with wild species is diploid ssp. *falcata* or tetraploid ssp. *falcata* with much reduced heterochromatin. Diploid ssp. *falcata*, which does not have heterochromatin on its chromosome arms, will provide many more opportunities for chiasmata formation; thus, exchange of genetic material can occur by crossing over. Because ssp. *falcata* readily crosses with ssp. *sativa*, it may be used as a bridge species between ssp. *sativa* and wild species of *Medicago*. Chromosome banding can also be used to identify the introgression of individual chromosomes, arms, and/or segments from the wild species into alfalfa as well as the exchange of genetic materials through crossing over, which has been done extensively in wheat for its improvement (Friebe et al. 1991).

Based on pachytene analysis, Clement and Stanford (1963) pointed out that the short arms of diploid alfalfa are heterochromatic; thus, chiasmata and crossing over are limited or restricted on the short arms of the chromosomes. The existence of highly heterochromatinized short arms has been shown to be true based on C- and N-banding of ssp. *coerulea*, ssp. *sativa*, and tetraploid ssp.

*falcata (*Bauchan and Hossain 1997, 1998a*)*. These findings support the conclusion of Stanford et al. (1972) regarding the fact that autotetraploid alfalfa forms very few multivalents, probably due to the reduction of chiasmata because of heterochromatin on the short arms of the chromosomes. There could also be undescribed gene(s) responsible for chromosome pairing. Banding studies of the meiotic chromosomes could answer this question through the identification of individual chromosomes and their pairing behavior.

The existence of large blocks of heterochromatic DNA on the short arms of the chromosomes of ssp. *coerulea,* ssp. *sativa,* and tetraploid ssp. *falcata* could explain the difficulty that molecular geneticists are having when mapping the genome of alfalfa. Heterochromatic DNA interferes with chiasmata formation and, thus, crossing over near these regions is reduced (Kaltsikes and Gustafson 1984). This could explain the phenomenon of segregation distortion that has been observed (Brummer, Bouton, and Kochert 1993; Brouwer and Osborn 1999; Sledge, Ray, and Jiang 2005; Robins et al. 2007a, b).

The development of an aneuploid series of diploid alfalfa would be especially helpful in the development of a chromosomal gene map. Kasha and McLennan (1967) were able to isolate several primary trisomics ($2n = 2x + 1 = 17$). However, only trisomics for chromosomes 1, 4, 6, 7, and 8 were identified using pachytene analysis (Gillies 1977). McCoy and Echt (1992) presented an example of how chromosome addition lines (trisomics) can be used to locate genes on specific chromosomes of alfalfa. A triploid ($3x$) hybrid would be produced from a cross between diploid *M. sativa* ssp. *sativa* ($2x$) and a chromosome doubled diploid *M. papillosa* ($4x$). This triploid hybrid would be backcrossed to diploid *M. papillosa* and the progeny screened for trisomics, which would contain two genomes of *M. papillosa;* the one extra chromosome would be from *M. sativa.* These trisomic addition lines could then be irradiated to induce chromosome breakage of the *M. sativa* chromosome to produce smaller pieces of the chromosome that could be analyzed to determine what genes are located on these chromosome fragments. The use of addition lines in this manner is being used in oat–corn addition lines to map the genome of corn (Ananiev et al. 1997).

2.6.7 Molecular Cytogenetics

The advent of the fluorescence *in situ* hybridization (FISH) technique has opened up a wide area to be explored in alfalfa. Schaff et al. (1990), Falistocco et al. (2002), and Calderini et al. (1996, 1997) showed that it is possible to tag a known gene sequence and identify the location of the gene(s) on the chromosome. Schaff et al. (1990) utilized molecular cytogenetic techniques on alfalfa chromosomes. They used enzymatic techniques (strepavidin–horseradish peroxidase complex) to label a specific gene (β-tubulin) and *in situ* hybridize the gene to alfalfa chromosomes to identify the location of the gene on two chromosomes. Calderini et al. (1996) used fluorescent stains [4′, 6-diamidion-2-phenylindole (DAPI) and chromomycin A_3 (CMA_3)] to band the chromosomes of alfalfa, showing that the NORs were high in GC content. Cluster et al. (1996) and Calderini et al. (1996) have used dual fluorescent staining (rhodamine and DAPI) for FISH technique to label the 18S gene of rDNA and identify the number and location of the genes on the chromosomes. Triple FISH labeling (DAPI, flourescein isothiocyanate [FITC], and steptavidin-Cy3) has been used to determine the probe sequence in genome mapping of *M. truncatula* (Kulikova et al. 2001).

It is possible to label multiple probes and several fluorechromes with contrasting fluorescent stains in order to identify whether the probes are linked or unlinked (Cluster et al. 1996; Calderini et al. 1996; Cerbah et al. 1999; Kulikova et al. 2001, 2004; Choi et al. 2004). FISH has been used in the physical mapping in *M. truncatula* (Kulikova et al. 2001; Choi et al. 2004). It was necessary to use FISH analysis to resolve ambiguities in the genetic map of *M. truncatula* (Choi et al. 2004). Due to the enormous focus on the mapping and sequencing of the genome of *M. truncatula,* these technologies have not been carried over to *M. sativa;* however, due to the similarity of the two genomes, they will be very useful in the improvement of alfalfa.

Genomic *in situ* hybridization (GISH) has been used to determine the relationship between two annual *Medicago* species: *M. murex* and *M. lesinsii*. Total DNA was extracted from the individual species, fluorescently labeled, and applied to the chromosome spreads of the other species (Falistocco, Torricelli, and Falcinelli 2002). This technique could be used to answer questions on the evolution of the *M. sativa* complex, the relationship between the tetraploid annual medic species and the diploid species, the incorporation of alien gene sequences, and chromosomal rearrangements.

2.6.8 Genetic Mapping

Molecular genetic maps of diploid or tetraploid alfalfa have been constructed based on amplified fragment length polymorphisms (AFLP), random amplification of polymorphic DNA (RAPD), and/ or RFLP (Barcaccia et al. 1999; Brouwer and Osborn 1999; Brummer et al. 1993, 1999; Echt et al. 1993; Kaló et al. 2000; Kiss et al. 1993; Robins, Bauchan, and Brummer 2007; Robins, Luth, et al. 2007). The maps consist of eight linkage groups corresponding to the base number of the *Medicago* species. The maps have been used to locate genes controlling flower color, dwarfism, sticky leaves (Kiss et al. 1993), a unifoliate leaf, and cauliflower head mutation (Brouwer and Osborn 1999). Diwan et al. (1997) were the first to show that simple sequence repeat (SSR) can be used to map tetraploid alfalfa, and thus extensive mapping programs have been developed using SSR (Julier et al. 2003; Sledge et al. 2005; Choi et al. 2004; Robins, Bauchan, et al. 2007; Robins, Luth, et al. 2007). The development of software programs JoinMap3.0 (Van Ooijen and Voorrips 2001), MapChart (Voorrips 2002), TetraploidMap (Hackett and Lou 2003), and MAPQTL 5.0 (Van Ooijen 2004) have made it possible to analyze mapping data from autotetraploid species. These molecular maps have been utilized to identify aluminum tolerance (Sledge et al. 2002; Narasimhamoorthy et al. 2007), forage yield, plant height and regrowth (Robins, Bauchan, et al. 2007a), and biomass production using quantitative trait locus (QTL) analysis (Robins, Luth, et al. 2007b).

There has been a concerted effort to map and sequence the entire genome of *M. truncatula*. The genome size of *M. truncatula* is ~454–526 Mbp, which is about half the size of alfalfa. The *M. truncatula* sequencing project was initiated with a generous grant from the Samuel Roberts Noble Foundation to the University of Oklahoma. Beginning in 2003, the National Science Foundation and the European Union's Sixth Framework Program provided funding to complete sequencing of the remaining euchromatic gene space. As of 2007 they have mapped 2,665 BAC clones 42,358 total genes with 22.8 genes per 100 kb, and have sequenced 251 million base pairs (www.medicago. org). Choi et al. (2004) used primers designed against *M. truncatula* exon sequences to map genes in *M. sativa*. They developed a comparative map of the two species and concluded that the two species are highly similar and proposed the development of a composite map for *Medicago*. The initial genome sequencing project was completed in 2006 with additional sequencing of the *Medicago* genome continuing into 2008 (www.medicago.org). With the advancements in the genome discovery in *M. truncatula*, its similarity to the alfalfa genome, alfalfa genomics, and improvement will greatly benefit from these discoveries.

2.7 TISSUE CULTURE AND GENETIC ENGINEERING

Alfalfa was one of the first major crop plants to be regenerated from tissue culture (Saunders and Bingham 1972). Alfalfa plants were initially regenerated from callus derived from anthers, ovaries, internodes, and seedling hypocotyls (Saunders and Bingham 1972). Bingham et al. (1975) developed RegenS alfalfa, which readily regenerated from tissue culture. Since that time, researchers have been able to regenerate plants from leaf, petiole, sepal, petal, cotyledon, root, and immature embryos (Bingham and McCoy 1988; McCoy and Walker 1984; Mroginski and Kartha 1984.). Plants can be regenerated from suspension cultured cells and leaf, cotyledon, root cell suspension protoplasts (dos Santos et al. 1980; Johnson et al. 1981; Pezzotti, Arcioni, and Mariotti 1984), and after protoplast fusion (Teoule 1983; Tian et al. 2002; Mizukami et al. 2006). Regeneration of alfalfa from tissue culture is recognized to be genetically inherited (Reisch and Bingham 1980; Hernandez-Fernandez and Christie 1989). Bingham and McCoy (1988) noted that most alfalfa cultivars can be regenerated, although some are regenerated at a low level that may require selection of regenerable types within the cultivar. Regeneration is primarily through somatic embryogenesis (Walker, Wendeln, and Jaworski 1979).

Genes can be inserted into alfalfa using *Agrobacterium tumumefaciens* (Pezzotti et al. 1991; Desgagnes et al. 1995). These same tissue culture and gene insertion technologies are also available for *M. truncatula* (Trinh et al. 1998), which has led to many fundamental studies on the genomics of *Medicago* (http://medicago.org/genome). The development and recent release of the first genetically engineered alfalfa in 2006 (Roundup Ready alfalfa) by Monsanto and Forage Genetics International follow in the path of Roundup Ready corn and soybeans. The Consortium for Alfalfa Improvement, composed of researchers from Forage Genetics International, the Samuel Roberts Noble Foundation, and the USDA-ARS Dairy Forage Research Center in Madison, Wisconsin, is planning the release of alfalfa with improved forage quality through the introduction of genes for reduced lignin content, which improves fiber digestibility and the efficiency of protein utilization.

2.8 CONCLUSION

Alfalfa will continue to rule as "queen of the forages" in the near future. Its high feed value and high biomass production—along with its ease of establishment; resistance to major diseases, insects, and nematodes; and tolerance to a wide range of environments—make this forage crop the plant of choice. Continued exploration for new germplasm sources and the exploitation of current germplasm collections will need to persist in order for researchers to advance new uses of alfalfa as a biofuel, for phytoremediation, and/or pharmaceutical factory, as well as toward any new diseases and insect pests that may arise. Utilization of genetic, cytogenetic, breeding, and biotechnology techniques will continue to play a huge role in the improvement of alfalfa. Cooperative research among the federal government, universities, and private industry will need to be continued and even strengthened in order to prolong the advancements that have been made, in the face of shrinking budgets, dwindling grant funding, and a reduced number of scientists conducting research on alfalfa. With the advent of molecular tools to study the *Medicago* genome, alfalfa stands as a model crop for improvement of leguminous plants.

REFERENCES

Agarwal, K., and P. K. Gupta. 1983. Cytological studies in the genus *Medicago* L. *Cytologia* 48:781–793.

Ananiev, E. V., O. Riera-Lizarazu, H. W. Rines, and R. L. Phillips. 1997. Oat-maize chromosome addition lines: A new system for mapping the maize genome. *Proceedings of the National Academy of Science USA* 94:3524–3529.

Arcioni, S. A., F. Damiani, M. Piccirilli, and F. Pupilli. 1993. Embryo rescue and somatic hybridization for the production of interspecific hybrids in the genus *Medicago*. *Proceedings of XVII International Grasslands Congress* 1041–1042.

Barcaccia, G., E. Albertini, S. Tavoletti, M. Falcinelli, and F. Veronesi. 1999. AFLP fingerprinting in *Medicago* spp.: Its development and application in linkage mapping. *Plant Breeding* 118:335–340.

Barker, D. G., S. Bianchi, F. Laondon, Y. Dattee, G. Duc, S. Essad, P. Flament, P. Gallusei, G. Genier, P. Guy, et al. 1990. *Medicago truncatula*: A model plant for studying the molecular genetics of the Rhizobium-legume symbiosis. *Plant Molecular Biology Reporter* 8: 40–49.

Barnes, D. K., E. T. Bingham, R. P. Murphy, O. J. Hunt, D. F. Beard, W. H. Skrdla, and L. R. Teuber. 1977. Alfalfa germplasm in the United States: Genetic vulnerability, use, improvement and maintenance. USDA Tech Bill. 1571. U.S. Government Print Office, Washington, D.C.

Barnes, D. K., B. P. Goplen, and J. E. Baylor. 1988. Highlights in the USA and Canada. In *Alfalfa and alfalfa improvement,* ed. A. A. Hanson, D. K. Barnes, and R. R. Hill, 1–24. Agronomy Monographs, 29. Madison, WI: ASA, CSSA and SSSA.

Basigalup, D. H., D. K. Barnes, and R. E. Stucker. 1995. Development of a core collection for perennial *Medicago* plant introductions. *Crop Science* 35:1163–1168.

Bauchan, G. R. 1987. Embryo culture of *Medicago scutellata* and *M. sativa. Plant Cell, Tissue and Organ Culture* 10:21–29.

Bauchan, G. R., and T. A. Campbell. 1994. Use of an image analysis system to karyotype diploid alfalfa (*Medicago sativa* L.). *Journal of Heredity* 85:18–22.

Bauchan, G. R., T. A. Campbell, and M. A. Hossain. 2002. Chromosomal polymorphism as detected by C-banding patterns in Chilean alfalfa germplasm. *Crop Science* 42:1291–1297.

———. 2003. Comparative chromosome banding of nondormant alfalfa germplasm. *Crop Science* 43:2037–2042.

Bauchan, G. R., and J. H. Elgin, Jr. 1984. A new chromosome number for the genus *Medicago. Crop Science* 24:193–195.

Bauchan, G. R., and S. L. Greene. 2002. Status of the *Medicago* germplasm collection in the United States. *Plant Genetics Research Newsletter* 129:1–8.

Bauchan, G. R., and M. A. Hossain. 1996. Karyotype analysis of alfalfa using C- and N-banding techniques. *Proceedings of the North American Alfalfa Improvement Conference* 35:4. Oklahoma City, OK, July 16–20, 1996.

———. 1997. Karyotypic analysis of C-banded chromosomes of diploid alfalfa: *Medicago sativa* ssp. *coerulea* and ssp. *falcata* and their hybrid. *Journal of Heredity* 88:533–537.

———. 1998a. Karyotypic analysis of N-banded chromosomes of diploid alfalfa: *Medicago sativa* ssp. *Caerulea* and ssp. *falcata* and their hybrid. *Journal of Heredity* 89:191–193.

———. 1998b. Cytogenetic studies of the nine germplasm sources of alfalfa. *Proceedings of the North American Alfalfa Improvement Conference* 36:21. Bozeman, MT, Aug. 2–6, 1998.

———. 1999a. Constitutive heterochromatin DNA polymorphisms in diploid *Medicago sativa* ssp. *falcata. Genome* 42:930–935.

———. 1999b. Detection of chromosome variations in *Medicago sativa. Proceedings of EUCARPIA Medicago species conference,* Perugia, Italy, Sept. 13–16, 1999.

———. 2001a. A computerized image analysis system to characterize small plant chromosomes. *Microscopy Analysis* 48:9–11.

———. 2001b. Distribution and characterization of heterochromatic DNA in the tetraploid African population alfalfa genome. *Crop Science* 41:1921–1926.

Bickmore, W., and J. Craig. 1997. *Molecular biology intelligence unit: Chromosome bands: Patterns in the genome.* Austin, TX: R. G. Landes Co.

Bickoff, E. M., G. O. Kohler, and D. Smith. 1972. Chemical composition of herbage. In *Alfalfa science and technology,* ed. C. H. Hanson. Agronomy Monograph 15. Madison, WI: ASA, CSSA, and SSSA.

Bingham, E. T., L. V. Hurley, D. M. Kaatz, and J. W. Saunders. 1975. Breeding alfalfa which regenerates from callus tissue in culture. *Crop Science* 15:719–721.

Bingham, E. T., and T. J. McCoy. 1979. Cultivated alfalfa at the diploid level: Origin, reproductive stability, and yield of seed and forage. *Crop Science* 19:97–100.

———. 1986. Somaclonal variation in alfalfa. *Plant Breeding Reviews* 4:123–152.

———. 1988. Alfalfa tissue culture. 903–929. In *Alfalfa and alfalfa improvement,* ed. A. A. Hanson, D. K. Barnes, and R. R. Hill, 903–929. Agronomy Monograph 29. Madison, WI: ASA, CSSA, and SSSA.

Bingham, E. T., and J. W. Saunders. 1974. Chromosome manipulations in alfalfa: Scaling the cultivated tetraploid to seven ploidy levels. *Crop Science* 14:474–477.

Blondon, F., D. Marie, S. Brown, and A. Kondorosi. 1994. Genome size and base composition in *Medicago sativa* and *M. truncatula* species. *Genome* 37:264–270.

Bolton, J. L. 1962. *Alfalfa botany, cultivation and utilization.* London: Leonard Hill, World Crops Books.

Bolton, J. L., and J. E. R. Greenshields. 1950. A diploid form of *Medicago sativa* L. *Science* 112:275–277.

Brouwer, D. J., and T. C. Osborn. 1999. A molecular marker linkage map of tetraploid alfalfa (*Medicago sativa* L.). *Theoretical and Applied Genetics* 99:1194–1200.

Brown, A. H. D. 1989. Core collection: A practical approach to genetic resources management. *Genome* 31:818–824.

Brummer, E. C., J. H. Bouton, and G. Kochert. 1993. Development of an RFLP map in diploid alfalfa. *Theoretical and Applied Genetics* 86:329–332.

Brummer, E. C., P. M. Cazcarro, and D. Luth. 1999. Ploidy determination of alfalfa germplasm accessions using flow cytometry. *Crop Science* 39:1202–1207.

Buss, G. R., and R. W. Cleveland. 1968. Karyotype of a diploid *Medicago sativa* L. analyzed from sporophytic and gemetophytic mitosis. *Crop Science* 8:713–716.

Calderini, O., F. P. D. Pupilli, A. Cluster, A. Mariani, and S. Arcioni. 1996. Cytological studies of the nucleolus organizing regions in the *Medicago* complex: *sativa-coerulea-falcata*. *Genome* 39:914–920.

Calderini, O., F. Pupilli, F. Paolocci, and S. Arcioni. 1997. A repetitive and species-specific sequence as a tool for detecting the genome contribution in somatic hybrids of the genus *Medicago*. *Theoretical and Applied Genetics* 95:734–740.

Cerbah, M., Z. Kevel, S. Siljak-Yakovlev, E. Kondorosi, A. Kondorosi, and T. H. Trinh. 1999. FISH chromosome mapping allowing karyotype analysis in *Medicago truncatula* lines Jemalong J5 and R-108-1. *Molecular Plant–Microbe Interactions* 12:947–950.

Choi, H-K., D. Kim, T. Uhm, E. Limpens, H. Lim, J. H. Mun, P. Kalo, et al. 2004. A sequence-based genetic map of *Medicago truncatula* and comparison of marker co-linearity with *M. sativa*. *Genetics* 166:1463–1502.

Classen, D., C. Nozzolillo, and E. Small. 1982. A phenolic-taxonomic study of *Medicago* (Leguminosae). *Canadian Journal of Botany* 60:2477–2495.

Clement, W. M., Jr., and E. H. Stanford. 1963. Pachytene studies at the diploid level in *Medicago*. *Crop Science* 3:147–150.

Cleveland, R. W., and E. H. Stanford. 1959. Chromosome pairing in hybrids between tetraploid *Medicago sativa* L. and diploid *Medicago falcata* L. *Agronomy Journal* 51:488–492.

Cluster, P. D., O. Calderini, F. Pupilli, F. Crea, F. Damiani, and S. Arcioni. 1996. The fate of ribosomal genes in three interspecific somatic hybrids of *Medicago sativa:* Three different outcomes including the rapid amplification of new spacer-length variants. *Theoretical and Applied Genetics* 93:801–808.

Cook, D. R. 1999. *Medicago truncatula*—A model in the making! *Current Opinion in Plant Biology* 2:301–304.

Cook, D. R., K. VandenBosch, F. J. De Bruijn, and T. Huguet. 1997. Model legumes get the nod. *Plant Cell* 9:275–281.

Desgagnes, R., S. Laberge, G. Allard, H. Khoudi, Y. Castonguay, J. Lapointe, R. Michaud, and L-P. Vezina. 1995. Genetic transformation of commercial breeding lines of alfalfa (*Medicago sativa*). *Plant Cell, Tissue and Organ Culture* 42:129–140.

Diwan, N., G. R. Bauchan, and M. S. McIntosh. 1994. A core collection for the United States annual *Medicago* germplasm collection. *Crop Science* 34:279–285.

Diwan, N., M. S. McIntosh, and G. R. Bauchan. 1995. Methods of developing a core collection of annual *Medicago* species. *Theoretical and Applied Genetics* 90:755–761.

Diwan, N., A. A. Bhagwat, G. R. Bauchan, and P. B. Cregan. 1997. Simple sequence repeat DNA markers in alfalfa and perennial and annual *Medicago* species. *Genome* 40:887–895.

dos Santos, A. V. P., D. E. Outka, E. C. Cocking, and M. R. Favey. 1980. Organogenesis and somatic embryogenesis in tissues derived from leaf protoplasts and leaf explants of *Medicago sativa*. *Zeitschrift für Pflanzenphysiologie* 99:261–270.

Dyer, A. F. 1964. Heterochromatin in American and Japanese species of *Trillium*. III. Chiasma frequency and distribution and the effect on it of heterochromatin. *Cytologia* 29:263–279.

Echt, C. S., K. K. Kidwell, S. J. Knapp, T. C. Osborn, and T. J. McCoy. 1993. Linkage mapping in diploid alfalfa (*Medicago sativa*). *Genome* 37:61–71.

Falistocco, E. 1987. Cytogenetic investigations and karyological relationships of two *Medicago: M. sativa* L. (alfalfa) and *M. arborea* L. *Caryologia* 40:339–346.

Falistocco, E., and M. Falcinelli. 1991. Cytological and morphological studies in *Medicago hispida* Gaertner (=*M. polymorpha* L.). *Annali Di Botanica* 49:13–25.

———. 1993. Karyotype and C-banding in *Medicago nöana* Boiss. Leguminosae. *Cytologia* 58:151–154.

Falistocco, E., M. Falcinelli, and F. Veronesi. 1995. Karyotype and C-banding pattern of mitotic chromosomes in alfalfa, *Medicago sativa* L. *Plant Breeding* 114:451–453.

Falistocco, E., R. Torricelli, and M. Falcinelli. 2002. Genomic relationships between *Medicago murex* Willd. and *Medicago lesinsii* E. Small. investigated by *in situ* hybridization. *Theoretical and Applied Genetics* 105:829–833.

Friebe, B., Y. Mukai, H. S. Dhaliwal, T. J. Martin, and B. S. Gill. 1991. Identification of alien chromatin specifying resistance to wheat streak mosaic and greenbug in wheat germplasm by C-banding and *in situ* hybridization. *Theoretical and Applied Genetics* 81:381–389.

Gillies, C. B. 1968. The pachytene chromosomes of a diploid *Medicago sativa*. *Canadian Journal of Genetics and Cytology* 10:788–793.

———. 1970a. Alfalfa chromosomes: I. Pachytene karyotype of a diploid *Medicago falcata* L. and its relationship to *M. sativa*. L. *Crop Science* 10:169–171.

———. 1970b. Alfalfa chromosomes: II. Pachytene karyotype of a tetraploid *Medicago sativa* L. *Crop Science* 10:172–175.

———. 1977. Identification of trisomics in diploid lucerne. *Australian Journal of Research* 28:309–317.

Gillies, C. B., and E. T. Bingham. 1971. Pachytene karyotypes of 2X haploids derived from tetraploid alfalfa (*Medicago sativa*)—Evidence for autotetraploidy. *Canadian Journal of Genetics and Cytology* 13:397–403.

Hackett, C. A., and Z. W. Lou. 2003. TetraploidMap: Construction of a linkage map in autotetraploid species. *Journal of Heredity* 94:358–359.

Harlan, J. R., and J. M. J. de Wet. 1971. Toward a rational classification of cultivated plants. *Taxon* 20:509–517.

Heichel, G. H., and D. K. Barnes. 1984. Opportunities for meeting crop nitrogen needs from symbiotic nitrogen fixation. In *Organic farming: Current technology and its role in a sustainable agriculture,* ed. D. Bezdicek and J. Power, 49–59. Spec. Pub. 46, Madison, WI: American Society of Agronomy.

Hernandez-Fernandez, M. M., and B. R. Christie. 1989. Inheritance of somatic embryogenesis in alfalfa (*Medicago sativa* L.). *Genome* 32:318–321.

Heyn, C. C. 1963. The annual species of *Medicago. Scripta Hierosolymithana XII.* Jerusalem: Magnes Press, the Hebrew University.

Hill, R. R., Jr., J. S. Shenk, and R. F Barnes. 1988. Breeding for yield and quality. In *Alfalfa and alfalfa improvement,* ed. A. A. Hanson, D. K. Barnes, and R. R. Hill, 809–825. Agronomy Monograph 29. Madison, WI: ASA, CSSA, and SSSA.

Ho, K. M., and K. J. Kasha. 1972. Chromosome homology at pachytene in diploid *Medicago sativa, M. falcata,* and their hybrids. *Canadian Journal of Genetics and Cytology* 14:829–838.

Hossain, M. A., and G. R. Bauchan. 1999. Identification of B chromosomes using Giemsa banding in *Medicago. Journal of Heredity* 90:428–429.

Ivanov, A. I. 1977. History, origin and evolution of the genus *Medicago,* subgenus Falcago. *Bulletin of Applied Botany and Genetic Selection* 59:3–40. (Trudy po prikladnoy botanique, genetike i selektsii. Translation by the Multilingual Services Division Department, Secretary of State, Canada.)

John, B., and K. R. Lewis. 1965. *The meiotic system. Protoplasmatologia.* New York: Springer–Verlag.

Johnson, L. B., D. L. Stuteville, R. K. Higgins, and D. Z. Skinner. 1981. Regeneration of alfalfa plants from protoplasts of selected Regen S clones. *Plant Science Letters* 20:297–304.

Johnson, L. B., D. L. Stuteville, S. E. Schlarbaum, and D. Z. Skinner. 1984. Variation in phenology and chromosome number in alfalfa protoclones regenerated from nonmutagenized calli. *Crop Science* 24:948–951.

Julier, G., S. Flajoulot, P. Barre, G. Cardinet, S. Santoni, T. Huguet, and C. Huyghe. 2003. Construction of tw genetic linkage maps in cultivated tetraploid alfalfa (*Medicago sativa*) using microsatellite and AFLP markers. BMC. *Plant Biology* Vol. 3:9 ww.biomedcentral.com/1471-2229/3/9.

Kaló, P., G. Endre, L. Zamányi, G. Csanádi, and G. B. Kiss. 2000. Construction of an improved linkage map of diploid alfalfa (*Medicago sativa*). *Theoretical and Applied Genetics* 100:641–657.

Kaltsikes, P. J., and J. P. Gustafson. 1984. The heterochromatin story in *Triticale. Proceedings of the EUCARPIA Genetics and Breeding of Triticale* meeting, Clermont-Ferrand, France, July 2–5, 1985.

Kasha, K. J., K. C. Armstrong, G. R. Buss, R. W. Cleveland, C. B. Gillies, K. Lesins, and E. H. Stanford. 1970. Report of the committee on chromosome numbering in alfalfa. *Twenty-second Alfalfa Improvement Conference.* USDA Agr. Res. Serv. Rept. CR-61-70, pp. 66–71.

Kasha, K. J., and H. A. McLennan. 1967. Trisomics in diploid alfalfa. I. Production, fertility, and transmission. *Chromosoma* 21:232–242.

Kiss, G. B., G. Csanádi, K. Kálmán, P. Kaló, and L. Ökrész. 1993. Construction of a basic genetic map for alfalfa using RFLP, RAPD, isozyme and morphological markers. *Molecular and General Genetics* 38:129–137.

Kulikova, O., G. Gualtieri, R. Geurts, D-J. Kim, D. R. Cook, T. Huguet, J. H. de Jong, P. F. Fransz, and T. Bisseling. 2001. Integration of the FISH pachytene and genetic maps of *Medicago truncatula*. *Plant Journal* 27:49–58.

Lesins, K., and C. B. Gillies. 1972. Taxonomy and cytogenetics of *Medicago*. In *Alfalfa science and technology*, ed. C. H. Hanson, 53–86. Agronomy Monograph 15. Madison, WI: ASA, CSSA, and SSSA.

Lesins, K., and I. Lesins. 1961. Some little-known *Medicago* species and their chromosome complements. *Canadian Journal of Genetics and Cytology* 3:7–9.

———. 1979. Genus *Medicago* (Leguminosae). *A taxogenetic study.* The Hague, the Netherlands: Junk. 228.

Mariani, A., and E. Falistocco. 1990. Chromosome studies in $2n = 14$ and $2n = 16$ types of *Medicago murex*. *Genome* 33:159–163.

———. 1991. Cytogenetic analysis of *Medicago rugosa* and *Medicago scutellata*. *Journal of Genetics and Breeding* 45:111–116.

Mariani, A., F. Pupilli, and O. Calderini. 1996. Cytological and molecular analysis of annual species of the genus *Medicago*. *Canadian Journal of Botany* 74:299–307.

Masoud, S. A., B. S. Gill, and L. B. Johnson. 1991. C-banding of alfalfa chromosomes: Standard karyotype and analysis of a somaclonal variant. *Journal of Heredity* 82:335–338.

McCoy, T. J. 1982. The inheritance of 2n pollen formation in diploid alfalfa *Medicago sativa*. *Canadian Journal of Genetics and Cytology* 24:315–323.

McCoy, T. J., and E. T. Bingham. 1988. Cytology and cytogenetics of alfalfa. In *Alfalfa and alfalfa improvement*, ed. A. A. Hanson, D. K. Barnes, and R. R. Hill, 737–776. Agronomy Monograph 29. Madison, WI: ASA, CSSA, and SSSA.

———. 1991 Alfalfa cytogenetics. In *Chromosome engineering in plants*, ed. P. K. Gupta and T. Tsuchiya, 399–418. New York: Elsevier Pub.

McCoy, T. J., and C. S. Echt. 1992. Chromosome manipulations and genetic analysis in *Medicago*. In *Plant Breeding Reviews*, 169–197. New York: John Wiley & Sons, Inc.

McCoy, T. J., and L. Y. Smith 1986. Interspecific hybridization of perennial *Medicago* species using ovule-embryo culture. *Theoretical and Applied Genetics* 71:772–783.

McCoy, T. J., and K. Walker. 1984. Alfalfa. In *Handbook of plant cell culture*, vol. 3, ed. P. V. Ammirato et al., 171–192. New York: Macmillian Publishing Co.

Michaud, R., W. F. Lehman, and M. D. Rumbaugh. 1988. World distribution and historical development. In *Alfalfa and alfalfa improvement*, ed. A. A. Hanson, D. K. Barnes, and R. R. Hill, 25–91. Agronomy Monograph 29. Madison, WI: ASA, CSSA, and SSSA.

Mizukami, Y., M. Kato, T. Takamizo, M. Kambe, S. Inami, and K. Hattoru. 2006. Interspecific hybrids between *Medicago sativa* L. and annual *Medicago* containing alfalfa weevil resistance. *Plant Cell, Tissue and Organ Culture* 84:79–88.

Mroginski, L. A., and K. K. Kartha. 1984. Tissue culture of legumes for crop improvement. *Plant Breeding Reviews* 2:215–264.

Nagarajan, P., and P. D. Walton. 1987. A comparison of somatic chromosomal instability in tissue culture regenerants from *Medicago media*. Pers. *Plant Cell Reports* 6:109–113.

Narasimhamoorthy, B., J. H. Bouton, K. M. Olsen, and M. K. Sledge, 2007. Quantitative trait loci and candidate gene mapping of aluminum tolerance in diploid alfalfa. *Theoretical and Applied Genetics* 114:901–913.

Nenz, E. F. Pupilli, F. Damiani, and S. Arcioni. 1996. Somatic hybrid plants between the forage legumes *Medicago sativa* and *Medicago arborea*. L. *Theoretical and Applied Genetics* 93:183–189.

North American Alfalfa Improvement Conference. 1999. Standard tests to characterize alfalfa cultivars. Beltsville, MD.

Pezzotti, M., S. Arcioni, and D. Mariotti. 1984. Plant regeneration from mesophyll root and cell suspension protoplasts of *Medicago sativa* cv. Adriana. *Genetic Agriculture* 38:195–208.

Pezzotti, M., F. Pupilli, F. Daniani, and S. Arcioni. 1991. Transformation of *Medicago sativa* L. using a *Ti* plasmid derived vector. *Plant Breeding* 106:39–46.

Piccirilli, M., and S. Arcioni. 1992. New interspecific hybrids in the genus *Medicago* through *in vitro* culture of fertilized ovules. In: *Angiosperm pollen and ovules.* ed. E. Ottaviano, M. Sari Goria and G. Bergamini Mulcahy, 325–330, Springer-Verlag, New York.

Pupilli, F., S. Businelli, M. E. Caceres, F. Damiani, and S. Arcioni. 1995. Molecular, cytological and morpho-agronomical characterization of hexaploid somatic hybrids in *Medicago. Theoretical and Applied Genetics* 90:347–355.

Quiros, C. F., and G. R. Bauchan. 1988. The genus *Medicago* and the origin of the *Medicago sativa* complex. In *Alfalfa and alfalfa improvement,* ed. A. A. Hanson, D. K. Barnes, and R. R. Hill, 93–124. Agronomy Monograph 29. Madison, WI: ASA, CSSA, and SSSA.

Reiger, R., A. Michaelis, and M. M. Green. 1991. *Glossary of genetics: Classical and molecular,* 5th ed. New York: Springer–Verlag.

Reisch, B., and E. T. Bingham. 1980. The genetic control of bud formation from callus cultures of diploid alfalfa. *Plant Science Letters* 20:71–77.

Robins, J. G., G. R. Bauchan, and E. C. Brummer. 2007. Genetic mapping forage yield, plant height, and regrowth at multiple harvests in tetraploid alfalfa (*Medicago sativa* L.). *Crop Science* 47:11–18.

Robins, J. G., D. Luth, T. A. Campbell, G. R. Bauchan, C. He, D. R. Viands, J. L. Hansen, and E. C. Brummer. 2007. Genetic mapping of biomass production in tetraploid alfalfa. *Crop Science* 47:1–10.

Rose, R. J., K. E. Nolan, and L. Bicego. 1999. The development of the highly regenerable seed line Jemalong 2HA for transformation of *Medicago truncatula*—Implications for regenerability via somatic embryogenesis. *Journal of Plant Physiology* 155:788–791.

Russelle, M. P. 2001. Alfalfa: After an 8,000-year journey, the "queen of the forages" stands poised to enjoy renewed popularity. *American Scientist* 89:252–261.

Sangduen, N., E. L. Sorensen, and G. H. Liang. 1982. A perennial × annual *Medicago* cross. *Canadian Journal of Genetics and Cytology* 24:361–365.

Saunders, J. W., and E. T. Bingham. 1972. Production of alfalfa plants from callus tissue. *Crop Science* 12:804–808.

Schaff, D. A., S. M. Koehler, B. F. Matthews, and G. R. Bauchan. 1990. *In situ* hybridization of β-tubulin to alfalfa chromosomes. *Journal of Heredity* 81:480–483.

Schlarbaum, S. E., L. B. Johnson, and D. L. Stuteville. 1988. Characterization of somatic chromosomes morphology in alfalfa, *Medicago sativa* L.: Comparison of donor plant with regenerated protoclone. *Cytologia* 53:499–507.

Schlarbaum, S. E., E. Small, and L. B. Johnson. 1984. Karyotypic evolution, morphological variability and phylogny in *Medicago* sect. *Intertextae. Plant System Evolution* 145:203–222.

Sinskaya, E. N. 1950. Flora of cultivated plants of the U.S.S.R. XII. Perennial leguminous plants. Part I. Medic, sweetclover, fenugreek. Translated by Israel Program for scientific translations, Jerusalem, 1961.

Skinner, D. Z., G. R. Bauchan, G. Auricht, and S. Hughes. 1999. A method for the efficient management and utilization of large germplasm collections. *Crop Science* 39:1237–1242.

Simon, J. P. 1976. Relationship in annual species of *Medicago*. V. Analysis of phenolics by means of one-dimensional chromatographic techniques. *Australian Journal of Botany* 15:83–93.

Sledge, M. K., I. M Ray, and G. Jiang. 2005. An expressed sequence tag SSR map of tetraploid alfalfa (*Medicago sativa* L.). *Theoretical and Applied Genetics* 111:980–992.

Sledge, M. K., J. H. Bouton, M. Dall'Agnoll, W. A. Parrott, and G. Kochert. 2002. Identification and confirmation of aluminum tolerance QTL in diploid *Medicago sativa* ssbsp. *coerulea. Crop Sciences* 42:1121–1128.

Small, E., and M. Jomphe. 1989. A synopsis of the genus *Medicago* (Leguminosae). *Canadian Journal of Botany* 67:3260–3294.

Smith, S. E., A. Al-Doss, and M. Warburton. 1991. Morphological and agronomic variation in North African and Arabian alfalfas. *Crop Science* 31:1159–1163.

Stanford, E. H. 1951. Tetrasomic inheritance in alfalfa. *Agronomy Journal* 43:222–225.

Stanford, E. H., W. M. Clement, Jr., and E. T. Bingham. 1972. Cytology and evolution of the *Medicago sativa-falcata* complex. In *Alfalfa science and technology,* ed. C. H. Hanson, 87–101. Agronomy Monograph 15. Madison, WI: ASA, CSSA, and SSSA.

Sybenga, J. 1975. *Meiotic configurations.* New York: Springer–Verlag.

Teoule, E. 1983. Somatic hybridization between *Medicago satiava* L. and *Medicago falcata* L. *Comptes Rendus Academy of Sciences,* Paris Ser. III. 297:13–16.

Tesar, M. B., and V. L. Marble. 1988. Alfalfa establishment. In *Alfalfa and alfalfa improvement,* ed. A. A. Hanson, D. K. Barnes, and R. R. Hill, 303–332. Agronomy Monograph 29. Madison, WI: ASA, CSSA, and SSSA.

Thoquet, P., M. Gherardi, E. P. Journet, and A. Kereszt. 2002. The molecular genetic linkage map of the model legume *Medicago truncatula:* An essential tool for comparative legume genomics and the isolation of agronomically important genes. *BMC Plant Biology* 2:doi:10.1186/1471-2229-2-1.

Tian, D., C. Niu, and R. J. Rose. 2002. DNA transfer by highly asymmetric somatic hybridization in *Medicago truncatula* (+) *Medicago rugosa,* and *Medicago truncatula* (+) *Medicago scutellata. Theoretical and Applied Genetics* 104:9–16.

Trinh, T. H., P. Ratet, E. Kondorosi, P. Durand, K. Kamate, P. Baur, and A. Kondorosi. 1998. Rapid and efficient transformation of diploid *Medicago truncatula* and *Medicago sativa* ssp. *falcata* lines improved in somatic embryogenesis. *Plant Cell Reports* 17:345–355.

Van Ooijen, J. W. 2004. Software for the calculation of QTL positions on genetic maps, MAPQTL 5.0. Plant Research International, Wageningen, the Netherlands.

Van Ooijen, J. W., and R. E. Voorrips. 2001. JoinMap 3.0 software for the calculation of genetic linkage maps. Plant Research International, Wageningen, the Netherlands.

Voorrips, R. E. 2002. MapChart: Software for the graphical presentation of linkage maps and QTLs. *Journal of Heredity* 93:77–78.

Walker, K. A., M. L. Wendeln, and E. G. Jaworski. 1979. Organogenesis in callus tissue of *Medicago sativa.* The temporal separation of induction processes from differentiation processes. *Plant Science Letters* 16:23–30.

Zhu, H., H-K. Choi, D. R. Cook, and R. C. Shoemaker. 2005. Bridging model and crop legumes through comparative genomics. *Plant Physiology* 137:1189–1196.

CHAPTER 3

Wheatgrass and Wildrye Grasses (Triticeae)

Richard R.-C. Wang and Kevin B. Jensen

CONTENTS

3.1 INTRODUCTION

Wheatgrass and wildrye grasses are valued throughout the temperate regions of the world as forage and habitat for livestock and wildlife as well as for other qualities relating to aesthetics, soil stabilization, weed control, and watershed management in semiarid environments (Asay and Jensen 1996a, 1996b). These perennial grasses are members of the Triticeae tribe, which also includes the cultivated cereal crops including wheat (*Triticum* spp.), barley (*Hordeum* spp.), and rye (*Secale cereale* L.); they are often hybridized and used as genetic sources for disease resistance, salinity tolerance, and other traits.

Depending on the taxonomic treatment, between 200 and 250 wheatgrass and wildrye species have been described worldwide. More than two thirds are native to Eurasia; from 22 to 30 are considered native to North America and are distributed throughout the vast prairies of the northern Great Plains of the United States and Canada (Rogler 1973; Cronquist et al. 1977). Relatively few species are found in South America, New Zealand, Australia, and Africa. Wheatgrass and wildrye grasses are generally adapted from subhumid to arid climatic conditions in steppe or desert regions. Both native and introduced forms are used in North American range improvement programs. The most important of these are listed in Table 3.1. Some species, such as intermediate and tall wheatgrass, are found in the arid to semi-irrigated pastures often associated with legumes such as alfalfa. In North America, the wheatgrass and wildrye grasses are most prevalent in the northern Great Plains as well as on the semiarid to arid rangelands of the intermountain and Great Basin regions. In their natural setting, wheatgrass and wildrye grasses are most often found in association with other grasses, sedges, forbs, and shrubs (Asay and Knowles 1985).

Wheatgrasses and wildryes have been previously reviewed by Asay and Jensen (1996a, 1996b), so this chapter concentrates on literature published after 1996.

3.2 DESCRIPTION AND CROP USE

3.2.1 World Production Area and Utilization

China has 392.8 million ha of grasslands and rangelands in which more than 6,000 plant species grow. Chinese wildrye (*Leymus chinensis* (Trin.) Tzvelev) is a predominant species grown on grasslands and rangelands of Inner Mongolia and northeastern China. Thus, most research on wildryes in China has focused on this species for its utilization in animal grazing and land reclamation. Vast hectares of arid to semiarid grasslands and rangelands are scattered throughout the former Soviet Union and are characterized by diverse stands of grasses, legumes, and forbs.

Table 3.1 Names and Genome Constitutions of Wheatgrasses, Wildryes, and Other Triticeae Grasses, Including Weeds

Common Name	Scientific Name	Synonymous Name	Statue	Genome Constitutions	Origin
Fairway crested wheatgrass	*Agropyron cristatum* (L.) J. Gaertner	*Triticum cristatum* (L.) Schreber	C, R	PP, PPPP, PPPPPP	I
Siberian crested wheatgrass	*Agropyron fragile* (Roth) Candargy	*Agropyron sibiricum* (Willd.) P. Beauv.	C	PP	I
Standard crested wheatgrass	*Agropyron desertorum* (Fischer ex Link) Schultes	*Triticum desertorum* Fischer ex Link	C	PPPP	I
Bluebunch wheatgrass	*Psuedoroegneria spicata* (Pursh) A. Löve	*Agropyron spicatum* (Pursh) Scribn. & Smith	C, R	StSt	N
Beardless wheatgrass	*Psuedoroegneria spicata* (Pursh) A. Löve	*Agropyron inermis* (Scribn. & Smith) Rydg.	C, R	StSt	N
Bearded wheatgrass	*Elymus caninus* (L.) L.	*Agropyron caninum* (L.) P. Beauv.	C	StStHH	I
Arizona wheatgrass	*Elymus arizonicus* (Scribn. & J.G. Sm.) Gould	*Agropyron arizonicum* Scribner & J.G. Smith	C	StStHH	N
Alaskan wheatgrass	*Elymus alaskanus* (Scribner & Merr.) A. Löve	*Agropyron alaskanum* Scribner & Merr.	C, R	StStHH	N
Baker's wheatgrass	*Elymus bakeri* (E.E. Nelson) A. Löve	*Agropyron bakeri* E.E. Nelson	C	StStHH	N
Green wheatgrass	*Elymus hoffmannii* K.B. Jensen & Asay		R	StStHH	I, M
Montana wheatgrass	*Elymus albicans* (Scribn. & J.G. Sm.) A. Löve	*Agropyron albicans* Scribn. & J.G. Sm.	R	StStHH	N
Scribner's wheatgrass	*Elymus scribneri* (Vasey) M.E. Jones	*Agropyron scribneri* Vasey	C	StStHH	N
Sierra wheatgrass	*Elymus sierrus* Gould	*Agropyron pringlei* (Scribner & Smith) Hitchc.	C	StStHH	N
Slender wheatgrass	*Elymus trachycaulus* (Link) Gould ex Shinners	*Agropyron trachycaulum* (Link) Malte ex H. F. Lewis	C, R	StStHH	N
Snake River wheatgrass	*Elymus wawawaiensis* J.R. Carlson & Barkworth	*Elymus lanceolatus* (Scribn. & Smith) Gould	C, R	StStHH	N

Continued

Table 3.1   Names and Genome Constitutions of Wheatgrasses, Wildryes, and Other Triticeae Grasses, Including Weeds (Continued)

Common Name	Scientific Name	Synonymous Name	Statue	Genome Constitutions	Origin
Streambank wheatgrass	Elymus lanceolatus ssp. riparius (Scribn. & Smith) Barkworth	Agropyron riparium Scribn. & Smith ex Piper	R	StStHH	N
Thickspike wheatgrass	Elymus lanceolatus (Scribn. & Smith) Gould ssp. lanceolatus	Agropyron dasystachyum (Hook.) Scribn.	R	StStHH	N
Northern wheatgrass	Elymus macrourus (Turcz. Ex Steud.) Tzvelev	Agropyron macrourum (Turcz.) Drobov	C	StStHH	N
Arctic wheatgrass	Elymus violaceus (Hornem.) Feilberg	Agropyron violaceum (Hornem.) Lange	C	StStHH	N
Stebbin's wheatgrass	Elymus stebbinsii Gould	Agropyron parishii Scribner & Smith	C, R	StStHH	N
Intermedium wheatgrass	Thinopyrum intermedium (Host) Barkworth & D.R. Dewey	Agropyron intermedium (Host) Beauv.	R	EEStSt(V-J-R)(V-J-R)	I
Pubescent wheatgrass	Thinopyrum intermedium (Host) Barkworth & D.R. Dewey	Agropyrum trichophorum (Link) Richt.	R	EEJJStSt=EEEEStSt	I
Russian wheatgrass	Thinopyrum junceum (L.) A. Löve	Agropyron junceum (L.) P. Beauv.	R	EEEEEE	I
Tall wheatgrass, rush wheatgrass	Thinopyrum ponticum (Podp.) Z.W. Liu & R.R.C. Wang	Elytigia pontica (Podp.) Holub	C	EEEEEEEEEE	I
Western wheatgrass	Pascopyrum smithii (Rydb.) A. Löve	Agropyron smithii Rydb.	R	StStHHNsNsXmXm	N
Russian wildrye	Psathyrostachys juncea (Fisch.) Nevski	Elymus junceus Fisch.	C	NsNs	I
Altai wildrye	Leymus angustus (Trin.) Pilger	Elymus angustus Trin.	C, R	(NsXm)x6	I
Beardless wildrye	Leymus triticoides (Buckl.) Pilger	Elymus triticoides Buckley	R	NsNsXmXm	N
Beach wildrye	Leymus mollis (Trin.) Pilger and L. arenarius (L.) Hochst.	Elymus mollis Trin.	R	NsNsXmXm	N
Great Basin wildrye	Leymus cinereus (Scribn. & Merr.) A. Löve	Elymus cinereus Scribn. & Merr.	C	NsNsXmXm	N
Giant wildrye	Leymus condensatus (K. Presl) A. Löve	Elymus condensatus K. Presl	C	(NsXm)x2 or x4	N

Table 3.1 Names and Genome Constitutions of Wheatgrasses, Wildryes, and Other Triticeae Grasses, Including Weeds (*Continued*)

Common Name	Scientific Name	Synonymous Name	Statue	Genome Constitutions	Origin
Mammoth wildrye	*Leymus racemosus* (Lam.) Tzvelev	*Elymus gigantis* Vahl	R	NsNsXmXm	I
Pacific wildrye	*Leymus pacificus* (Gould) D.R. Dewey	*Elymus pacificus* Gould	R	NsNsXmXm	N
Alkali wildrye	*Leymus simplex* (Scribn. & T.A. Williams) D.R. Dewey	*Elymus simplex* Scribner & T.A. Williams	R	NsNsXmXm	N
Many-stem wildrye	*Leymus multicaulis* (Kar. & Kir.) Tzvelev	*Elymus multicaulis* Kar. & Kir.	C, R	NsNsNsXmXmXm	I
Many-flowered wildrye	*Leymus x multiflorus* (Gould) Barkworth & R.J. Atkins	*Elymus triticoides* ssp. *multiflorus* Gould	C, R	NsNsNsXmXmXm	N
Salina wildrye	*Leymus salina* (M.E. Jones) Barkworth subsp. *salina*	*Elymus salina* M.E. Jones subsp. *salina*	C	NsNsXmXm	N
Colorado wildrye	*Leymus ambiguus* (Vasey & Scribn.) D.R. Dewey	*Elymus ambiguus* Vasey & Scribner	C, R	NsNsXmXm	N
Boreal wildrye	*Leymus innovatus* (Beal) Pilg.	*Elymus innovatus* Beal	C, R	(NsXm)x2 or x4	N
Yellow wildrye	*Leymus flavescens* (Scribn. & J.G. Sm.) Pilg.	*Elymus flavescens* Scribner & J.G. Smith	C, R	NsNsXmXm	N
Blue wildrye	*Elymus glaucus* Buckley	*Elymus americanus* Vasey & Scribner	C, R	StStHH	N
Great Plains wildrye; Canada wildrye	*Elymus canadensis* L.	*Elymus philadelphicus* L.	C, R	StStHH	N
Dahurian wildrye	*Elymus dahuricus* Turcz. ex Grieseb.	*Elymus cylindricus* (Franch.) Honda	C	StStHHYY	I
Siberian wildrye	*Elymus sibiricus* L.	*Elymus tener* L.	C, R	StStHH	I
Virginia wildrye	*Elymus virginicus* L.	*Elymus striatus* Willd.	C	StStHH	N
Glumeless wildrye	*Elymus hystrix* L.	*Hystrix patula* Moench	C	StStHH	N
European dunegrass; lymegrass	*Leymus arenarius* (L.) Hochst.	*Elymus arenarius* L.	C, R	(NsXm)x4	I
American dunegrass; sea lymegrass	*Leymus mollis* (Trin.) Pilg.	*Elymus mollis* Trin.	R	NsNsXmXm	N

Continued

Table 3.1 Names and Genome Constitutions of Wheatgrasses, Wildryes, and Other Triticeae Grasses, Including Weeds (*Continued*)

Common Name	Scientific Name	Synonymous Name	Statue	Genome Constitutions	Origin
California bottlebrush	*Leymus californicus* (Bol. Ex Thurb.) Barkworth	*Hystrix californica* (Bolander) Kuntze	C, R	(NsXm)x4	I
Tick quackgrass	*Thinopyrum pycnanthum* (Godr.) Barkworth	*Elytrigia pycnantha* (Gordon.) A. Löve	R	EEEEPP	I
Bottlebrush or common squirreltail	*Elymus elymoides* (Rafin.) Sweezey	*Elymus sitanion* Schult.	C	StStHH	N
Big squirreltail	*Elymus multisetus* (J.G. Smith) Burtt Davy	*Sitanion multisetum* J.G. Smith	C	StStHH	N
Medusahead	*Taeniatherum caput-medusae* (L.) Nevski	*Elymus caput-medusae* L.	C	TaTa	I
Quackgrass, couchgrass	*Elymus repens* (L.) Gould	*Elytrigia repens* (L.) Nevski	R	StStStStHH	I

Notes: C = cespitose, R = rhizomatous, N = native to North America, I = introduced into North America, M = man-made hybrids.

Disturbances by wildfires, livestock, and wildlife and varied impacts by humans, including recreational activities, have contributed to degraded conditions on much of America's 800 million acres of rangeland. Undesirable weeds have invaded many ecosystems, and their continuing spread to new areas increasingly threatens most rangelands. Soil erosion has increased on weed-dominated lands and valuable watersheds have been seriously impaired; wildfire frequency has increased and biological diversity has diminished. Seeding wheatgrass and wildrye grasses on rangelands is a valuable tool for restoring burned and disturbed areas and managing noxious weeds (Asay et al. 2001). Due to their rapid establishment characteristics, persistence, and drought tolerance, many of the wheatgrasses and wildryes are components in frequently used seed mixtures on rangeland restoration and revegetation projects (Renault, Qualizza, and MacKinnon 2004; Jones and Larson 2005; Aschenbach 2006).

Expenses associated with producing and feeding stored forage during winter reduce profit margins of livestock operations in North America. On western rangelands where summer regrowth is limited by the availability of water, use of stockpiled wheatgrasses and wildryes as a source of fall and winter forage is gaining in popularity. Additional uses of wheatgrass include use as a low-maintenance turfgrass in areas where irrigation water is limited and maintenance is reduced (Hanks et al. 2005; Robins et al. 2006). With biofuel becoming an important alternative energy source, grasses including the wheatgrasses and wildryes are being evaluated as potential high-yielding plant materials for use in the biofuel industry (Jefferson et al. 2004).

3.2.2 Botany

3.2.2.1 Reproductive System

The wheatgrasses and wildryes cover the full spectrum from complete self-fertility to complete self-sterility (Jensen, Zhang, and Dewey 1990), and from sexual to apomictic (Murphy 2003).

Within this group, there are many more self-fertile species than self-sterile species. About two thirds of the approximately 250 wheatgrasses and wildryes are self-fertilizing. The greatest majority of the self-pollinating species are in *Elymus* and a few are in *Thinopyrum* and *Leymus*. Species within *Agropyron, Pseudoroegneria, Psathyrostachys, Pascopyrum,* and the *E. lanceolatus* (Scribner & Smith) Gould complex are predominately cross-pollinating (Jensen et al. 1990). The long-awned *Elymus rectisetus* (Nees in Lehm.) Á. Löve & Connor is apomictic, the shorter awned *E. scaber* var. *scaber* (R. Br.) Á. Löve varies from facultative apomictic to sexual, and the very short-awned *E. multiflorus* (Banks & Solander ex Hook f.) Á. Löve & Connor, and *E. scaber* (R. Br.) Á. Löve var. *plurinervis* are sexual.

3.2.2.2 Growth Habit and Plant Structure

Growth habit is an important, adaptive trait in the wheatgrasses and wildryes. In general, grasses with caespitose and rhizomatous growth habit are better adapted to xeric and mesic environments, respectively (Barkworth et al. 2007). Caespitose grasses form a compact plant of upright aerial tiller stem branches, whereas sod-forming grasses typically spread by underground stems defined as rhizomes. Despite differences in growth habit, aerial tiller and subterranean rhizome branches are homologous in that they develop from the same type of axillary meristem and eventually form the same basic phytomer organization and structure. The axillary meristems of caespitose grasses grow upwards, emerging within the leaf sheath; the axillary meristems of rhizomatous grasses cut through the leaf sheath and grow outwards. Most *Leymus* species are rhizomatous; however, rhizomatous plants also occur in the following genera: *Agropyron, Pseudoroegneria, Psathyrostachys, Thinopyrum, Pascopyrum,* and *Elymus* (Barkworth et al. 2007). There is a direct correlation between rhizome growth habit and the plant's competitive ability against invasive annual species. Rhizomes provide protection from herbivory and trampling, storage tissues for vegetative propagation, and dispersal.

3.3 ORIGIN, DOMESTICATION, AND DISPERSION

The majority of Triticeae species originated from Eurasia and/or Asia (Tables 3.1 and 3.2); between 22 and 30 are native to North America (Table 3.2). The wheatgrasses and wildryes are broadly adapted to western U.S. rangelands and the eastern Central Great Plains (Vogel and Jensen 2001). Wheatgrasses and wildrye adapted to the eastern Central Great Plains were crested

Table 3.2   Origin and Distribution of Triticeae Species

Genus	Haplome Combinations	Number of Species	Center(s) of Origin	Number of Species in N.A.
Agropyron	P; PP; PPP.	2 to 15	Eurasia	2 introduced
Pseudoroegneria	St; StSt.	15 to 20	Eurasia or Asia	1 native
Elymus	StH; StY; StHY; StPY; StWY.	About 150	Eurasia and Australia	32 native and 7 introduced
Pascopyrum	StHNsXm	Only 1	North America	1 native
Leymus	NsXm; NsNsXmXm.	About 50	East Asia and North America	11 native and 4 introduced
Psathyrostachys	Ns	Only 8	Central Asia	1 introduced
Thinopyrum	E; EE; ESt; EEE; EESt; EEEEE.	About 10	Mediterranean to western Asia	4 introduced

Source: Summarized from Barkworth, M. E. et al. 2007. *Flora of North America North of Mexico,* vol. 24. New York: Oxford University Press.

wheatgrass, Russian wildrye, intermediate and tall wheatgrass, western wheatgrass, and species with *Elymus* and *Leymus*. Species with only the H genome (*Hordeum*) and the St genome (Psuedoroegneria) were not adapted to this region because of poor survival and low forage yield (Vogel and Jensen 2001).

In the intermountain semiarid area of the United States, introduced wheatgrass species (*Agropyron cristatum* (L.) Gaertn.), *A. fragile* (Roth) Candargy, *Thinopyrum intermedium* (Host) Barkworth & D. R. Dewey, and *A. cristatum × A. desertorum* (Fisch. ex Link) Shultes)) outperformed native grasses (*Pseudoroegneria spicata* (Pursh) Á. Löve, *Elymus lanceolatus*, *E. wawawaiensis* J. Carlson & Barkworth, and *Pascopyrum smithii* (Rydb.) Á. Löve) at sites where water was limited (<300 mm). Western wheatgrass stands increased during the seasons after establishment, even though it was difficult (Asay et al. 2001).

3.4 TAXONOMY AND GERMPLASM RESOURCES

The taxonomy and nomenclature for the tribe, which includes the wheatgrasses and wildryes, have been and still are in a state of confusion. Löve (1984) concluded that the taxonomic confusion in the tribe is not only at the species level, but also at the genus level. Possible solutions to the taxonomic problems of the wheatgrasses and wildryes range from recommendations that all species be treated in a single genus to Löve's (1984) treatment in which he recognized 38 genera.

3.4.1 Center of Diversity and Taxonomy

Based on N. I. Vavilov, there are two centers of origin for the wheatgrasses and wildryes: Asia Minor and the Mediterranean region. These centers were in the past associated with the taming of the main domestic animals and the development of husbandry (Vavilov 1987).

The wheatgrasses (Table 3.1) traditionally have been included in the genus *Agropyron,* and the wildryes have been treated as species in the genus *Elymus* (Bowden 1965; Hitchcock 1971). More recently, however, taxonomic realignments were proposed based on genomic or biological relationships as well as plant morphology (Tzvelev 1976; Barkworth, Dewey, and Atkins 1983; Dewey 1983, 1984; Löve 1984; Yen, Yang, and Yen 2005). Wheatgrasses encompasses six genera of Triticeae (*Agropyron, Pseudoroegneria, Thinopyrum, Elytrigia, Elymus,* and *Pascopyrum*) and wildryes belong to three genera (*Psathyrostachys, Leymus,* and *Elymus*) (Table 3.1). A brief description of genera containing the wheatgrasses and wildryes follows. For individual species descriptions, readers are referred to the review of Asay and Jensen (1996a, 1996b). Additional reference material and taxonomic keys for the wheatgrasses and wildryes can be found in the *Flora of North America North of Mexico,* vol. 24 (Barkworth et al. 2007).

3.4.1.1 Wheatgrasses

Agropyron in its restricted genomic sense consists of just the crested wheatgrasses, with no more than 10 species, all from Eurasia (Tzvelev 1976). The three common species are *A. cristatum* (broad-spiked taxa), *A. desertorum* (cylindrical-spiked taxa), and *A. fragile* (narrow-spiked taxa). The remaining species listed by Tzvelev are rather uncommon endemics. *Agropyron* species occur at three ploidy levels: diploid ($2n = 2x = 14$), tetraploid ($2n = 4x = 28$), and hexaploid ($2n = 6x = 42$). However, all of these taxa are based on one basic genome (**P**), which means that the polyploidy races are autoploid or near autoploid (Dewey and Asay 1982).

Pseudoroegneria consists of about 15 species that were previously included in *Agropyron* and *Elytrigia* (Löve 1984). The genus consists of diploid ($2n = 2x = 14$) and tetraploid ($2n = 4x = 28$) taxa, all of which contain the **St** genome or some variation of it. As with *Agropyron,* the polyploid races of *Pseudoroegneria* are autoploids or near autoploids (Stebbins and Pun 1953; Dewey 1970). In addition to being the foundation of *Pseudoroegneria*, the **St** genome is a component of all the species of *Elymus*. Within *Pseudoroegneria*, there is a group of species *P. deweyi* K. B. Jensen, S. L. Hatch, and J. K. Wipff and *P. tauri* (Bioss. & Bal.) Á. Löve that are composed of the **St** and **P** genomes (Jensen, Hatch, and Wipff 1992).

Thinopyrum is a genus erected by Löve (1984) that, according to Dewey's (1984) definition, encompasses about 20 species that had previously been in traditional *Agropyron* or *Elytrigia* (Tzvelev 1976). As treated by Dewey (1984), *Thinopyrum* is morphologically and genomically diverse. Löve (1984), in the "Conspectus of the Triticeae," recognized within Dewey's (1984) *Thinopyrum* the following genera: *Thinopyrum, Lophopyrum,* and *Elytrigia. Thinopyrum* as treated herein consists of three species complexes: (1) *T. junceum* (L.) A. Löve; *Th. elongatum* (Host) D. R. Dewey; and *T. intermedium.* Species in this genus possess the **J** or **E** genome (which Dewey, 1984, designated as **J=E**) and sometimes contain the **St** genome (Liu and Wang 1993a; Kishii, Wang, and Tsujimoto 2005). *Thinopyrum* consists of diploids, segmental allotetraploids, segmental allohexaploids, and genomically complex octoploids and decaploids.

Elytrigia as defined genomically is a small genus of about five species with chromosome numbers of $2n = 42$ or 56. All species are complex segmental autoallopolyploids. *Elytrigia repens* L. ($2n = 42$), the type species, is known to contain two genomes (St_1St_2) from *Pseudoroegneria* and an **H** genome from the small seeded *Hordeum* species (Assadi and Runemark 1995).

Elymus is by far the largest genus of the wheatgrasses and wildryes (Dewey 1984). This genus contains about 150 species with various genome combinations based on the **St** genome, in combination with one or more of **H, Y, W,** or **P** genomes (Wang et al. 1995). Yen et al. (2005) divided *Elymus* into six genera strictly based on genome compositions: *Douglasdeweya* (**PPStSt**), *Roegneria* (**StStYY**), *Australoroegneria* (**StStWWYY**), *Kengylia* (**PPStStYY**), *Campeiostachys* (**HHStStYY**), and *Elymus* (**StStHH, StStStHH,** and **StStHHHH**). Octoploids ($2n = 56$) are rare in this genus.

Pascopyrum is a monotypic genus consisting of *P. smithii* and is always octoploid ($2n = 8x = 56$) (Dewey 1984). It is a true allooctoploid with a genomic formula of **StStHHNsNsXmXm** (Wang et al. 1995). Intermediate parents came from *Elymus* (*E. lanceolatus;* **StStHH**) and *Leymus* (*L. triticoides* (Buckley) Pilg.; **NsNsXmXm**), and its diploid ancestors came from *Pseudoroegneria* (**StSt**), *Hordeum* (**HH**), *Psathyrostacys* (**NsNs**), and an unknown diploid (**XmXm**). Based on chloroplast DNA, its maternal lineage traces through *E. lanceolatus* to a diploid *Pseudoroegneria* species (Jones, Redinbaugh, and Zhang 2000).

3.4.1.2 Wildryes

Psathyrostachys is a small genus of about 10 species, all of which contain the basic N genome. The genus was erected by S. A. Nevski (1934) and consists of species previously in traditional *Elymus* or *Hordeum and* composed of the **Ns** genomes. With the exception of Russian wildrye (*Psathyrostachys juncea* (Fisch.) Nevski), which has $2n = 14$ and $2n = 28$ chromosome numbers, all other species studied behave cytologically as diploids ($2n = 14$). The **Ns** genome of *Psathyrostachys* is one of the progenitors to the polyploid genus *Leymus* (Dewey 1984).

Leymus is a polyploid genus of about 30 species worldwide (Dewey 1984). All species of *Leymus* are based on the **Ns** genome of the genus *Psathyrostachys* and the **Xm** genome of an unknown origin (Wang et al. 1995; Jensen and Wang 1997; Zhang et al. 2006) (Figure 3.1). There is some controversy as to the origin of the **Xm** genome. It has been proposed that the genomic composition of *Leymus* is **Ns_1Ns_2** (Zhang and Dvorák 1991; Anamthawat-Jónsson and Bödvarsdóttir 2001). More than half of the *Leymus* species are allotetraploids ($2n = 28$) represented genomically as **NsNsXmXm.** The higher polyploidy species ($2n = 42–84$) are complex autoallopolyploids.

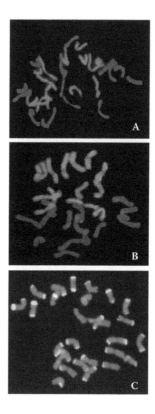

Figure 3.1 (See color insert following page 274.) Molecular cytogenetics revealing the **NsXm** genome constitution in *Leymus* species. (A) 14 chromosomes of *Hystrix duthiei* ssp. *longearistata* (now a *Leymus* species) show red fluorescence when being probed with **Ns**[h]-genome DNA of *Psathyrostachys huashanica* and blocked with **E**[e]-genome DNA of *Thinopyrum elongatum*. (B) 14 chromosomes of *Hystrix duthiei* ssp. *duthiei* (now a *Leymus* species) show red fluorescence when being probed with **Ns**[h]-genome DNA of *Psathyrostachys huashanica* and blocked with **E**[e]-genome DNA of *Thinopyrum elongatum*. (C) Fluorescence *in situ* hybridization (FISH) sites of the pLrTail-1 (red) and pLrPstI-1 (green) repetitive sequences in somatic chromosomes of *Leymus racemosus* accessions PI 502402. The pLrPstI-1 repetitive sequence is not found in four **Ns**-genome *Psathyrostachys* species, making it an **Xm**-originated repetitive sequence. (A and B are provided by Dr. Rui-wu Yang of Sichuan Agricultural University, China. C is one that has not been used in Wang, R. R.-C. et al. 2006. *Genome* 49:511–519.)

3.4.2 Germplasm Resources

3.4.2.1 Ex Situ *Collections*

Wheatgrasses and wildryes are being preserved as seed in many gene banks (Table 3.3). Of these *ex situ* collections, the National Plant Germplasm System (NPGS) located in Pullman, Washington, holds the largest number of accessions and more detailed database information on each accession for the wheatgrasses and wildryes. NPGS lists approximately 380 accessions of *A. cristatum*. Most of the *A. cristatum* accessions were collected from regions of the former Soviet Union (140), Iran (66), China (33), Turkey (31), and Mongolia (20). *Agropyron desertorum* is represented by 107 accessions in GRIN, of which 98 originated from within regions of the former Soviet Union. There are currently 124 accessions of Siberian wheatgrass *A. fragile* in the NPGS, and all but four were collected from the former Soviet Union. There are nearly 1,000 *Elymus* accessions constituting more than 300 species/hybrids within the NPGS collection. The genus *Elytrigia* has only 17 accessions (three species) listed. The *Pseudoroegneria* NPGS collection consists of 229 accessions across eight species. There are 534 accessions of *Thinopyrum* representing 12 species.

Table 3.3 *Ex Situ* Collections of Wheatgrasses and Wildryes

Organization	Location	Genus	No. Species	No. Accessions
USDA-ARS National Plant Germplasm System (http://www.ars-grin.gov/)	Pullman, Washington	Agropyron	17	807
		Elymus	137	2135
		Elytrigia (including Thinopyrum)	26	923
		Leymus	34	850
		Pascopyrum	1	73
		Pseudoroegneria	15	268
		Psathyrostachys	6	223
	Total	7	236	5279
Nordic Gene Bank (http://www.nordgen.org/ngb/)	Arslev, Sweden	Agropyron	1	2
		Elymus	50	273
		Leymus	2	27
		Thinopyrum	1	2
		Pseudoroegneria	2	3
		Psathyrostachys	1	3
	Total	6	57	310
PGRDEU (http://www.genres.de/pgrdeu/)	Gatersleben, Germany	Agropyron (including some other genera)	18	87
		Elymus	21	266
		Leymus	7	59
		Psathyrostachys	2	5
		Pascopyrum	1	1
		Roegneria	?	94
	Total	6	>50	512
CAAS-ICGR (http://icgr.caas.net.cn/)	Beijing, China	Agropyron (including other genera)	?	354
		Elymus	10	28
		Thinopyrum	3	50
		Leymus	3	4
		Psathyrostachys	2	30
		Roegneria	2	2
	Total	6		468

The Forage and Range Research Unit at Logan, Utah, maintains a seed collection of wheatgrasses and wildryes that contains more than 400 accessions of *Agropyron* (4 species), 1,188 accessions of *Elymus* (118 species), 106 accessions of *Elytrigia* (4 species), 450 accessions of *Leymus* (28 species), 31 accessions of *Pascopyrum* (1 species), 86 accessions of *Pasthyrostacys* (4 species), 349 accessions of *Pseudoroegneria* (11 species), and 220 accessions of *Thinopryum* (15 species). Seed requests should be made through the NPGS at Pullman, Washington.

3.4.2.2 *Wild and Weedy Relatives*

All undeveloped accessions of wheatgrass or wildrye species can be treated as wild or weedy relatives of named cultivars of wheatgrass or wildrye. The most fitting example is the quackgrass weed in relation to the NewHy RS-wheatgrass cultivar; both of them have the same genome constitution, **StStStStHH,** and the former is a parent for the latter.

3.4.2.3 Core Collections

Because there are much fewer cultivars of wheatgrass and wildrye than of major grain crops, the need to develop core collections for these two forage crops is not urgent at the present.

3.4.2.4 Special Genetic Stocks

Special genetic stocks developed by cytogeneticists aim at transferring useful genes from wild Triticeae species to bread wheat (*Triticum aestivum* L.). Derived from hybrids between wheat and wild Triticeae grasses, these genetic stocks are in the forms of amphidiploids, addition lines, substitution lines, and translocation lines (Table 3.4).

Two genetic stocks derived from the $E^bE^bE^eE^e$ (=**JJEE**) amphiploids of *Thinopyrum bessarabicum* (Savul & Rayass) Löve × *Th. elongatum* (Host) D. R. Dewey have been released (Wang 2006). The original diploid hybrid and amphidiploid plants had been used to assess the genome relationship between E^b (=**J**) and E^e (=**E**) (Wang 1985; Jauhur 1988; Wang and Hsiao 1989). Scientists generally accept the closeness between the **J** and **E** genomes and recognize them as different versions of the same basic genome designated as E^b and E^e or J^b and J^e. Germplasms TBTE001 and TBTE002 were intended for scientists who want to study leaf glaucousness or to isolate the gene controlling this trait (Wang 2006). These genetic stocks are also useful in transferring genes from the two diploid species to tetraploid *Thinopyrum* species having the same genome constitution, such as *Th. junceiforme* (Löve & Löve) Löve and *Th. sartorii* (Bioss. & Helder.) Á. Löve (Liu and Wang 1992). A trisomic series within the **P** genome of crested wheatgrass was developed by Imanywhoha and Jensen (1994).

3.5 CYTOGENETICS

3.5.1 Genome and Phylogenetic Relationships among Species

3.5.1.1 Genome Designation

At the First International Triticeae Symposium, the Committee on Genome Designation was formed. This committee made recommendations on "rules for genome designation and current genome designations in Triticeae" at the Second International Triticeae Symposium (Wang et al. 1995). To avoid the confusion caused by the use of the same genome symbols in both annual and perennial Triticeae species, the symbols of the latter were changed to two letters. Thus, the **S** of *Pseudoroegneria* was changed to **St** and **N** of *Psathyrostachys* was changed to **Ns.** The unknown genome in *Leymus* species was designated as **Xm,** whereas the genomes in *Hordeum marinum* and *H. murinum* were designated as **Xa** and **Xu,** respectively. The symbols **J** and **E** of *Thinopyrum* and *Lophopyrum* were changed to E^b and E^e, respectively. Thereafter, perennial Triticeae genera classified by Dewey (1984) and Löve (1984) based on their genome constitutions are *Agropyron* (**P** genome), *Pseudoroegneria* (**St**), *Psathyrostachys* (**Ns**), *Australopyrum* (**W**), *Hordeum* (**H, I, Xa, Xu**), *Thinopyrum* (**E, ESt, EE, EESt, EEEEE,** etc.), *Elymus* (**StH, StY, StP, StHY, StPY, StWY,** etc.), *Leymus* (**NsXm, NsNsXmXm,** etc.), and *Pascopyrum* (**StHNsXm**). The unknown genome **Y** in *Elymus* has been suggested to share a common ancestral genome with the **St** genome of *Pseudoroegneria* (Liu et al. 2006).

Table 3.4 Special Genetic Stocks Having Chromosomes of Wheatgrass or Wildrye Grasses

Institution or Person Holding the Genetic Stocks	Genetic Stock	No. Lines
Adam Lukaszewski's collection, Univ. of California, Riverside	*Thinopyrum* transfers of ER Sears:	
	Bread wheat 3D.3Ag translocation lines	17
	Bread wheat 7D.7Ag translocation lines	14
Wheat Genetic Resource Center, Kansas State University	Alien addition lines	373
	Alien substitution lines	248
	Amphiploids	124
	Bread wheat/*Thinopyrum* translocation lines	22
The John Innes Center, UK	Bread wheat/intermedium wheatgrass partial amphiploid	1
	Bread wheat/intermedium wheatgrass addition lines	10
	Bread wheat/intermedium wheatgrass substitution lines	5
National BioResource Project, Japan	Bread wheat/*Leymus racemosus* addition lines	15
	Bread wheat/*Leymus racemosus* substitution lines	2
	Durum wheat/intermedium wheatgrass amphidiploid	1
	Bread wheat/intermedium wheatgrass partial amphiploid	2
	Bread wheat/intermedium wheatgrass addition lines	12
	Bread wheat/*Th. elongatum* addition lines	13
	Bread wheat/*Leymus mollis* addition lines	3
	Bread wheat/*Psathyrostachys huashanica* addition lines	5
	Bread wheat/*Elymus trachycaulus* addition lines	25
	Bread wheat/*Elymus trachycaulus* translocation lines	1
	Bread wheat/*Elymus ciliaris* addition lines	7
	T. durum/*Th. elongatum* amphidiploid	1
Richard R.-C. Wang, USDA-ARS FRRL, Logan, Utah	Bread wheat/*Elymus rectisetus* addition lines	4
	Bread wheat/*Elymus rectisetus* substitution lines	3
	Lines of A. Charpentier (France):	
	Bread wheat/*Thinopyrum junceum* amphiploids	9
	Bread wheat/*Thinopyrum junceum* addition lines	13

3.5.1.2 *Crossing Affinity*

Natural interspecific and intergeneric hybrids in Triticeae are frequently found (Jensen et al. 1999; Barkworth et al. 2007), attesting to the crossability between and promiscuity of different species. However, hybridization between diploid species of perennial Triticeae is difficult without the aid of embryo rescue. Diploid intergeneric hybrids have been synthesized for genome analysis using chromosome pairing (Wang 1986a, 1986b, 1987a, 1987b, 1987c, 1988b, 1990, 1992a, 1992b).

3.5.1.3 Chromosome Pairing

Kimber (1984) reviewed the different methods proposed for assessing biological and phylogenetic relationships. The fundamental principle of classical genome analysis is that like (homologous) chromosomes pair during meiosis and unlike (nonhomologous) chromosomes do not pair. Homologous chromosomes are defined as having the same nucleotide base sequence; homeologous chromosomes have only residual homology from homologous chromosomes (Rieger, Michaelis, and Green 1968). Stebbins (1971) concluded that the most important factor determining the nature of chromosome pairing within species and hybrids is the structural and chemical similarity, or homology, of the chromosomes. The level of chromosome pairing in a species hybrid reflects the degree of biological relationships between parental species (Dewey 1982; Sadasiviah and Weifer 1981; Wang 1989a, 1990; Wang and Berdahl 1990; Liu and Wang 1992, 1993a, 1993b; Jones, Wang, and Li 1995). Chromosome pairing patterns in diploid and triploid intergeneric hybrids having different genome combinations were analyzed to elucidate genome relationships among genomes of perennial Triticeae (Wang 1992a, 1992b).

3.5.1.4 Molecular Methods

Various molecular markers, such as restriction fragment length polymorphism (RFLP) and random amplified polymorphic DNA (RAPD), have been employed to study genome and phylogenetic relationships among species of Triticeae (Wang and Wei 1995; Wei and Wang 1995; Svitashev et al. 1998). The nucleotide sequence of the internal transcribed spacer (ITS) of nuclear rDNA has been utilized for studying phylogenetic relationships of the monogenomic species of Triticeae (Hsiao et al. 1995). Lately, the high-throughput amplified fragment length polymorphism (AFLP) technique has been used to study 46 *Elymus elymoides* (Rafin.) Sweezy and 13 *E. multisetus* (J. G. Smith) Davy accessions along with 9 other *Elymus* species (Larson, Palazzo, and Jensen 2003b). In that study, *E. sibiricus* L., *E. mutabilis* (Drobov) Tzvelev, and *E. caninus* (L.) L. grouped in one cluster. *Elymus wawawaiensis* and *E. lanceolatus* formed one cluster, and *E. canadensis* L. was grouped with *E. hystrix* L. in another cluster. *E. glaucus* Buckl. and *E. trachycaulus* (Link) Gould ex Shinners were singly situated between the two clusters. *Elymus elymoides* was more closely related to *E. multisetus* than to other *Elymus* species, supporting the taxonomic treatment of *Elymus* sect. *Sitanion* (Raf.) Á. Löve comprises these two species.

Even EST-simple sequence repeat (SSR) markers from tall fescue were useful in phylogenic analysis of a wide range of cool-season forage grasses (Mian et al. 2005). Barley expressed sequence tag-simple sequence repeat (EST-SSR) markers have also been utilized to construct a dendrogram depicting relationships among selected annual and perennial Triticeae species (Hagras et al. 2005). A single copy gene, encoding plastid acetyl-CoA carboxylase, was used to elucidate relationships among *Hystrix, Leymus,* and their relatives (Fan et al. 2007). Fan and colleagues found that *Hystrix coreana* (Honda) Ohwi (formerly *Elymus coreana* Honda and now *Leymus coreanus*), *Hy. duthiei* (=*Elymus duthi* (Stapf)), and *Hy. duthiei* ssp. *longearistata* (=*Hystrix longeristata* (Hack.) Honda) are closely related to the species of *Leymus* by having the **NsXm** genomes. They also concluded that the **Xm** genome in *Leymus* is different from the **E**[b] and **E**[e] genome of *Th. bessarabicum* and *Th. elongatum,* respectively, which are related to the **St** genome of *Pseudoroegneria* species. These results are consistent with those obtained from morphological and cytological studies (Jensen and Wang 1997; Zhang et al. 2006).

Chloroplast DNA data are useful in phylogenetic analysis to identify the maternal genome donor of allopolyploid species (Redinbaugh, Jones, and Zhang 2000; Mason-Gamer, Orme, and Anderson 2002). These studies demonstrated that all North American *Elymus* wheatgrass taxa originated from intergeneric hybrids having a *Pseudoroegneria* species as the maternal parent.

Molecular cytogenetic techniques, such as genomic *in situ* hybridization (GISH), fluorescence *in situ* hybridization (FISH), and multicolor FISH (MC-FISH), are additional tools for genome analysis to ascertain genome relationships (Zhang, Dong, and Wang 1996; Kishii, Wang, and Tsujimoto 2003, 2005). Employing both FISH and RFLP of 18S-26S rDNA, Anamthawat-Jónsson and Bódvarsdóttir (2001) confirmed the close relationship between *Psathyrostachys* and *Leymus* but suggested complex relationships among geographically distinct *Leymus* species rather than having a simple $(NsXm)_n$ formula for all *Leymus* species.

3.5.2 Polyploid Complexes

The perennial Triticeae grasses have many examples of autopolyploid, allopolyploid, and autoallopolyploid complexes. Crested wheatgrass (*Agropyron* spp.) complex is composed of diploid, tetraploid, and hexaploid species made up of only one genome, **P**, and should be treated as a common gene pool (Jensen et al. 2006a). Although most *Psathyrostachys* species are diploid, tetraploid cytotypes have been found in Russian wildrye, *Ps. Juncea* (Fisch.) Nevski. Additionally, artificial tetraploid Russian wildrye has been produced and used in plant breeding (Jensen et al. 2005a).

3.5.3 Chromosomal Aberrations—Structural and Numerical Changes

Structural and numerical changes in chromosomes of wheatgrass and wildrye grasses have not been thoroughly investigated; thus, few published papers are available. Structural changes in Russian wildrye were detected using chromosomal C-banding techniques (Wei, Campbell, and Wang 1995). Aneuploids are rare in diploid species of wheatgrass and wildrye. Trisomics of crested wheatgrass and Russian wildrye have been reported (Imanywhoha and Jensen 1994; Wei et al. 1995). In seed lots of open-pollinated polyploid species, however, odd-numbered ploidy (such as triploid and pentaploid) plants are often found. This could be the result of outcrossing between polyploid species and diploids/polyploids of the same or different genera, which is the basis of reticulate evolution in Triticeae (Kellogg 1989; Kellogg, Appels, and Mason-Gamer 1996; Mason-Gamer 2004). On the other hand, haploidization through chromosome elimination and polyploidization through unreduced gametes (usually female gametes) are common mechanisms leading to numerical changes of chromosomes in wheatgrass and wildrye grasses (Wang 1987b, 1988a, 1988c; Jain, Sopory, and Veilleux 1996; Refoufi, Jahier, and Esnault 2001).

3.5.4 Linkage Mapping

3.5.4.1 Chromosome Map

Chromosome karyotypes in 22 diploid species of perennial Triticeae have been studied to establish the genome relationships among **P, St, J** (=E), **H, I, Ns, W**, and **R** genomes (Hsiao, Wang, and Dewey 1986). Almost without exception, karyotypes of species within a genus manifest a unique pattern specific to the basic genome for that genus. Earlier, C-banding patterns were studied in 10 diploid species encompassing five basic genomes: **P, Ns, J** (=E), **St**, and **W** (Endo and Gill 1984). Based on C-banding patterns, Endo and Gill questioned the equivalence of **J** and **E**. Studying 10 accessions of diploid Russian wildrye (**Ns**), Wei et al. (1995) observed variations in C-banding patterns not only between accessions but also between homologues within an accession. Therefore, difference in C-banding patterns alone is not an indication that the genomes do not belong to the same basic genome. Using C-banded karyotypes, Linde-Laursen, von Bothmer, and Jacobsen (1992) elucidated relationships in the genus *Hordeum,* which includes the four basic genomes **H, I, Xa,** and **Xu.**

3.5.4.2 Genetic and Molecular Map

Molecular investigations on wheatgrass and wildrye grasses lagged far behind those on cereals. As a result, polymerase chain reaction (PCR)-based molecular markers such as AFLP and SSR markers, rather than RFLP markers, were commonly used in mapping studies of perennial grasses. Comparative genomics and synteny facilitate the construction of molecular maps of wheatgrass and wildrye grasses. Mullan et al. (2005) mapped 22 SSR markers to chromosomes of *Thinopyrum elongatum;* nine of them were mapped to homoeologous chromosome locations compared to wheat chromosomes; the remaining markers were mapped to nonhomoeologous chromosomes due to rearrangements of chromosomes during speciation.

Polymorphism and linkage mapping of the gene for 6-SFT in many cool-season grasses have been investigated (Wei et al. 2000). A molecular linkage map has been constructed for 14 chromosomes in the interspecific hybrid *Leymus* mapping populations (Wu et al. 2003). Building on this molecular map of *Leymus,* additional SSRs, sequence tagged site (STS) markers, and quantitative traits loci (QTLs) for many traits were mapped (Hu et al. 2005; Larson et al. 2006; Larson and Mayland 2007). QTLs controlling spring regrowth (in terms of new tillers, new leaves, leaf length, and dry matter weight) and metabolite accumulation (percent soluble carbohydrate and anthocyanin coloration) were mapped using AFLP markers to 14 linkage groups in the two mapping populations (Hu et al. 2005). QTLs for circumference of rhizome spread, reproductive tillers, anthesis date, and plant height were mapped in the *L. cinereus* x *L. triticoides* mapping populations (Larson et al. 2006). Two of the four rhizome QTLs are in homoeologous regions of linkage groups 3a and 3b in both mapping families. A major plant height QTL may correspond with dwarfing mutations on barley 2H and wheat 2A chromosomes. Fiber, protein, and mineral content QTLs were comparatively mapped (Larson and Mayland 2007) (Figure 3.2). All these mapping studies revealed 14 linkage groups that represented two different genomes, **Ns** and **Xm.**

3.6 GERMPLASM ENHANCEMENT AND CONVENTIONAL BREEDING

3.6.1 Cultivar Development

3.6.1.1 Wheatgrasses

Since its introduction from Asia in the early 1900s, crested wheatgrass (*Agropyron cristatum* (L.) Gaertn., *A. desertorum* (Fisch. ex Link) Schultes) has become the major cool-season grass used to improve semiarid rangelands of western North America. This widely adapted cool-season perennial grass is a complex of diploid ($2n = 14$), tetraploid ($2n = 28$), and hexaploid ($2n = 42$) species. On the basis of chromosome-pairing relationships in hybrids among species in the complex, Dewey (1969, 1974) concluded that the same basic genome **P,** modified by structural rearrangements, occurred at the three ploidy levels and that the crested wheatgrasses should be treated as a single gene pool.

Cultivars of crested wheatgrass within the different ploidy levels have been released in North America (Asay and Jensen 1996a). However, until recently, genetic improvement in these cultivars has been restricted to selection and hybridization within ploidy levels. The most progress from interploidy breeding to date has been achieved at the 4x level. The cultivar Hycrest, which was developed from a hybrid between induced tetraploid *A. cristatum* and natural tetraploid *A. desertorum,* was released in 1984 (Asay et al. 1985a). Hybridization schemes involving 6x–2x, 6x–4x, and 4x–2x have shown potential for expanding the genetic resources of 4x breeding populations (Asay and Dewey 1979; Dewey 1969, 1971, 1974; Dewey and Pendse 1968; Jensen et al. 2006a). Dewey

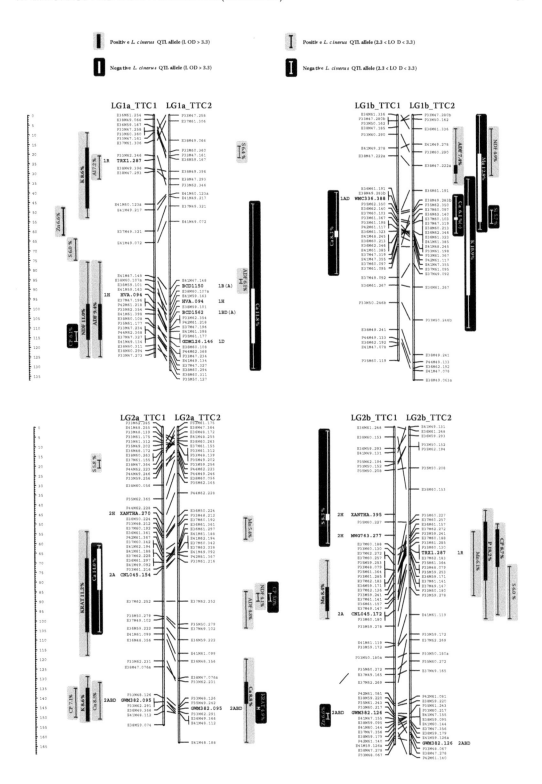

Figure 3.2 Molecular map of fiber, protein, and mineral content QTLs in two full-sib families of *Leymus* interspecific hybrids. (Taken from Larson, S. R., and H. F. Mayland. 2007. *Molecular Breeding* DOI 10.1007/s11032-007-9095-9.)

Continued

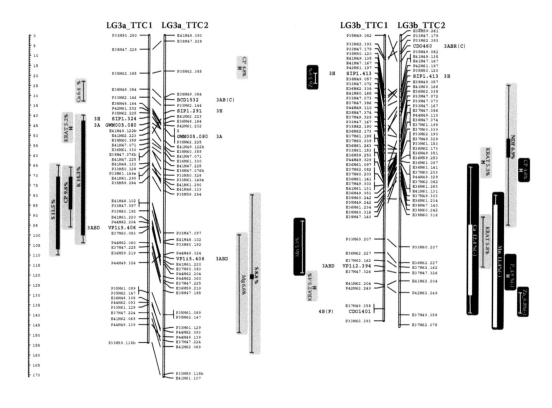

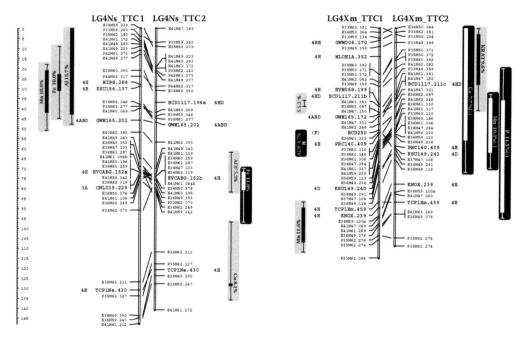

Figure 3.2 *Continued.*

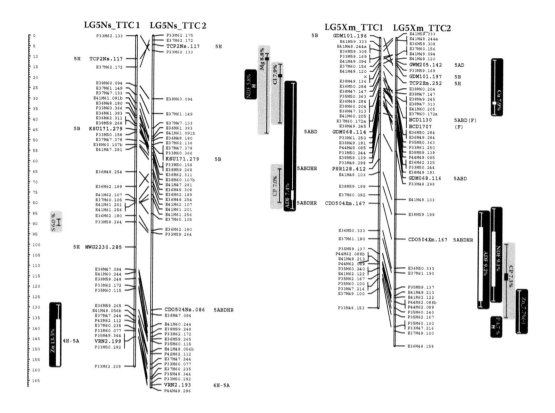

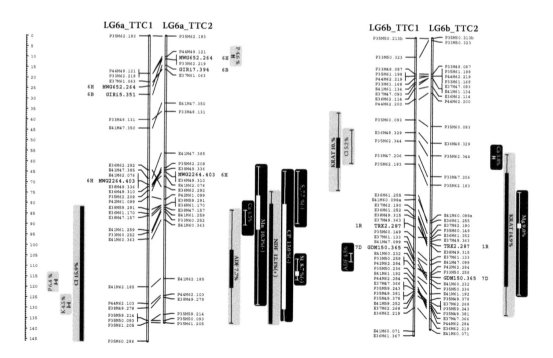

Figure 3.2 *Continued.*

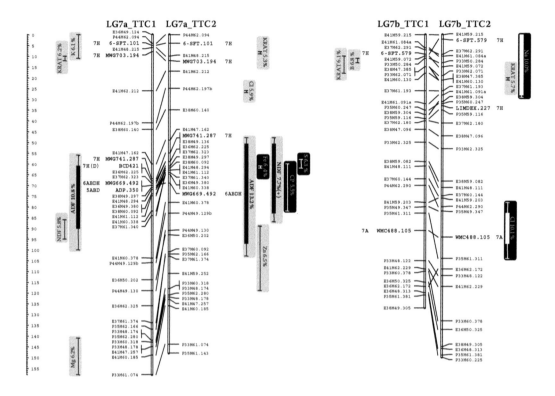

Figure 3.2 *Continued.*

and Pendse's (1968) data suggested that selection for improved fertility would be effective in these interploidy hybrids.

The hexaploid accession—(6x–BL), PI 406442, from the former Soviet Union—is characterized by exceptionally broad leaves that maintain their green color 2–3 weeks longer in the growing season than typical crested wheatgrass. A breeding program is in progress to transfer the broadleaf and color-retention characters into the genetic background of the tetraploid cultivar Hycrest. Newly released crested and Siberian wheatgrass cultivars include Hycrest II and Vavilov II, respectively.

Intermediate wheatgrass plant introduction PI 98568 was released as the cultivar Ree and has subsequently provided parental germplasm for the cultivars Chief, Greenar, Nebraska 50, Oahe, and Slate (Asay and Jensen 1996a). The cultivars Reliant (Berdahl et al. 1992) and Mansaka (Berdahl et al. 1993) were released for their improved persistence, forage quality, animal performance, and forage and seed yields. Rush intermediate wheatgrass (released in 1996) is a selection from a population received as PI 281863 that was selected for its superior seedling emergence and high forage production on dryland sites receiving less than 12 inches of precipitation. More recently released intermediate wheatgrass cultivars include Beefmaker (Vogel et al. 2005a) and Haymaker (Vogel et al. 2005b). Haymaker intermediate wheatgrass is a very broadly adapted high-yielding intermediate wheatgrass. Beefmaker intermediate wheatgrass is a cultivar developed for improved forage quality. See Alderson and Sharp (1994) for a more detailed description of cultivars released.

Commercially available tall wheatgrass cultivars have originated from a very narrow genetic base—predominately from PI 98526 introduced from the former Soviet Union in 1932 (Alderson and Sharp 1994). The first cultivar, Largo, was released in 1937 (Alderson and Sharp 1994). Cultivars Alkar and Jose were released in 1958 and 1965, respectively, with selections originating from PI 98526 and PI 150123, respectively (Alderson and Sharp 1994). Other released cultivars—Orbit, Tyrrell, and Platte—originated from PI 98526.

The first commercial cultivar of bluebunch wheatgrass was Whitmar, which originated from a population collected in the Palouse of eastern Washington in the 1940s. The cultivar Goldar originated from material collected on Malley Ridge, Umatilla Forest, in Asotin, Washington, and was released in 1989 (Alderson and Sharp 1994). More recent releases include P7 in 2001, which is described as a multiple-origin polycross generated by intermating 23 open-pollinated, native-site collections, and two cultivars from Washington, Oregon, Nevada, Utah, Idaho, Montana, and British Columbia (Jones et al. 2002). The cultivar Secar was originally released as a cultivar of bluebunch wheatgrass; however, based on cytogenetic data, it is more closely related to the thick-spike wheatgrass complex in the genus *Elymus*. Secar was derived from germplasm collected near Lewiston, Idaho (Morrison and Kelly 1981). Recently, the Forage and Range Research Lab released the Snake River wheatgrass cultivar Discovery (Jones, 2008).

Early seedings of western wheatgrass were made with seed harvested from native stands. The cultivars Barton, Rosana, Arriba, and Flintlock were released in the 1970s (Alderson and Sharp 1994). The cultivar Walsh is a 20-clone synthetic developed from 468 ecotypes and was selected for increased forage and seed yields (Smoliak and Johnston 1984). Selection within a native collection in the Missouri river bottoms near Mandan resulted in the release of the cultivar Rodan (Barker, Berdahl, and Jacobson 1984).

Slender wheatgrass cultivars Primar, Revenue (Crowle 1970), Pryor, and San Luis all originated from single collections. Recently, the cultivar FirstStrike, with increased establishment characteristics and combining collections from Wyoming and Colorado, was released (Jensen et al. 2007).

3.6.1.2 Wildryes

Russian wildrye is native to the steppe and desert regions of Russia, Kazahkstan, and China. Its range of adaptation is similar to crested wheatgrass, and it performs well with 200–400 mm of annual precipitation. The first cultivar of Russian wildrye, Vinall, was released in 1960 as a five-clone synthetic selected for increased seed yield (Alderson and Sharp 1994). Diploid cultivars Cabree, Mayak, and Swift were released by Agriculture Canada between 1971 and 1978 and selected for seed retention, high forage and seed yield, and seedling vigor, respectively (Alderson and Sharp 1994). The only tetraploid ($2n = 28$) cultivar Tetracan was released in 1988 based on increased spring vigor, leafiness, seed production, and reduced lodging (Alderson and Sharp 1994).

Bozoisky-Select is a winter-hardy, drought-resistant bunchgrass that is widely adapted to semi-arid rangelands. It was derived from germplasm obtained form the former Soviet Union (Asay et al. 1985b). During its development, selection was based primarily on traits associated with seedling and vegetative vigor, tolerance to abiotic and biotic stresses, leafiness, and seed yield. Bozoisky-Select is an excellent source of forage during the spring and summer; because its leaves cure relatively well during the fall, it is often used for grazing during the late fall and winter. Although it has improved seedling vigor compared to other diploid cultivars, it is more difficult to establish than grasses such as crested wheatgrass.

Bozoisky-II Russian wildrye was selected for seedling vigor (emergence from a deep planting depth), seed mass, seed yield, vegetative vigor, total dry matter production, and response to drought. Bozoisky-II has a much broader genetic base than other Russian wildrye cultivars and has been evaluated extensively on rangeland sites in the western United States. Seedling establishment of Bozoisky-II has been equal to or greater than that of commercially available cultivars (Jensen et al. 2006b).

Altai wildrye (*Leymus angustus* (Trin.) Pilger) was introduced from Asia but relatively little breeding work has been done on it. Altai wildrye is productive and has considerable potential for grazing during the fall and winter, but it is difficult to establish on range sites. Mustang Altai wildrye (Jensen, Larson, and Waldron 2005) was developed from a selected Asian collection after successive cycles of selection for characters related to stand establishment.

There are only two cultivars of basin wildrye: Magnar and Trailhead (Alderson and Sharp 1994). Magnar is octoploid, blue-green in color, and grows tall, with broad course leaves; it is frequently used for winter forage and wind barriers. Trailhead was selected from 125 collections of basin wildrye in Montana and Wyoming. This tetraploid cultivar is more drought tolerant than Magnar and is used as forage early in the spring and winter (Asay and Jensen 1996b). Washoe germplasm basin wildrye is a prevarietal selected class release from a naturally occurring germplasm originating near Anaconda, Montana. An induced octoploid cultivar, Continental, was recently released by the Forage and Range Research Lab for its increased seedling establishment characteristics.

Cultivar development in beardless wildrye has been restricted to evaluation of ecotypes. The cultivar Shoshone originated in 1958 from a stand at the Riverton, Wyoming, fairgrounds. It is an exceptionally leafy, fine-stemmed, high-forage producer; the rhizomes are vigorous. The most recent released cultivar, Rio, was collected in 1973 from a native stand in Stratford, Kings County, California. It demonstrated superior seed viability and initial sod establishment in comparison with about 12 other California native collections.

Two cultivars of beach wildrye (*L. mollis*)—Benson and Reeve—were developed by the Alaska Plant Materials Center (Wright 1994a, 1994b). Benson was derived from a native accession collected near Narrow Cape, Kodiak, Alaska. Reeve originated from a Eurasian collection of *L. arenarius* (PI 345978) to provide a grass of coastal reclamation and dune stabilization.

To date, one cultivar each of Canada wildrye and Virginia wildrye has been released or developed. Mandan Canada wildrye was developed from collections made near Mandan, North Dakota (Alderson and Sharp 1994). Omaha wildrye is a privately developed Virginia wildrye cultivar produced by Stock Seed Farms, Murdock, Nebraska. It originates from plant material collected in eastern Nebraska (Vogel et al. 2006).

Only two cultivars of Dahurian wildrye—James and Arthur—have been released. Both are reported to have excellent stand-establishment vigor, high forage yield during establishment, and good seed yield (Lawrence, Jefferson, and Ratzlaff 1990).

The Agriculture & Agri-Food of Canada (AAFC) scientists in Saskatoon, Saskatchewan, and Lethbridge, Alberta, have been active in breeding wheatgrass and wildrye grasses. They have released wheatgrass cultivars AEC Highlander, AEC Hillcrest, AEC Montaineer, Goliath, and AC Saltlander (Darroch and Acharya 1995, 1996; Darroch, Acharya, and Woosaree 2001; Coulman 2006; Steppuhn et al. 2006). Ducks Unlimited Canada (DUC) undertook the Native Plants Solution Ecovar® program to address the persistent shortage of Canadian-sourced native plant materials in western Canada. Several ecovars have been released from this program (DUC 2003; Ferdinandez, Coulman, and Fu 2005), including AC Pintail and AC Sprig awned wheatgrass (*Elymus trachycaulus*), AC Polar northern wheatgrass (*Elymus lanceolatus*), and W. R. Poole western wheatgrass (*Pascopyrum smithii*).

In Australia, a tall wheatgrass cultivar, Dundas, was developed by Agriculture Victoria and CSIRO and then released by Wrightson Seeds (Australia) Pty Ltd. (Smith and Kelman 2000). It is a replacement for Tyrrell (Oram 1990) in southern Australia in the reclamation of land with high water tables and salinity.

3.6.2 Germplasm Enhancement

In addition to cultivar release, FRRL released germplasms of wheatgrasses and wildryes, such as SL-1 hybrid (Asay et al. 1991b), RWR-Tetra-1 tetraploid Russian wildrye (Jensen et al. 1998), *Leymus* Hybrid-1 (Jensen et al. 2002), and RS-H hybrid wheatgrass germplasm (Jensen, Asay, and Waldron 2003). Breeding populations derived from interspecific hybrids of *L. cinereus* and *L. triticoides* have been developed for plant improvement. They have excellent potential for high biomass production, reduced susceptibility to grazing or harvest, and better regrowth.

For prairie restoration, erosion control, and wildlife habitat enhancement, USDA's Natural Resources Conservation Service (USDA-NRCS), with cooperating Indiana Association of Soil and Water Conservation Districts and Indiana Department of Natural Resources, released in 2004 a germplasm line of Canada wildrye (*Elymus canadensis* L.) Icy Blue for the Midwest and Great Lakes regions of the United States (Durling, Leif, and Burgdorf 2006). Anatone germplasm is a selection from a native plant collection made in Asotin County, Washington, in 1988 by the USDA Forest Service, Rocky Mountain Research Station, Provo, Utah.

3.6.3 Gene Pools

Dewey's work on genome analysis of Triticeae perennial grasses has significantly contributed to our current strategies of germplasm enhancement (Dewey 1984). Based on the knowledge of genome constitutions of Triticeae species, geneticists and breeders can correctly search for and then select appropriate plant materials from primary, secondary, and tertiary gene pools (GP-1, GP-2, and GP-3, respectively) for desirable traits to be incorporated into the target germplasm. Because of the complex genome relationships among wheatgrass and wildrye grasses, each genome combination has its own gene pool (Table 3.5). The best example of using GP-2 for germplasm enhancement is the release of RS hybrids (Asay and Dewey 1981), which are hexaploid **StStStHH** derivatives of the cross between hexaploid quackgrass (*Elymus repens*; **StStStStHH**) and tetraploid bluebunch wheatgrass (*Pseudoroegneria spicata*; **StStStSt**). These germplasm populations eventually led to the release of the cultivar NewHy (Asay et al. 1991a).

3.6.4 Yield Potential and Stability

In natural plant communities, wheatgrass and wildrye coexist with local plants of various taxa. The persistence of wheatgrass or wildrye in an ecosystem depends on the competitiveness of coexisting plants in the area under particular environmental and climatic conditions. Invasiveness of introduced species has caused great concerns recently, especially to land managers. Waldron et al. (2005) evaluated the invasiveness of Russian wildrye and crested and Siberian wheatgrass mixed with seven North American native perennial grasses. Introduced species, such as crested and Siberian wheatgrass, were so competitive due to fast germination and rapid root growth in spring that they established well and suppressed the land cover by both native grasses and weeds. However, the introduced Russian wildrye had slower germination with increased ground cover by native grasses and weeds the first year. Later, however, all introduced wheatgrass and wildrye were able to coexist with the native grasses while maintaining adequate weed suppression.

Within the wheatgrasses and wildryes, forage yield is more a function of available moisture through annual precipitation than other environmental factors. Mean dry matter yields for Russian wildrye and Siberian and crested wheatgrass at two dryland research sites in northern Utah were 1,290 kg ha^{-1}; 2,337 kg ha^{-1}; and 2,849 kg ha^{-1}, respectively (Asay et al. 2001). Selection for carbon isotope discrimination (delta, Δ) is thought to have potential for improving water-use efficiency (WUE) and subsequent forage yields in cool-season grasses. These two measures were found to be independent in several wheatgrass and wildrye grasses, suggesting that breeding efforts to improve cool-season grasses should involve simultaneous selection for dry matter yield and Δ (Johnson, Asay, and Jensen 2003).

3.6.5 Improved Resistance to Biotic Constraints

3.6.5.1 Diseases

The wheatgrasses and wildryes are generally resistant to many of the common plant diseases (Wang, Dong, and Zhou 1993; Cox, Murphy, and Jones 2002); however, Chang et al. (2001) and

Table 3.5   Primary, Secondary, and Tertiary Gene Pools (GP-1, GP-2, and GP-3) for Wheatgrass and Wildrye

Genus	Species	Genome Composition	GP-1	GP-2	GP-3
Agropyron	cristatum	PP	PP	PPPP, PPPPP	StStPP, StStPPYY
		PPPP	PPPP	PP, PPPPP	StStPP, StStPPYY
		PPPPP	PPPPP	PP, PPPP	StStPP, StStPPYY
Psathyrostachys	juncea	NsNs	NsNs	NsNsNsNs	NsNsXmXm
		NsNsNsNs	NsNsNsNs	NsNs	NsNsXmXm
Pseudoroegneria	spicata	StSt	StSt	StStStSt	StStHH, StStPP, StStYY, StStEE
		StStStSt	StStStSt	StStStStHH	StStHH, StStPP, StStYY, StStEE
Elymus	trachycaulus	StStHH	StStHH	StStStHH	StStStSt, HHHH
	wawawaiensis	StStHH	StStHH	StStStHH	StStStSt, HHHH
	lanceolatus	StStHH	StStHH	StStStHH	StStStSt, HHHH
	dahuricus	StStHHYY	StStHHYY	StStHH, StStYY	StStPPYY, StStWWYY
	repens	StStStStHH	StStStStHH	StStStSt, StStHH	StSt, HHHH
Leymus	angustus	(NsNsNsXmXmXm)x2	(NsNsNsXmXmXm)x2	NsNsNsNsXmXmXmXm	StStHHNsNsXmXm
	cinereus	NsNsXmXm	NsNsXmXm	NsNsNsNsXmXmXmXm	StStHHNsNsXmXm
	triticoides	NsNsXmXm	NsNsXmXm	NsNsNsNsXmXmXmXm	StStHHNsNsXmXm
Thinopyrum	intermedium	EEEEStSt	EEEEStSt	EEEE, StStStSt	EEStSt, StStHH, StStPP, StStYY
	ponticum	(EEEE)x2	(EEEE)x2	EEEE	EEStSt
Pascopyrum	smithii	StStHHNsNsXmXm	StStHHNsNsXmXm	StStHH, NsNsXmXm	None

Mohan, Bijman, and St. John (2001) have reported new disease races affecting intermediate wheat-grass. Recurrent selection for resistance to *Bipolaris sorokiniana* and *Fusarium graminearum* in intermediate wheatgrass has resulted in genetic gains in disease resistance (Krupinsky and Berdahl 2000). Selection for disease resistance in wheatgrass and wildrye grasses is very limited.

Many of the Triticeae grasses have been used as a gene reservoir for cereal crop disease improvement. Disease resistance genes *Lr19, Lr24, Lr29,* and *Lr38* for leaf rust (caused by *Puccinia recondite* Rob. Ex Desm.), as well as *Sr24, Sr25, Sr26,* and *Sr43* for stem rust (*P. graminis* Pers.), have been successfully transferred from *Thinopyrum* species into wheat (McIntosh et al. 1998). Intermediate and tall wheatgrasses are resistant to tan spot (caused by *Pyrenophora tritici-repentis*), barley yellow dwarf virus, and wheat streak mosaic virus (Cox et al. 2005). *Thinopyrum elongatum* ($2n = 14$, **EE**) was found to be resistant to *Fusarium* head blight (FHB) or scab disease caused by *Fusarium graminearum* Schwabe, which can cause devastating yield losses in both wheat and bar-ley. The type II FHB resistance gene in *Th. elongatum* is located on the long arm of the 7E chromo-some (Shen and Ohm 2006). From derivatives of *Thinopyrum* (Podpera) Liu and Wang and wheat, a type II FHB resistance QTL, designated *Qfhs.pur-7EL*, was also located in the distal region of the long arm of 7el2 and delimited with flanking markers *XBE445653* and *Xcfa2240* (Shen and Ohm 2007). Using wheat-alien addition lines, Oliver et al. (2008) identified individual chromosomes of perennial Triticeae carrying genes for resistance to tan spot and *Stagonospora nodorum* blotch.

3.6.5.2 Viruses

Intermediate and tall wheatgrasses are known to be resistant to barley yellow dwarf virus and wheat streak mosaic virus (Fedak and Han 2005; Zhang et al. 1996). Intensive efforts have been made to transfer resistance to these two viruses into wheat (Wang and Zhang 1996; Crasta et al. 2000). Attempts to isolate genes for resistance to BYDV on a *Th. intermedium* chromosome have been made using two different approaches (Jiang et al. 2004, 2005).

3.6.5.3 Pests

Most of the wheatgrasses comprising the St and H genomes are susceptible to bluegrass billbug (*Sphenophorus parvulus*). The black grass bug (*Labops hesperius* Uhler) damages many of the wil-dryes and wheatgrasses on the rangelands (Higgins, Browns, and Haws 1977). Different varieties of bluegrass exhibit a range in resistance to bluegrass billbug. The common varieties tend to be more resistant. Billbug resistance also occurs in many perennial ryegrass cultivars, particularly those that contain endophytic fungi. Sequential effects of grass bugs (*Irbisia pacifica*) and subsequent drought stress on crested wheatgrass were investigated under a controlled environment (Nowak, Hansen, and Nowak 2003). No true resistance to grass bugs was detected.

Fifteen diploid species within the tribe Triticeae were evaluated for the Columbia root-knot nematode *Meloidogyne chitwoodi* race 2 for gall and reproductive indexes. Species from the genus *Thinopyrum* (*Thinopyrum bessarabicum;* **J** genome) and *Psathyrostachys* (*P. juncea;* **Ns** genome) expressed more resistance to *M. chitwoodi* than species within the genus *Agropyron* (*Agropyron cristatum;* **P** genome), *Pseudoroegneria* (*Pseudoroegneria spicata;* **St** genome), and *Hordeum* (*Hordeum bogdanii* L.; **H** genome). The variation among genera and within species indicates that it would be possible to select Triticeae grasses for resistance to *M. chitwoodi* in order to identify and introgress genes for resistance into cultivated cereals (Jensen and Griffin 1994, 1997).

Within the Triticeae grasses, resistance to the Russian wheat aphid *Diuraphis noxia* (Mordvilko) can be grouped into three categories: (1) moderately resistant: *Leymus* and *Elytrigia;* (2) tolerant to moderately susceptible: *Agropyron, Pseudoroegneria, Elymus,* and *Pascopyrum;* and (3) suscep-tible: *Hordeum* and *Thinopyrum*. Within the tribe Triticeae, the genera *Leymus* and *Elytrigia* are the best sources of resistance. However, within each species, extensive genetic variation for resistant types exists (Kindler and Jensen 1993, 1999; Kindler, Springer, and Jensen 1995).

3.6.6 Improved Tolerance of Abiotic Constraints

Some wheatgrass and wildrye species are tolerant of low temperatures and/or poor soil conditions. *Agropyron desertorum* is more cold and drought tolerant than *Thinopyrum elongatum,* whereas the opposite is true for salt tolerance (Tabaei-Aghdaei, Harrison, and Pearce 2000). *Agropyron fragile* is more persistent on sandy soils under severe water limitation than crested wheatgrass.

3.6.6.1 Salinity Tolerance

Many wild annual and perennial Triticeae species are highly tolerant of soil salinity. The most salt-tolerant species belong to the genus *Thinopyrum,* composed of the diploid *Th. bessarabicum* and *Th. elongatum,* tetraploid *Th. junceiforme,* hexaploid *Th. junceum* (L.) A. Löve and *Th. intermedium,* octoploid *Th. runemarkii* A. Löve, and decaploid *Th. ponticum* (Dewey 1960; McGuire and Dvorák 1981; Forster, Miller, and Law 1988). Some of these species have been crossed to wheat, and studies on derived addition or substitution lines showed that salt tolerance in these perennial grasses is controlled by multiple genes on several chromosomes (Forster et al. 1988; Dvorák, Edge, and Ross 1988; Dubcovsky, Galvez, and Dvorák 1994; Zhong and Dvorák 1995; Colmer, Flowers, and Munns 2006). Therefore, both improving salt tolerance in the wheatgrass or wildrye forage species and transfer of salt tolerance by introducing alien genes into wheat are difficult and more complicated than transfer of pest resistance, which is usually controlled by a single gene. However, progress has been achieved in both efforts.

After several cycles of selection at increased salinity levels, increased plant persistence in the cultivar NewHy was reported (Jensen et al. 2005a). Equaling or exceeding the salt tolerance of tall wheatgrass and NewHy RS-wheatgrass, AC Saltlander (a variety of green RS-wheatgrass) has been released for commercialization and seed increase (Steppuhn and Asay 2005).

Using the Ph^I gene to induce homoeologous recombination in the hybrid of a wheat-*Thinopyrum* addition line (AJDAj5) and the Ph^I line, Wang et al. (2003b) successfully transferred the salt tolerance from a *Th. junceum* chromosome into wheat chromosomes. Two recombinant lines, W4909 and W4910, have been released as germplasm lines (Wang et al. 2003b) for wheat improvement and gene discovery. Microarray analysis of these two lines along with their parental lines and Chinese Spring (CS) confirmed that both genetic materials of AJDAj5 and Ph^I had been integrated into wheat chromosomes (Mott and Wang 2007). Introgression of salt tolerance from *Th. ponticum* into wheat was also reported by Chen et al. (2004) using UV-induced asymmetric somatic hybridization between the two species. However, the chromosome number in the two wheat–*Th. ponticum* introgression lines has not been stabilized even in the F_5 families.

Sodium exclusion, controlled by *Kna1* on chromosome 4D and *Nax1* on chromosome 2A of wheat, is the main mechanism found in perennial Triticeae. The ability to exclude Na+ was strongest when 3E of *Th. elongatum* or 5J of *Th. bessarabicum* was added to Chinese Spring, although the enhanced Na+ exclusion in wheat resulting from 5J appears to be much less than that from 3E (Colmer et al. 2006). However, W4909, W4910, and Ph^I had increasing sodium ion concentrations in leaf tissue but maintained a relatively stable K+/Na+ ratio (ranging from 10 to 3). CS and AJDAj5 had lower sodium concentrations but a widely variable K+/Na+ ratio (ranging from 44 to 4) under treatments of increasing salinity levels from electrical conductivity (EC) values 3–22 (Mott and Wang 2007). Therefore, some accessions of *Aegilops speltoides* must have the tissue tolerance, which was not found in Chinese Spring and AJDAj5 termed by Colmer et al. (2006). This mechanism is probably also present in tall wheatgrass varieties Tyrrell and Dundas that were compared with *Puccinellia ciliata* Bor var. 'Menemen' (Zhang et al. 2005).

3.7 MOLECULAR VARIATION

Genetic diversity within and/or between species of Triticeae has been investigated using various allozymes and molecular markers. Russian wildrye accessions have been analyzed with allozymes (Wei, Campbell, and Wang 1996) and RAPD (Wei, Campbell, and Wang 1997). More recently, the AFLP technique has been used to describe genetic diversity between ecotypes and cultivars in bluebunch and beardless wheatgrass (Larson et al. 2000; Larson, Jones, and Jensen 2004), *Agropyron* species (Mellish, Coulman, and Ferdinandez 2002), and western wheatgrass (Larson, Palazzo, and Jensen 2003a). AFLP markers were used to monitor genetic changes in AC Pintail, a composite population, and a cultivar AEC Hillcrest of slender wheatgrass following seed increase (Ferdinandez et al. 2005).

The polycross population P-7 was developed by intermating 25 source populations collected from geographically dispersed locations. It is more genetically diverse than Whitmar and Goldar, which originated from two plant communities fewer than 83 km apart and were released in 1946 and 1989 for awnless and awned spike morphology, respectively. Despite this morphological difference, no true breeding AFLP markers distinguished Goldar from Whitmar (Larson et al. 2000). There was a high correlation ($r = 0.58$) in bluebunch wheatgrass accessions and collection origin (Larson et al. 2004).

Based on AFLP analysis, cultivars and NPGS accessions of western wheatgrass formed three groups that were correlated with geographic origins: northern Great Plains, northern Rocky Mountains, and southern Rocky Mountains (Larson et al. 2003a). The correlation of DNA polymorphism and geographic distance among western wheatgrass localities was $r = 0.66$. Results of the AFLP study on crested wheatgrass (Mellish et al. 2002) showed that (1) Fairway, Parkway, and S9240 are likely the true *A. cristatum* populations; (2) Hycrest, CD-II, and Kirk grouped loosely between *A. cristatum* and *A. desertorum,* suggesting their hybrid nature and the misclassification of Kirk; (3) *A. desertorum* cv. 'Nordan' sat between *A. cristatum* and *A. mongolicum,* confirming that *A. desertorum* is a hybrid derivative of *A. cristatum* and *A. mongolicum;* and (4) cultivars Vavilov, Douglas, and Ephraim clustered with *A. desertorum.*

Refoufi and Esnault (2006) studied genetic variation and population structure of 15 populations of *Elytrigia pycnantha* (Godron) A. Löve collected in Mont Saint-Michel Bay using five microsatellite or simple sequence repeat (SSR) loci. Only one population clearly separated from the remaining populations and little genetic differentiation among populations was detected. Sun and Salomon (2003) used SSR markers to study variability in Alaskan wheatgrass (*Elymus alaskanus* (Scribner & Merr.) A. Löve) complex and detected a significant heterozygote deficiency in seven populations from Canada, Greenland, and the United States.

Russian wildrye, its relative species in *Psathyrostachys,* and some *Leymus* species have been analyzed for the abundance of two repetitive sequences using PCR amplification and FISH (Wang et al. 2006). Based on the survey results, *Leymus* species are highly variable for the abundance of pLrTaiI-1 sequence, which is present only in *Psathyrostachys juncea* and *Ps. lanuginose* (Trin.) Nevski—thus making these two species the likely donor species of the **Ns** genome to *Leymus* species. None of the four species of *Psathyrostachys* contains the pLrPstI-1 sites detectable by FISH, so the source of this repetitive sequence (and thus the genome "Xm") in *Leymus* remains unknown (Figure 3.1C).

3.8 TISSUE CULTURE AND GENETIC TRANSFORMATION

Tissue culture is a prerequisite of genetic transformation of plants. Advances of these methods for wheatgrass and wildrye grasses lagged behind those for cereal grain crops. The generation of haploids via anther culture was not successful in Triticeae grasses (Marburger and Wang 1988;

Chekurov and Razmakhnin 1999). Inflorescence culture was used to produce amphidiploids of rare diploid intergeneric hybrids (Wang, Marburger, and Hu 1991). An efficient plant regeneration system was developed for different wheatgrass species, including tall wheatgrass, intermediate wheatgrass, crested wheatgrass, and western wheatgrass (Wang, Bell, and Hopkins 2003). Efficient plant regeneration obtained from embryogenic suspensions with regeneration frequencies ranged from 20 to 65% in tall wheatgrass, 21 to 40% in intermediate wheatgrass, 32 to 51% in crested wheatgrass, and 25 to 48% in western wheatgrass. Plant transformations were successful in hybrid crested wheatgrass only recently (*Agropyron cristatum* × *A. desertorum* cv. 'Mengnong') (Huo et al. 2004). The phosphinothricin acetyltransferase (bar) gene was transformed into the wheatgrass by the particle bombardment method. Transgenic plants were recovered from calli resistant to glufosinate and later confirmed by PCR and Southern analysis of digested genomic DNA.

Polyploid species of Triticeae can easily cross-hybridize with other Triticeae species in nature. Production of herbicide-resistant transgenics of wheatgrass and wildrye grasses should not be encouraged. Otherwise, herbicide resistance genes could readily spread to both annual cereal crops and aggressive weeds. However, transgenics for other traits such as turf quality (Zhang, Lomba, and Altpeter 2007) using the *ATHB16* gene (a repressor of cell expansion) to limit leaf and tiller growth could still be exploited.

3.9 INTERSPECIFIC AND INTERGENERIC HYBRIDIZATION

Interspecific and intergeneric hybrids in Triticeae have been previously reviewed by Dewey (1984) and Wang (1989b). Interspecific and intergeneric hybrids (Tables 3.6 and 3.7) are useful in both basic research, such as genome analysis, and applied science, including plant improvement through gene introgression and linkage mapping of molecular markers and agronomically important traits. By crossing tetraploid plants of parental species, a fertile amphiploid was obtained from the crosses *Hordeum chilense* × *Agropyron cristatum* and *Aegilops tauschii* × *A. cristatum* (Martin, Rubiales, and Cabrera 1999; Martin et al. 1999).

Jensen (2005) reported on advanced populations of *Elymus lanceolatus* × *Elymus caninus* hybrids. Hybrids and amphiploid derivatives of *Pseudoroegneria spicata* (**StSt**) × *Elymus lanceolatus* (**StStHH**) led to stable autoallohexaploid plants having the genome constitution **St₁St₁St₂St₂HH** (Jensen, Maughan, and Asay 2006). The genes of the two **St** genomes could recombine through multivalent formation during meiotic chromosome pairing, leading to gene exchanges.

3.10 FUTURE DIRECTION

Learning from the experience of researchers working on other plant species, molecular scientists working on wheatgrass and wildrye species are rapidly catching up in molecular studies. FRRL recently led a cooperative effort with the W. M. Keck Center for Comparative and Functional Genomics, University of Illinois, to develop expressed sequence tag (EST) libraries for bluebunch wheatgrass (*Pseudoroegneria spicata* seedlings and salt-/drought-stressed whole plants), thickspike and Snake River wheatgrasses (*Elymus lanceolatus* and *E. wawawaiensis* rhizome and tiller buds), and basin and beardless wildryes (*Leymus cinereus* and *L. triticoides* rhizome and tiller buds). A BAC library has also been constructed for the wildryes in cooperation with Dr. H. B. Zhang of Texas A&M University. These tools are being utilized on available relevant genetic stocks (Table 3.4) and mapping populations to develop molecular and physical maps that will be useful in breeding of wheatgrass and wildrye cultivars. Molecular markers will be increasingly developed and utilized to select important agronomic and physiological traits needed for forage and biofuel production from wheatgrass and wildrye species.

Table 3.6   Interspecific Hybrids of Wheatgrass and Wildrye Species Made by FRRL Scientists

Maternal Parent	Paternal Parent	Status
Elymus canadensis	*Elymus albicans*	Amphiploids
Elymus canadensis	*Elymus caninus*	Amphiploids
Elymus canadensis	*Elymus glaucus*	Amphiploids
Elymus canadensis	*Elymus lanceolatus*	Amphiploids
Elymus canadensis	*Elymus oschensis*	Amphiploids
Elymus canadensis	*Elymus subsecundus*	Amphiploids
Elymus canadensis	*Elymus semicostatus*	Amphiploids
Elymus arizonicus	*Elymus canadensis*	Amphiploids
Elymus donianus	*Elymus caninus*	Amphiploids
Elymus donianus	*Elymus subsecundus*	Amphiploids
Elymus lanceolatus	*Elymus caninus*	Amphiploids, F8
Elymus lanceolatus	*Elymus elymoides*	Amphiploids, F5
Elymus lanceolatus	*Elymus praecaespitosus*	Amphiploids
Elymus lanceolatus	*Elymus trachycaulus*	Amphiploids
Elymus lanceolatus	*Elymus glaucus*	F8
Elymus caninus	*Elymus praecaespitosus*	Amphiploids
Elymus trachycaulus	*Elymus elymoides*	Amphiploids
Elymus fibrosus	*Elymus trachycaulus*	Amphiploids
Elymus glaucissimus	*Elymus ugamicus*	Amphiploids
Elymus tilcarensis	*Elymus lanceolatus*	Amphiploids, F4
Elymus ugamicus	*Elymus praecaespitosus*	Amphiploids
Elymus scribneri	*Elymus angustiglumis*	Amphiploids
Elymus alatavicus	*Elymus batalinii*	F3
Elymus breviaristatus	*Elymus nutans*	F2
Elymus kengii	*Elymus grandiglumis*	F2
Elymus grandiglumis	*Elymus kengii*	F2
Elymus kokonoricus	*Elymus kengii*	F2
Elymus kengii	*Elymus thoraldianus*	F2
Elymus thoraldianus	*Elymus kengii*	F2
Elymus laxiflorus	*Elymus trachycaulus*	F2
Elymus wawawaiensis	*Elymus lanceolatus*	F2
Elymus sibiricus	*Elymus nutans*	F2
Elymus sibiricus	*Elymus submuticus*	F1
Elymus tilcarensis	*Elymus trachycaulus*	F4
Elymus tschimganicus	*Elymus ugamicus*	F5
Agropyron mongolicum	*Agropyron cristatum*	Amphiploids, F3
Thinopyrum bessarabicum	*Thinopyrum elongatum*	Amphiploids
Thinopyrum junceiforme	*Thinopyrum bessarabicum*	Amphiploids
Thinopyrum curvifolium	*Thinopyrum gentryi*	Amphiploids
Thinopyrum curvifolium	*Thinopyrum scythicum*	Amphiploids
Thinopyrum intermedium	*Thinopyrum junceum*	Amphiploids
Thinopyrum acutum	*Thinopyrum intermedium*	F2

Continued

Table 3.6   Interspecific Hybrids of Wheatgrass and Wildrye Species
            Made by FRRL Scientists (*Continued*)

Maternal Parent	Paternal Parent	Status
Thinopyrum intermedium	*Thinopyrum podperae*	Fx
Leymus arenarius	*Leymus racemosus*	Amphiploids
Leymus multicaulis	*Leymus cinereus*	Amphiploids
Leymus multicaulis	*Leymus karataviensis*	Amphiploids
Leymus multicaulis	*Leymus triticoides*	Amphiploids
Leymus secalinus	*Leymus cinereus*	Amphiploids
Leymus secalinus	*Leymus triticoides*	Amphiploids
Leymus secalinus	*Leymus racemosus*	F2
Leymus innovatus	*Leymus secalinus*	Amphiploids
Leymus racemosus	*Leymus mollis*	Amphiploids
Leymus cinereus	*Leymus angustus*	F5
Leymus racemosus	*Leymus angustus*	F5
Leymus simplex	*Leymus cinereus*	F4
Leymus triticoides	*Leymus cinereus*	F3, BC1

Table 3.7   Intergeneric Hybrids of Wheatgrass and Wildrye Species Made
by FRRL Scientists

Maternal Parent	Paternal Parent	Status
Elymus canadensis	*Pseudoroegneria libanotica*	Amphiploid
Elymus canadensis	*Pseudoroegneria spicata*	Amphiploid
Elymus tilcarensis	*Pseudoroegneria libanotica*	Amphiploid, F6
Pseudoroegneria libanotica	*Elymus caninus*	Amphiploid
Pseudoroegneria libanotica	*Elymus trachycaulus*	Amphiploid
Pseudoroegneria libanotica	*Elymus elymoides*	Amphiploid
Pseudoroegneria libanotica	*Elymus sibiricus*	Amphiploid
Pseudoroegneria libanotica	*Elymus lanceolatus*	Amphiploid
Pseudoroegneria libanotica	*Elymus praecaespitosus*	Amphiploid
Pseudoroegneria libanotica	*Elymus drobovii*	Amphiploid
Pseudoroegneria libanotica	*Elymus yezoensis*	Amphiploid
Pseudoroegneria spicata	*Elymus lanceolatus*	Amphiploid
Pseudoroegneria spicata	*Elymus trachycaulus*	Amphiploid
Pseudoroegneria spicata-4x	*Elymus caninus*	Amphiploid
Pseudoroegneria spicata-4x	*Elymus lanceolatus*	Amphiploid
Pseudoroegneria spicata-4x	*Elymus elymoides*	Amphiploid
Pseudoroegneria ferganensis	*Elymus patagonicus*	Amphiploid
Pseudoroegneria spicata	*Hordeum violaceum*	Amphiploid
Pseudoroegneria inermis	*Hordeum violaceum*	Amphiploid
Elymus canadensis	*Hordeum bogdanii*	Amphiploid
Hordeum jubatum	*Elymus trachycaulus*	Amphiploid
Elymus canadensis	*Leymus secalinus*	Amphiploid
Elymus repens	*Agropyron cristatum*	Amphiploid
Elymus repens	*Agropyron desertorum*	Amphiploid
Elymus repens	*Thinopyrum curvifolium*	Amphiploid
Thinopyrum scirpeum	*Elytrigia pycnantha*	Amphiploid
Thinopyrum scythicum	*Elymus repens*	Amphiploid
Leymus karataviensis	*Psathyrostachys juncea*	Amphiploid
Leymus secalinus	*Psathyrostachys juncea*	Amphiploid
Elymus repens	*Pseudoroegneria spicata-4x*	F10
Elymus repens	*Pseudoroegneria stipifolia-4x*	F7
Thinopyrum elongatum	*Pseudoroegneria spicata*	Amphiploid
Thinopyrum elongatum	*Pseudoroegneria inermis*	Amphiploid
Thinopyrum elongatum	*Psathyrostachys juncea*	Amphiploid
Thinopyrum elongatum	*Psathyrostachys fragilis*	Amphiploid
Hordeum californicum	*Secale montanum*	Amphiploid
Pseudoroegneria inermis	*Psathyrostachys juncea*	Amphiploid
Pseudoroegneria spicata	*Secale montanum*	Amphiploid

REFERENCES

Alderson, J., and W. C. Sharp. 1994. Grass varieties in the United States. *Agriculture handbook* No. 170. Soil Conservation Service, USDA. Washington, D.C.: U.S. Government Printing Office.

Anamthawat-Jónsson, K., and S. Bödvarsdóttir. 2001. Genomic and genetic relationships among species of *Leymus* (Poaceae: Triticeae) inferred from 18S-26S ribosomal genes. *American Journal of Botany* 88:553–559.

Asay, K. H., and D. R. Dewey, 1979. Bridging ploidy differences in crested wheatgrass with hexaploid x diploid hybrids. *Crop Science* 19:519–523.

———. 1981. Registration of *Agropyron repens* × *A. spicatum* germplasms RS-1 and RS-2. *Crop Science* 21:351.

Asay, K. H., and K. B. Jensen. 1996a. Wheatgrasses. In *Cool-season forage grasses,* ed. L. E. Moser, D. R. Buxton, and M. D. Casler, 691–724. Agronomy Monograph no. 34. Madison, WI: ASA-CSSA-SSSA.

———. 1996b. Wildryes. In *Cool-season forage grasses,* ed. L. E. Moser, D. R. Buxton, and M. D. Casler, 725–748. Agronomy Monograph no. 34. Madison, WI: ASA-CSSA-SSSA.

Asay, K. H., and R. P. Knowles. 1985. The wheatgrasses. In *Forages: The science of grassland agriculture,* ed. R. F. Barnes et al., 166–176. Ames: Iowa State Univ. Press.

Assay, K. H. et al. 1985a. Registration of 'Hycrest' wheatgrass. *Crop Science* 25:368–369.

Assay, K. H. et al. 1985b. Registration of 'Bozoisky-Select' Russian wildrye. *Crop Science* 25:575–576.

Asay, K. H. et al. 1991a. Registration of NewHy RS hybrid wheatgrass. *Crop Science* 31:1384–1385.

———. 1991b. Registration of *Pseudoroegneria spicata* × *Elymus lanceolatus* hybrid germplasm SL-1. *Crop Science* 31:1391.

———. 1995a. Registration of Vavilov Siberian crested wheatgrass. *Crop Science* 35:1510.

———. 1995b. Registration of Douglas crested wheatgrass. *Crop Science* 35:1510–1511.

———. 1997. Registration of CD-II crested wheatgrass (Reg. No. CV-24). *Crop Science* 37:1023.

———. 1999. Registration of RoadCrest crested wheatgrass. *Crop Science* 39:1535.

———. 2001. Merits of native and introduced Triticeae grasses on semiarid rangelands. *Canadian Journal of Plant Science* 81:45–52.

Aschenbach, T. A. 2006. Variation in growth rates under saline conditions of *Pascopyrum smithii* (western wheatgrass) and *Distichlis spicata* (inland saltgrass) from different source populations in Kansas and Nebraska: Implications for the restoration of salt-affected plant communities. *Restoration Ecology* 14:21–27.

Assadi, M., and H. Runemark, 1995. Hybridization, genomic constitution and generic delimitation in *Elymus* s.l. (Poaceae, Triticeae). *Plant Systematics and Evolution* 194:189–205.

Barker, R. E., J. D. Berdahl, and E. T. Jacobson. 1984. Registration of Rodan western wheatgrass. *Crop Science* 24:1215–1216.

Barkworth, M. E., D. R. Dewey, and R. J. Atkins, 1983. New generic concepts in the Triticeae of the intermountain region: Key and comments. *Great Basin Nature* 43:561–572.

Barkworth, M. E. et al. 2007. Magnoliophyta: Commelinidae (in part): Poaceae, part 1, *Flora of North America north of Mexico,* vol. 24. New York: Oxford University Press.

Berdahl, J. D. et al. 1992. Registration of Reliant intermediate wheatgrass. *Crop Science* 32:1072.

———. 1993. Registration of Manska pubescent intermediate wheatgrass. *Crop Science* 33:881.

Bowden, W. M. 1965. Cytotaxonomy of the species and interspecific hybrids of the genus *Agropyron* in Canada and neighboring areas. *Canadian Journal of Botany* 43:1421–1448.

Chang, K. F. et al. 2001. Stem smut (*Ustilago hypodytes*) on intermediate wheatgrass in Canada. *Plant Disease* 85:96.

Chekurov, V. M., and E. P. Razmakhnin. 1999. Effect of inbreeding and growth regulators on the in vitro androgenesis of wheatgrass, *Agropyron glaucum*. *Plant Breeding* 118:571–573.

Chen, S. Y. et al. 2004. Introgression of salt-tolerance from somatic hybrids between common wheat and *Thinopyrum ponticum*. *Plant Science* 167:773–779.

Colmer, T. D., T. J. Flowers, and R. Munns. 2006. Use of wild relatives to improve salt tolerance in wheat. *Journal of Experimental Botany* 57:1059–1078.

Coulman, B. 2006. Goliath crested wheatgrass. *Canadian Journal of Plant Science* 86:743–744.

Cox, C. M., T. D. Murphy, and S. S. Jones. 2002. Perennial wheat germplasm lines resistant to eyespot, *Cephalosporium* stripe, and wheat streak mosaic. *Plant Disease* 86:1043–1048.

Cox, C. M. et al. 2005. Reactions of perennial grain accessions to four major cereal pathogens of the Great Plains. *Plant Disease* 89:1235–1240.

Crasta, O. R. et al. 2000. Identification and characterization of wheat–wheatgrass translocation lines and localization of barley yellow dwarf virus resistance. *Genome* 43:698–706.

Cronquist, A. et al. 1977. *Intermountain flora, vascular plants of the intermountain west, U.S.A.*, vol. 6. New York: Columbia Univ. Press.

Crowle, W. L. 1970. Revenue slender wheatgrass. *Canadian Journal of Plant Science* 50:748–749.

Darroch, B. A., and S. N. Acharya. 1995. AEC Highlander slender wheatgrass. *Canadian Journal of Plant Science* 75:699–701.

———. 1996. AEC Hillcrest awned slender wheatgrass. *Canadian Journal of Plant Science* 76:345–347.

Darroch, B. A., S. N. Acharya, and J. Woosaree. 2001. AEC mountaineer broadglumed wheatgrass. *Canadian Journal of Plant Science* 81:745–747.

Dewey, D. R. 1960. Salt tolerance of twenty-five strains of *Agropyron*. *Agronomy Journal* 52:631–635.

———. 1969. Inbreeding depression in diploid and induced-autotetraploid crested wheatgrass. *Crop Science* 9:592–595.

———. 1970. Hybrids of South American *Elymus agropyroides* with *Agropyron caespitosum, Agropyron subsecundum,* and *Sitanion hystrix. Botanical Gazette* 131:210–216.

———. 1971. Reproduction in crested wheatgrass triploids. *Crop Science* 11:575–580.

———. 1974. Reproduction in crested wheatgrass pentaploids. *Crop Science* 14:867–872.

———. 1982. Genomic and phylogenetic relationships among North American perennial Triticeae grasses. In *Grasses and Grasslands: Systematics and Ecology.* ed. J. E. Estes et al., 51–80. Norman: Univ. of Oklahoma Press.

———. 1983. New nomenclatural combinations in the North American perennial Triticeae (Gramineae). *Brittonia* 35:30–33.

———. 1984. The genomic system of classification as a guide to intergeneric hybridization with the perennial Triticeae. In *Gene manipulation in plant improvement,* ed. J. P. Gustafson, 209–279. New York: Plenum.

Dewey, D. R., and K. H. Asay. 1982. Cytogenetic and taxonomic relationships among three diploid crested wheatgrasses. *Crop Science* 22:645–650.

Dewey, D. R., and P. C. Pendse. 1968. Hybrids between *Agropyron desertorum* and induced-tetraploid *Agropyron cristatum. Crop Science* 8:607–611.

Dubcovsky, J., A. F. Galvez, and J. Dvořák. 1994. Comparison of the genetic organization of the early salt-stress-response gene system in salt-tolerant *Lophopyrum elongatum* and salt-sensitive wheat. *Theoretical and Applied Genetics* 87:957–964.

DUC. 2003. W. R. Poole—Western wheatgrass. In *The protector,* Soil Conservation Council of Canada, p. 8, winter 2003 (http://www.soilcc.ca/newsletters/Protector_winter2003.pdf).

Durling, J. C., J. W. Leif, and D. W. Burgdorf. 2006. Registration of Icy Blue Canada wildrye germplasm. *Crop Science* 46:2330.

Dvořák, J., M. Edge, and K. Ross. 1988. On the evolution of the adaptation of *Lophopyrum elongatum* to growth in saline environments. *Proceedings of the National Academy of Sciences (USA)* 85:3805–3809.

Endo, T. R., and B. S. Gill. 1984. The heterochromatin distribution and genome evolution in diploid species of *Elymus* and *Agropyron. Canadian Journal of Genetics and Cytology* 26:669–678.

Fan, X. et al. 2007. Phylogenetic analysis among *Hystrix, Leymus,* and its affinitive genera (Poaceae: Triticeae) based on the sequences of a gene encoding plastid acetyl-CoA carboxylase. *Plant Science* 172:701–707.

Fedak, G., and F. Han. 2005. Characterization of derivatives from wheat-*Thinopyrum* wide crosses. *Cytogenetic and Genome Research* 109:360–367. (DOI: 10.1159/000082420).

Ferdinandez, Y. S. N., B. E. Coulman, and Y.-B. Fu. 2005. Detecting genetic changes over two generations of seed increase in an awned slender wheatgrass population using AFLP markers. *Crop Science* 45:1064–1068.

Forster, B. P., T. E. Miller, and C. N. Law. 1988. Salt tolerance of two wheat-*Agropyron junceum* disomic addition lines. *Genome* 30:559–564.

Hagras, A. A-A. et al. 2005. Extended application of barley EST markers for the analysis of alien chromosomes added to wheat genetic background. *Breeding Science* 55:335–341.

Hanks, J. D. et al. 2005. Breeding CWG-R crested wheatgrass for reduced-maintenance turf. *Crop Science* 45:524–528.

Higgins, K. M., J. E. Browns, and A. B. Haws. 1977. The black grass bug (*Labops hesperious* Uhler): Its effect on several native and introduced grasses. *Journal of Range Management* 30:380–384.

Hitchcock, A. S. 1971. *Manual of grasses of the United States,* 2nd ed. New York: Dover Publ. Inc.

Hsiao, C., R. R.-C. Wang, and D. R. Dewey. 1986. Karyotype analysis and genome relationships of 22 diploid species in the tribe Triticeae. *Canadian Journal of Genetics and Cytology* 28:109–120.

Hsiao, C. et al. 1995. Phylogenetic relationships of the monogenomic species of the wheat tribe, Triticeae (Poaceae), inferred from nuclear rDNA (internal transcribed spacer) sequences. *Genome* 38:211–223.

Hu, Z.-M. et al. 2005. Detection of linkage disequilibrium QTLs controlling low-temperature growth and metabolite accumulations in an admixed breeding population of *Leymus* wildryes. *Euphytica* 141:263–280.

Huo, X. W. et al. 2004. Plant regeneration and genetic transformation in wheatgrass (*Agropyron cristatum* × *A. desertorum* cv. 'Mengnon'). *Scientia Agricultura Sinica* 37:642–647.

Imanywhoha, J. and K. B. Jensen. 1994. Production and identification of primary trisomics in diploid *Agropyron cristatum* (crested wheatgrass). *Genome* 37:469–476.

Jain, S. M., S. K. Sopory, and R. E. Veilleux. 1996. *In vitro haploid production in higher plants,* vol. 1. *Fundamental aspects and methods.* Dordrecht, the Netherlands: Kluwer Academic Publishers.

Jauhur, P. P. 1988. A reassessment of genome relationships between *Thinopyrum bessarabicum* and *T. elongatum* of the Triticeae. *Genome* 30:903–914.

Jefferson, P. G. et al. 2004. Potential utilization of native prairie grasses from western Canada as ethanol feedstock. *Canadian Journal of Plant Science* 84:1067–1075.

Jensen, K. B. 2005. Cytology and fertility of advanced populations of *Elymus lanceolatus* (Scribn. & Smith) Gould × *Elymus caninus* (L.) L. hybrids. *Crop Science* 45:1211–1215.

Jensen, K. B., K. H. Asay, and B. L. Waldron. 2003. Registration of RS-H hybrid wheatgrass germplasm. *Crop Science* 43:1139–1140.

Jensen, K. B., and G. D. Griffin. 1994. Resistance of diploid Triticeae species and accessions to the Columbia root-knot nematode, *Meloidogyne chitwoodi. Annals of Applied Nematology* 26:635–639.

———. 1997. Resistance of auto- and allotetraploid Triticeae species and accessions to *Meloidogyne chitwoodi* based on genome composition. *Journal of Nematology* 29:104–111.

Jensen, K. B., S. L. Hatch, and J. K. Wipff. 1992. Cytology and morphology of *Psuedoroegneria deweyi* (Poaceae: Triticeae): A new species from the foothills of the Caucasus Mountains (Russia). *Canadian Journal of Botany* 70:900–909.

Jensen, K. B., S. R. Larson, and B. L. Waldron. 2005. Registration of 'Mustang' Altai wildrye. *Crop Science* 45:1168–1169

Jensen, K. B., K. W. Maughan, and K. H. Asay. 2006. Characterization of amphiploid hybrids between bluebunch and thickspike wheatgrasses. *Crop Science* 46:655–661.

Jensen, K. B., and R. R.-C. Wang. 1997. Cytological and molecular evidence for transferring *Elymus coreanus* and *Elymus californicus* from the genus *Elymus* to *Leymus* (Poaceae: Triticeae). *International Journal of Plant Science* 158:872–877.

Jensen, K. B., Y. F. Zhang, and D. R. Dewey. 1990. Mode of pollination of perennial species of the Triticeae in relation to genomically defined genera. *Canadian Journal of Plant Science* 70:215–225.

Jensen, K. B. et al. 1998. Registration of RWR-Tetra-1 tetraploid Russian wildrye germplasm. *Crop Science* 38:1405.

———. 1999. Natural hybrids of *Elymus elymoides* × *Leymus salinus* subsp. *salmonis* (Poaceae: Triticeae). *Crop Science* 39:976–982.

———. 2002. Registration of *Leymus* Hybrid-1 wildrye germplasm. *Crop Science* 42:675–676.

———. 2005a. Characterization of hybrids from induced and natural tetraploids of Russian wildrye. *Crop Science* 45:1305–1311.

———. 2005b. Persistence after three cycles of selection in NewHy RS-wheatgrass (*Elymus hoffmannii* K. B. Jensen & Asay) at increased salinity levels. *Crop Science* 45:1717–1720.

———. 2006a. Cytogenetic and molecular characterization of hybrids between 6*x,* 4*x,* and 2*x* ploidy levels in crested wheatgrass. *Crop Science* 46:105–112.

———. 2006b. Registration of Bozoisky-II Russian wildrye. *Crop Science* 46:986–987.

———. 2007. Registration of FirstStrike slender wheatgrass. *Journal of Plant Registrations* 1:24–25.

Jiang, S.-M. et al. 2004. Screening and analysis of differentially expressed genes from an alien addition line of wheat-*Thinopyrum intermedium* induced by barley yellow dwarf virus infection. *Genome* 47:1114–1121.

————. 2005. Cloning of resistance gene analogs located on the alien chromosome in an additional line of wheat-*Thinopyrum intermedium*. *Theoretical and Applied Genetics* 111:923–931.

Johnson, D. A., K. H. Asay, and K. B. Jensen. 2003. Carbon isotope discrimination and yield in 14 cool-season grasses. *Journal of Range Management* 56:654–659.

Jones, T. A. 2008. 'Discovery' Snake River wheatgrass. *Native Plants Journal* 9:99–101.

Jones, T. A., and S. R. Larson. 2005. Development of native Western North American Triticeae germplasm in a restoration context. In *Proceedings of the 5th International Triticeae Symposium*, Prague, June 6–10, 2005. ed. V. Hulubec, M. Barkworth, and R. von Bothmer, 108–111. *Czech Journal of Genetic Plant Breeding* 41:108–111.

Jones, T. A., M. G. Redinbaugh, and Y. Zhang. 2000. The western wheatgrass chloroplast genome originates in *Pseudoroegneria*. *Crop Science* 40:43–47.

Jones, T. A., R. R.-C. Wang, and L.-H. Li. 1995. Meiotic stability of intersubspecific hybrids of Snake River × thickspike wheatgrasses. *Crop Science* 35:962–964.

Jones, T. A. et al. 2002. Registration of P-7 bluebunch wheatgrass germplasm. *Crop Science* 42:1754–1755.

Kellogg, E. A. 1989. Comments on genomic genera in the Triticeae. *American Journal of Botany* 76:796–805.

Kellogg, E. A., R. Appels, and R. J. Mason-Gamer. 1996. When genes tell different stories: The diploid genera of Triticeae (Gramineae). *Systematic Botany* 21:312–347.

Kimber, G., 1984. Evolutionary relationships and their influence on plant breeding. In *Gene manipulation in plant improvement*, ed. J. P. Gustefson, 281–283. 16th Stadler Genetics Symposium. New York: Plenum Publishing Corp.

Kindler, S. D., and K. B. Jensen. 1993. An overview: Resistance to the Russian wheat aphid (Homoptera: Aphididae) within the perennial Triticeae. *Journal of Economic Entomology* 86:1609–1618.

————. 1999. Detection and characterization of mechanisms of resistance to Russian wheat aphid (Homoptera: Aphididae) in crested wheatgrass. *Journal of Agriculture and Urban Entomology* 16:129–139.

Kindler, S. D., T. L. Springer, and K. B. Jensen. 1995. Detection and characterization of the mechanisms of resistance to the Russian wheat aphid (Homoptera: Aphididae) in tall wheatgrass. *Journal of Economic Entomology* 88:1503–1509.

Kishii, M., R. R.-C. Wang, and H. Tsujimoto. 2003. Characteristics and behavior of the chromosomes of *Leymus mollis* and *L. racemosus* (Triticeae, Poaceae) during mitosis and meiosis. *Chromosome Research* 11:741–748.

————. 2005. GISH analysis revealed new aspect of genomic constitution of *Thinopyrum intermedium*. In *Proceedings of the 5th International Triticeae Symposium,* ed. V. Hulubec, M. Barkworth, and R. von Bothmer, Prague, June 6–10. *Czech Journal of Genetic Plant Breeding* 41:92–95.

Krupinsky, J. M., and J. D. Berdahl. 2000. Selecting resistance to *Bipolaris sorokinana* and *Fusarium graminearum* in intermediate wheatgrass. *Plant Disease* 84:1299–1302.

Larson, S. R., T. A. Jones, and K. B. Jensen. 2004. Population structure of *Pseudoroegneria spicata* (Poaceae: Triticeae) modeled by Bayesian clustering of AFLP genotypes. *American Journal of Botany* 91:1789–1801.

Larson, S. R., and H. F. Mayland. 2007. Comparative mapping of fiber, protein, and mineral content QTLs in two interspecific *Leymus* wildrye full-sib families. *Molecular Breeding* 20:331–347. DOI 10.1007/s11032-007-9095-9.

Larson, S. R., A. J. Palazzo, and K. B. Jensen. 2003a. Identification of western wheatgrass cultivars and accessions by DNA fingerprinting and geographic provenance. *Crop Science* 43:394–401.

————. 2003b. Amplified fragment length polymorphism in *Elymus elymoides, E. multisetus,* and other *Elymus* taxa. *Canadian Journal of Botany* 81:789–804.

Larson, S. R. et al. 2000. Genetic diversity of bluebunch wheatgrass cultivars and a multiple-origin polycross. *Crop Science* 40:1142–1147.

————. 2006. Comparative mapping of growth habit, plant height, and flowering QTLs in two interspecific families of *Leymus*. *Crop Science* 46:2526–2539.

Lawrence, T., P. G. Jefferson, and C. D. Ratzlaff. 1990. James and Arthur, two cultivars of Dahurian wildrye grass. *Canadian Journal of Plant Science* 48:75–84.

Linde-Laursen, I., R. von Bothmer, and N. Jacobsen. 1992. Relationships in the genus *Hordeum:* Giemsa C-banded karyotypes. *Hereditas* 116:111–116.

Liu, Q. et al. 2006. Phylogenetic relationship in *Elymus* (Poaceae: Triticeae) based on the nuclear ribosomal internal transcribed spacer and chloroplast *trnl*-F sequences. *New Phytologist* 170:411–420.

Liu, Z. -W., and R. R.-C. Wang. 1992. Genome analysis of *Thinopyrum junceiforme* and *T. sartorii*. *Genome* 35:758–764.

———. 1993a. Genome analysis of *Elytrigia caespitosa, Lophopyrum nodosum, Pseudoroegneria geniculata* ssp. *scythica,* and *Thinopyrum intermedium. Genome* 36:102–111.

———. 1993b. Genome constitutions of *Thinopyrum curvifolium, T. scirpeum, T. distichum,* and *T. junceum* (Triticeae: Gramineae). *Genome* 36:641–651.

Löve, A. 1984. Conspectus of the Triticeae. *Feddes Repertorium* 95:425–521.

Marburger, J. E., and R. R.-C. Wang. 1988. Anther culture of some perennial Triticeae. *Plant Cell Reports* 7:313–317.

Martin, A., D. Rubiales, and A. Cabrera. 1999. A fertile amphiploid between a wild barley (*Hordeum chilense*) and crested wheatgrass (*Agropyron cristatum*). *International Journal of Plant Science* 160:783–786.

Martin, A. et al. 1999. A fertile amphiploid between diploid wheat (*Triticum tauschii*) and crested wheatgrass (*Agropyron cristatum*). *Genome* 42:519–524.

Mason-Gamer, R. J. 2004. Reticulate evolution, introgression, and intertribal gene capture in an allohexaploid grass. *Systematic Biology* 53:25–37.

Mason-Gamer, R. J., N. L. Orme, and C. M. Anderson. 2002. Phylogenetic analysis of North American *Elymus* and the monogenomic Triticeae (Poaceae) using three chloroplast DNA data sets. *Genome* 45:991–1002.

McGuire, P. E., and J. Dvořák. 1981. High salt-tolerance potential in wheatgrasses. *Crop Science* 21:702–705.

McIntosh, R. A. et al. 1998. Catalogue of gene symbols for wheat. In *Proceedings of the 9th International Wheat Genetics Symposium*, vol. 5, Saskatoon, Saskatchewan, Canada, August 2–7, 1998, University Extension Press, Univ. Saskatchewan, Canada.

Mellish, A., B. E. Coulman, and Y. S. N. Ferdinandez. 2002. Genetic relationships among selected crested wheatgrass cultivars and species determined on the basis of AFLP markers. *Crop Science* 42:1662–1668.

Mian, M. A. R. et al. 2005. Use of tall fescue EST-SSR markers in phylogenetic analysis of cool-season forage grasses. *Genome* 48:637–647.

Mohan, S. K., V. P. Bijman, and L. St. John. 2001. Bacterial leaf stripe caused by *Xanthomonas translucens* pv. *cerealis* on intermediate wheatgrass in Idaho. *Plant Disease* 85:921.

Morrison, K. J., and C. A. Kelly. 1981. Secar bluebunch wheatgrass. Washington State Univ. Coop. Extension Bulletin 0991.

Mott, I. W., and R. R.-C. Wang. 2007. Comparative transcriptome analysis of salt-tolerant wheat germplasm lines using wheat genome arrays. *Plant Science* 173:327–339.

Mullan, D. J. et al. 2005. EST-derived SSR markers from defined regions of the wheat genome to identify *Lophopyrum elongatum* specific loci. *Genome* 48:811–822.

Murphy, M. A. 2003. Relationships among taxa of *Elymus* (Poaceae: Triticeae) in Australia: Reproductive biology. *Australian Systematic Botany* 16:633–642.

Nevski, S. A. 1934. Tribe XIV. Hordeae Benth. In *Flora of the U.S.S.R,* vol. II, ed. V. L. Komarov, 469–579. Israel Program for Science Translation, Jerusalem.

Nowak, R. S., J. D. Hansen, and C. L. Nowak. 2003. Effects of grass bug feeding and drought stress on selected lines of crested wheatgrass. *Western North American Naturalist* 63:67–177.

Oliver, R. E. et al. 2008. Resistance to tan spot and *Stagonospora nodorum* blotch derived from relatives of wheat. *Plant Disease* 92:150–157.

Oram, R. N. 1990. *Thinopyrum ponticum* (Podp.) Z.-W. Liu & R. R.-C. Wang (tall wheatgrass) cv. Tyrrell. In *Register of Australian herbage plant cultivars,* 3rd ed., 95–96. Melbourne, Australia: CSIRO.

Redinbaugh, M. G., T. A. Jones, and Y. T. Zhang. 2000. Ubiquity of the St chloroplast genome in St-containing Triticeae polyploids. *Genome* 43:846–852.

Refoufi, A., and M. A. Esnault. 2006. Genetic diversity and population structure of *Elytrigia pycnantha* (Godron) A. Löve (Triticeae) in Mont Saint-Michel Bay using microsatellite markers. *Plant Biology* (Stuttgart) 8:234–242.

Refoufi, A., J. Jahier, and M. A. Esnault. 2001. Genome analysis of a natural hybrid with $2n = 63$ chromosomes in the genus *Elytrigia* Desv. (Poaceae) using the GISH technique. *Plant Biology* (Stuttgart) 3:386–390.

Renault, S., C. Qualizza, and M. MacKinnon. 2004. Suitability of altai wildrye (*Elymus angustus*) and slender wheatgrass (*Agropyron trachycaulum*) for initial reclamation of saline composite tailings of oil sands. *Environmental Pollution* 128:339–349.

Rieger, R., A. Michaelis, and M. M. Green. 1968. *A glossary of genetics and cytogenetics,* 3rd ed., 225. New York: Springer–Verlag.

Robins, J. G. et al. 2006. Evaluation of crested wheatgrass managed as turfgrass. *Applied Turfgrass Science* 1: 1–6 (http://www.plantmanagementnetwork.org/ats/), May 23, 2006.

Rogler, G. A. 1973. The wheatgrasses. In *Forages: The science of grassland agriculture,* ed. M. E. Heath et al., 221–230. Ames: Iowa State Univ. Press.

Sadasiviah, R. S., and J. Weifer. 1981. Cytogenetics of some natural intergeneric hybrids between *Elymus* and *Agropyron* species. *Canadian Journal of Genetics and Cytology* 23:131–140.

Shen, X., and H. Ohm. 2006. Fusarium head blight resistance derived from *Lophopyrum elongatum* chromosome 7E and its augmentation with *Fhb1* in wheat. *Plant Breeding* 125:424–429.

———. 2007. Molecular mapping of *Thinopyrum*-derived *Fusarium* head blight resistance in common wheat. *Molecular Breeding* 20:131–140.

Smith, K. F., and W. M. Kelman. 2000. Register of Australian herbage plant cultivars. A. Grasses. 18. Wheatgrass. *Thinopyrum ponticum* (Podp.) Z.-W. Liu & R. R.-C. Wang (tall wheatgrass) cv. Dundas. *Australian Journal of Experimental Agriculture* 40:119–120.

Smoliak, S., and A. Johnston. 1984. Registration of Walsh western wheatgrass. *Crop Science* 24:1216.

Stebbins, G. L. 1971. *Chromosomal evolution in higher plants,* 44–48. Reading, MA: Addison–Wesley Publishing Co.

Stebbins, G. L., and F. T. Pun. 1953. Artificial and natural hybrids in the Gramineae, tribe Hordeae. V. Diploid hybrids of *Agropyron. American Journal of Botany* 40:444–449.

Steppuhn, H., and K. H. Asay. 2005. Emergence, height, and yield of tall, NewHy, and green wheatgrass forage crops grown in saline root zones. *Canadian Journal of Plant Science* 85:863–875.

Steppuhn, H. et al. 2006. AC Saltlander green wheatgrass. *Canadian Journal of Plant Science* 86:1161–1164.

Sun, G. L., and B. Salomon. 2003. Microsatellite variability and heterozygote deficiency in the arctic-alpine Alaskan wheatgrass (*Elymus alaskanus*) complex. *Genome* 46:729–737.

Svitashev, S. et al. 1998. Genome specific repetitive DNA and RAPD markers for genome identification in *Elymus* and *Hordelymus. Genome* 41:120–128.

Tabaei-Aghdaei, S. R., P. Harrison, and R. S. Pearce. 2000. Expression of dehydration-stress-related genes in the crowns of wheatgrass species (*Lohphopyrum elongatum* (Host) A. Löve and *Agropyron desertorum* (Fish. Ex Link.) Schult) having contrasting acclimation to salt, cold and drought. *Plant, Cell and Environment* 23:561–571.

Tzvelev, N. N. 1976. Tribe 3, Triticeae Dum. In *Poaceae,* ed. A. A. Fedorov, 105–206. Leningrad, USSR: Nauka Publ. House.

Vavilov, N. I. 1987. The phytogeographical basis for plant breeding. p. 360. In *Origin and geography of cultivated plants,* ed. V. F. Dorofeyev. (English translation 1992). Cambridge: Cambridge University Press.

Vogel, K. P. and K. B. Jensen. 2001. Adaptation of perennial Triticeae to the eastern Central Great Plains. *Journal of Range Management* 54:674–679.

Vogel, K. P. et al. 2005a. Registration of Beefmaker intermediate wheatgrass. *Crop Science* 45:414–415.

———. 2005b. Registration of Haymaker intermediate wheatgrass. *Crop Science* 45:415–416.

———. 2006. Genetic variation among Canada wildrye accessions from midwest USA remnant prairies for biomass yield and other traits. *Crop Science* 46:2348–2353.

Waldron, B. L. et al. 2005. Coexistence of native and introduced perennial grasses following simultaneous seeding. *Agronomy Journal* 97:990–996.

Wang, R. R.-C. 1985. Genome analysis of *Thinopyrum bessarabicum* and *T. elongatum. Canadian Journal of Genetics and Cytology* 27:722–728.

———. 1986a. Diploid perennial intergeneric hybrids in the tribe Triticeae. I. *Agropyron cristatum* x *Pseudoroegneria libanotica* and *Critesion violaceum* x *Psathyrostachys juncea. Crop Science* 26:75–78.

———. 1986b. Diploid perennial intergeneric hybrids in the tribe Triticeae. II. Hybrids of *Thinopyrum elongatum* with *Pseudoroegneria spicata* and *Critesion violaceum. Biologisches Zentralblatt* 105:361–368.

———. 1987a. Diploid perennial intergeneric hybrids in the tribe Triticeae. III. Hybrids among *Secale montanum, Pseudoroegneria spicata,* and *Agropyron mongolicum. Genome* 29:80–84.

———. 1987b. Progenies of *Thinopyrum elongatum* x *Agropyron mongolicum. Genome* 29:738–743.

———. 1987c. Synthetic and natural hybrids of *Psathyrostachys huashanica. Genome* 29:811–816.

———. 1988a. Cytological studies on a polyhaploid of *Critesion iranicum* obtained after hybridization with *C. bulbosum. Genetica* 76:225–228.

———. 1988b. Diploid perennial intergeneric hybrids in the tribe Triticeae. IV. Hybrids among *Thinopyrum bessarabicum, Pseudoroegneria spicata,* and *Secale montanum. Genome* 30:356–360.

———. 1988c. Coenocytism, ameiosis, and chromosome diminution in intergeneric hybrids in the perennial Triticeae. *Genome* 30:766–775.

———. 1989a. An assessment of genome analysis based on chromosome pairing in hybrids of perennial Triticeae. *Genome* 32:179–189.

———. 1989b. Intergeneric hybrids involving perennial Triticeae. *Genetics (Life Science Advances)* 8:57–64.

———. 1990. Intergeneric hybrids between *Thinopyrum* and *Psathyrostachys* (Triticeae). *Genome* 33:845–849.

———. 1992a. New intergeneric diploid hybrids among *Agropyron, Thinopyrum, Pseudoroegneria, Psathyrostachys, Hordeum,* and *Secale. Genome* 35:545–550.

———. 1992b. Genome relationships in the perennial Triticeae based on diploid hybrids and beyond. *Hereditas* 116:133–136.

———. 2006. Registration of TBTE001 and TBTE002 *Thinopyrum* amphidiploid genetic stocks differing for leaf glaucousness. *Crop Science* 46:1013–1014.

Wang, R. R.-C., and J. D. Berdahl. 1990. Meiotic associations at metaphase I in diploid, triploid, and tetraploid Russian wildrye. *Cytologia* 55:639–643.

Wang, R. R.-C., Y. Dong, and R. Zhou. 1993. Resistance to powdery mildew and/or barley yellow dwarf in perennial Triticeae species. *Genetic Resources and Crop Evolution* 40:171–176.

Wang, R. R.-C., and C. T. Hsiao. 1989. Genome relationship between *Thinopyrum bessarabicum* and *T. elongatum:* Revisited. *Genome* 32:802–809.

Wang, R. R.-C., J. E. Marburger, and C-J. Hu. 1991. Tissue-culture-facilitated-production of aneupolyhaploid *Thinopyrum ponticum* and amphidiploid of *Hordeum violaceum* x *H. bogdanii* and their uses in phylogenetic studies. *Theoretical and Applied Genetics* 81:151–156.

Wang, R. R.-C., and J-Z. Wei. 1995. Variations for two repetitive DNA sequences in several Triticeae genomes revealed by polymerase chain reaction and sequencing. *Genome* 38:1221–1229.

Wang, R. R.-C., and X-Y. Zhang. 1996. Characterization of the translocated chromosome using fluorescence *in situ* hybridization and random amplified polymorphic DNA on two *Triticum aestivum–Thinopyrum intermedium* translocation lines resistant to wheat streak mosaic or barley yellow dwarf virus. *Chromosome Research* 4:583–587.

Wang, R. R.-C. et al. 1995. Genome symbols in the Triticeae. In *Proceedings of the 2nd International Triticeae Symposium,* ed. R. R.-C. Wang, K. B. Jensen, and C. Jaussi, 29–34, Logan, UT, June 20–24, 1994. Logan: Utah State University Publication Design and Production.

———. 2003a. Development of salinity-tolerant wheat recombinant lines from a wheat disomic addition line carrying a *Thinopyrum junceum* chromosome. *International Journal of Plant Science* 164:25–33.

———. 2003b. Registration of W4909 and W4910 bread wheat germplasm lines with high salinity tolerance. *Crop Science* 43:746.

———. 2006. Variations in abundance of 2 repetitive sequences in *Leymus* and *Psathyrostachys* species. *Genome* 49:511–519.

Wang, Z-Y., J. Bell, and A. Hopkins. 2003. Establishment of a plant regeneration system for wheatgrasses (*Thinopyrum, Agropyron,* and *Pascopyrum*). *Plant Cell, Tissue and Organ Culture* 73:265–273.

Wei, J.-Z., W. F. Campbell, and R. R.-C. Wang. 1995. Standard giemsa C-banded karyotype of Russian wildrye (*Psathyrostachys juncea*) and its use in identification of a deletion-translocation heterozygote. *Genome* 38:1262–1270.

———. 1996. Allozyme variation in accessions of Russian wildrye. *Crop Science* 36:785–790.

———. 1997. Genetic variability in Russian wildrye (*Psathyrostachys juncea*) assessed by RAPD. *Genetic Resources and Crop Evolution* 44:117–125.

Wei, J-Z., and R. R.-C. Wang. 1995. Genome- and species-specific markers and genome relationships of diploid perennial species in Triticeae based on RAPD analyses. *Genome* 38:1230–1236.

Wei, J-Z. et al. 1995. Cytological identification of some trisomics of Russian wildrye (*Psathyrostachys juncea*). *Genome* 38:1271–1278.

———. 2000. Linkage mapping and nucleotide polymorphisms of the 6-SFT gene of cool-season grasses. *Genome* 43:931–938.

Wright, S. J. 1994a. Registration of Benson beach wildrye. *Crop Science* 34:537.

———. 1994b. Registration of Reeve beach wildrye. *Crop Science* 34:537.

Wu, X.-L. et al. 2003. Molecular genetic linkage maps for allotetraploid *Leymus* (Triticeae). *Genome* 46:627–646.

Yen, C., J.-L. Yang, and Y. Yen. 2005. Hitoshi Kihara, Áskell Löve and the modern genetic concept of the genera in the tribe Triticeae (Poaceae). *Acta Phytotaxonmica Sinica* 43:82–93.

Zhang, B. et al. 2005. Comparative studies on salt tolerance of seedlings for one cultivar of puccinellia (*Puccinellia ciliate*) and two cultivars of tall wheatgrass (*Thinopyrum ponticum*). *Australian Journal of Experimental Agriculture* 45:391–399.

Zhang, Hai-qin et al. 2006. Genome constitutions of *Hystrix patula, H. duthiei* ssp. *duthiei* and *H. duthiei* ssp. *longearistata* (Poaceae: Triticeae) revealed by meiotic pairing behavior and genomic *in-situ* hybridization. *Chromosome Research* 14:595–604.

Zhang, H. B., and J. Dvořák. 1991. The genome origin of tetraploid species of *Leymus* species (Poaceae: Triticeae) inferred from variation in repeated nucleotide sequences. *American Journal of Botany* 78:871–874.

Zhang, H.-N., P. Lomba, and F. Altpeter. 2007. Improved turf quality of transgenic bahiagrass (*Paspalum notatum* Flugge) constitutively expressing the *ATHB16* gene, a repressor of cell expansion. *Molecular Breeding* 20:415–423. DOI 10.1007/s11032-007-9101-2.

Zhang, X.-Y., Y. S. Dong, and R. R.-C. Wang. 1996. Characterization of genomes and chromosomes in partial amphiploids of the hybrid *Triticum aestivum x Thinopyrum ponticum* by in situ hybridization, isozyme analysis, and RAPD. *Genome* 39:1062–1071.

Zhang, X.-Y. et al. 1996. Molecular verification and characterization of BYDV-resistant germ plasms derived from hybrids of wheat with *Thinopyrum ponticum* and *Th. intermedium*. *Theoretical and Applied Genetics* 93:1033–1039.

Zhong, G.-Y., and J. Dvořák. 1995. Chromosomal control of the tolerance of gradually and suddenly imposed salt stress in the *Lophopyrum elongatum* and wheat, *Triticum aestivum* L., genomes. *Theoretical and Applied Genetics* 90:229–236.

CHAPTER 4

Bahiagrass

Ann R. Blount and Carlos A. Acuña

CONTENTS

Figure 4.1   Bahiagrass hay field located in Quincy, Florida.

4.1 INTRODUCTION

Paspalum notatum Flüggé, bahiagrass, is a warm-season, perennial grass native to the Americas. It has become widely distributed in almost all tropical and subtropical regions of the world, particularly in the Western Hemisphere. It is the primary constituent of native grassland in southern Brazil, Paraguay, Uruguay, and northeastern Argentina (Gates, Quarin, and Pedreira 2004). In the United States, bahiagrass is grown in the southeastern portion of the country, throughout Florida (Figure 4.1), and in the Coastal Plain and Gulf Coast regions (Chambliss and Adjei 2006). It has been planted extensively as a pasture grass and utility turf and for soil stabilization.

4.2 TAXONOMY

The taxonomy of the bahiagrass is as follows:

Order: Poales
Family: Poaceae
Subfamily: Pooideae
Tribe: Paniceae
Genus: *Paspalum*
Species: *notatum* Flüggé
Botanical name: *Paspalum notatum* Flüggé

4.2.1 General Botanical Description

Bahiagrass is a deep-rooted perennial grass that spreads by short, stout, woody horizontal rhizomes and by seed. Bahiagrass is considered to be one of the major grass species of the native grasslands in the New World (Chase 1929). As described by Chase (1929), bahiagrass is a perennial

Figure 4.2   (See color insert following page 274.) Bahiagrass inflorescence.

grass with characteristically strong, shallow rhizomes formed by short, stout internodes usually covered with old, dry leaf sheaths:

culms: simple, ascending, geniculate at the node between the first and the second elongated internodes, and otherwise erect (10–60 cm tall)

leaves: mostly crowded at the base with overlapped keeled sheaths, glabrous, or with ciliate margins mainly toward the summit

blades: typically flat or somewhat folded toward the base, linear lanceolate 3–30 cm long, 3–12 mm wide, usually glabrous or ciliate toward the base, and rarely pubescent throughout

inflorescences: subconjugate with a short, almost imperceptible common axis, having two racemes, rarely three, that are ascending or recurved divergent in some races, with racemes 3–14 cm long (Figure 4.2)

rachis: glabrous, flexuous, and green or purplish

spikelets: solitary in two rows on one side of the rachis, obovate or ovate, shining, glabrous, 2.5–4 mm long and 2–2.8 mm wide

anthers and stigmata: generally purple

fruit: oval, about 1.8 mm long and 1.2 mm wide

4.2.2 Botanical Varieties

Paspalum notatum is divided into two botanical types: the common form (*P. notatum*), which is tetraploid ($2n = 4x = 40$) and has broad leaves, with stout rhizomes, few inflorescences, large spikelets, and short internodes, and the 'Pensacola'-type (*P. notatum* var. *saurae*), which is diploid ($2n = 2x = 20$), with long, narrow leaves, smaller spikelets, longer internodes, and is generally taller than the common form. Hitchcock (1971) describes the two types as follows:

Paspalum notatum Flügge: Culms are typically 15–50 cm tall, ascending from short, stout, woody, horizontal rhizomes. Leaf blades are flat or folded with racemes recurved ascending, usually 4–7 cm long. Spikelets are typically ovate to obovate, about 3–3.5 mm long, smooth, and shining.

Paspalum notatum var. *saurae* Parodi: This is a hardier form, 40–70 cm tall, with blades to 35 cm long, two or three racemes (rarely to five), suberect, and spikelets 2.8–3 mm long. It has been called the Pensacola strain in the southeastern United States.

4.2.3 Common Names

Common names include bahiagrass (United States), Capii cabayu (Paraguay), Grama batatais (Brazil), and Pasto horqueta (Argentina).

4.3 ORIGIN AND DOMESTICATION

4.3.1 Geographic Origin

The genus *Paspalum* L. contains approximately 330 species (Zuloaga and Morrone 2005) and is distributed throughout tropical, subtropical, and some temperate regions of the New World; only a few species can be found in the Old World. The genus is common to Brazil, eastern Bolivia, Paraguay, and northeastern Argentina. *Paspalum notatum* Flüggé is commonly found in open areas, savannas, and cultivated pastures from sea level to 2000 m from central Mexico to Argentina and throughout the West Indies (Chase 1929). The bahiagrass tetraploid form is the primary grass species of many native pasturelands in southern Brazil, Paraguay, Uruguay, and northeastern Argentina (Gates et al. 2004). The original distribution of the variety *saurae* was confined to Corrientes, Entre Rios, and the eastern edge of Santa Fe provinces in Argentina. The diploid populations are infrequent and usually restricted to wet, sandy soils along rivers and flat, sandy islands in the Parana River (Burton 1967; Gates et al. 2004). Because the polyploid forms of the species are considered autopolyploid and because of the higher variability observed in tetraploid populations that were in contact with diploid populations (Daurelio et al. 2004), this region is considered the center of origin of this species.

4.3.2 Geographic Distribution

Bahiagrass was first introduced into the United States by the Bureau of Plant Industry (Scott 1920). Following its introduction into Florida, the variety *saurae* was brought into cultivation throughout the state, the Coastal Plain, and the Gulf Coast regions of the southern United States. It soon escaped from cultivation and rapidly became naturalized throughout that region (Gates et al. 2004). It can now be found growing from Texas to North Carolina, extending into Arkansas and Tennessee (Watson and Burson 1985). Several accessions of the tetraploid form were also introduced in the United States more than 50 years ago, and a few of them have persisted as cultivars. Naturalized populations of both Pensacola and tetraploid races occur in Australia (Gates et al. 2004). Bahiagrass is used as a pasture grass in Japan (Sugimoto, Hirata, and Ueno 1985), Taiwan (Jean and Juang 1979), and Zimbabwe (Mills and Boultwood 1978).

4.4 UTILIZATION

4.4.1 Forage for Livestock

As a forage species, bahiagrass provides adequate nutritional quality for many types of livestock, is satisfactory quality for mature cattle, and is less desirable for growing animals. In young, growing animals or highly productive animals such as lactating dairy cows, animal nutritional requirements are high. Grazing bahiagrass or feeding it as conserved forage without supplementation often is not enough to meet the nutritional requirements of the animal (Rollins and Hoveland 1960). In the southern Coastal Plain region of the United States, the beef cattle industry is predominantly a cow–calf operation, with bahiagrass as the primary pasture grass (Chambliss and Sollenberger 1991). The variable nutritive value of bahiagrass is well complemented by the large buffering capacity of the mature cow in this system. In studies with beef cattle in Florida, Sollenberger and co-workers (1998) found total-season average daily gains on yearling animals seldom exceeded 0.5 kg.

Summer digestibility of bahiagrass showed a decline in concentration of *in vitro* digestible organic matter (IVDOM) from 599 to 432 g kg^{-1} in hay when cutting intervals increased from 6 to 14 wk (Moore et al. 1971). In addition, nitrogen concentration declined between the two hay ages, from 12.2 to 10.1 g kg^{-1}, respectively, and lignin concentration and acid detergent fiber concentrations increased.

Cultivars may differ slightly in forage nutritive value due to the nature of their ploidy levels. In diploid bahiagrass, seasonal forage harvests from April to September had more of an influence on digestibility among cultivars than the effects of age of regrowth, cultivar, or cutting height (Gates, Hill, and Burton 1999). As diploid plants have been selected for increased forage production, some alteration in lowering lignin content in bahiagrass has been reported (Gates and Burton 1990). In tetraploid cultivars, there tends to be a higher leaf-to-stem ratio compared to diploid types. Argentine bahiagrass typically has more leaves and fewer inflorescences than diploid cultivars such as Pensacola and Tifton 9 (Cuomo et al. 1996). This germplasm is often associated with notably fewer inflorescences occurring in tetraploid than diploid bahiagrass species.

Bahiagrass is also characteristically known for its persistence under continuous stocking and intense defoliation (Beaty, Brown, and Morris 1970). Its adaptation to sandy, marginally fertile soils; its ability to persist on such soils; and its marked persistence under grazing make it unique among forage grasses. Persistence under intense grazing pressure is the main reason for the predominance of this plant species in its own native grasslands and its popularity for beef cattle in the southeastern United States.

Short days and cool temperatures limit the forage production of bahiagrass (Mislevy, Sinclair, and Ray 2001); however, through plant breeding, selection for better cold tolerance, winter survival, and long season growth have been achieved (Blount et al. 2008). Recently released cultivars such as Tifton 9, TifQuik, and UF-Riata have been improved through conventional plant breeding techniques to have better cold adaptation and produce more forage than the cultivars Argentine, Paraguay, Paraguay 22, and Pensacola.

4.4.2 Turf and Soil Stabilization

Bahiagrass has also been planted extensively as a utility turfgrass and for soil stabilization, particularly in the southern Coastal Plain region of the United States. Although cultivars used for turf or soil stabilization have been mostly limited to the cultivar Pensacola, other cultivars, such as Argentine and Paraguayan, have become popular for those uses (Kretschmer and Hood 1999). Bahiagrass may be established by seed, which makes it relatively easier and more economical than sprigging a utility turfgrass, although it may be sodded. Once established, it forms a dense sod that is very persistent and requires little fertilizer inputs. It is also tolerant of frequent, close mowing; prevents soil erosion; and has few known insect pests and disease problems. In the southeastern United States, bahiagrass is often planted or sodded along the roadways, including interstate highways, for these reasons. Argentine is preferred to Pensacola as a utility turfgrass (Busey 1989), mainly because it produces fewer inflorescences.

4.4.3 Crop Rotation

Bahiagrass is often used in crop rotations, particularly in rotation with row crops and various vegetable crops. Its establishment from seed makes bahiagrass more attractive than many vegetatively propagated grasses for use in rotations. Recent efforts in sod-based rotation production systems in the southeastern United States have reported a multitude of benefits from the use of bahiagrass in rotation with groundnut (peanut) (Norden et al. 1977; White, Sparrow, and Carter 1962; Katsvairo et al. 2006), soybean (Rodriguez-Kabana et al. 1989a, 1989b), and cotton (Katsvairo et al. 2006). In addition to higher yields of row crops, great reductions in insect, nematode, and disease problems associated with these crops have been linked to short-term crop rotation with bahiagrass (Katsvairo et al. 2006).

There have been few studies reporting differences comparing the effectiveness of the various bahiagrass cultivars on reducing pests and pathogens in crop production systems where it is used in rotation. A recent trial was established at the North Florida Research and Education Center, Marianna, in 2007 to compare pest suppression by bahiagrass cultivars; to date, cultivars appear to be equally effective.

4.4.4 Bioenergy

Bahiagrass also has potential for bioenergy production. Reports from various global locations show that annual productivity ranges from 2 to 37 MT/ha (Duke 1981). As cellulose conversion becomes more cost effective, the role of bahiagrass as a low-input bioenergy crop should evolve.

4.4.5 Seed Crop

Bahiagrass seed is of considerable value in the southeastern United States, and many livestock producers harvest seed of Argentine and Pensacola varieties for sale to retail seed distributors. In 2007, farm gate seed values in north Florida were approximately $5.00 kg^{-1} for Argentine and $2.75 kg^{-1} for Pensacola (U.S. currency). Typically, Argentine bahiagrass yields less seed than Pensacola, but seed yields of 60–150 kg ha^{-1} and greater may be expected. Care to manage the bahiagrass with adequate and timely N applications in the spring and adequate moisture ensures a good seed yield (Adjei, Mislevy, and Chason 2000). Argentine is susceptible to ergot (*Claviceps* spp.), and upwards of 30% loss of viable seed production in Florida has been attributed to ergot.

4.5 CHROMOSOME NUMBER, PLOIDY, AND REPRODUCTIVE CHARACTERISTICS

4.5.1 Chromosome Number and Ploidy

Bahiagrass is characterized by having a chromosome base number of $x = 10$ and an important range of ploidy levels. Diploids (Burton 1946), triploids (Tischler and Burson 1995), tetraploids (Burton 1940), and pentaploids (Tischler and Burson 1995) are the known ploidy levels. The tetraploid form is the most abundant in natural environments, followed by the diploid form (Tischler and Burson 1995; Pozzobon and Valls 1997). The diploid form is taller, spreads faster, and has longer, narrower leaves than the naturally occurring tetraploid form (Burton, Forbes, and Jackson 1970). Although triploids are rarely found in nature, they are suspected to play an important role in the evolution of the species (see Section 4.5.2).

Cytological and molecular evidence suggest that all polyploid forms of this species are autopolyploids. The considerably high number of multivalent chromosome associations observed during diakinesis and metaphase I of meiosis in induced autotetraploids, natural tetraploids, and induced × natural tetraploid hybrids is partial evidence of autopolyploidy (Forbes and Burton 1961; Quarin, Burson, and Burton 1984). Moreover, the high frequency of trivalent chromosome associations observed in a triploid hybrid generated by crossing diploid and tetraploid plants indicates a high degree of homology between the two genomes from the tetraploid parent and one from the diploid parent (Forbes and Burton 1961). In addition, Stein et al. (2004) reported molecular evidence of tetrasomic inheritance for most analyzed loci of the tetraploid plants.

4.5.2 Reproductive Characteristics

Bahiagrass diploid genotypes reproduce sexually (Burton 1955). In their ovules, the archesporial cell undergoes normal meiosis resulting in four megaspores. The megaspore at the chalazal end develops after several mitotic divisions into an embryo sac of the polygonum type, while the other three megaspores degenerate. The embryo sac has the egg apparatus, the egg cell and two synergids at the micropylar end, and the central cell with two polar nuclei next to the egg apparatus; there is a mass of antipodals at the chalazal end. Although there is a marked variation for self-fertility among diploid genotypes (Burton 1955; Acuña et al. 2007), cross-pollination predominates and results in highly heterogeneous and heterozygous natural and naturalized populations.

Based on results from field progeny tests, Burton (1948) suggested that tetraploid ecotypes reproduced by apomixis. In ovules from these genotypes, nucellar cells can develop into unreduced embryo sacs accompanying the normal megagametogenesis or completely replacing it. The egg cell develops without fertilization (parthenogenesis) into an embryo, which is a genetic copy of the mother plant. This type of apomixis, known as apospory, is common among warm-season grasses. Apomictic ecotypes are classified as pseudogamous because pollination and fertilization of the polar nuclei are required for endosperm development (Burton 1948; Quarin 1999). The male gametes of these plants are normally reduced because they are originated by a normal meiotic process.

Facultative apomictic plants—those that can reproduce by apomixis or sexuality—seem to predominate in natural environments (Martínez et al. 2001). However, the ecotypes that have been domesticated, such as Paraguay 22, Argentine, and Wilmington, are highly apomictic plants having multiple aposporous embryo sacs in most of their ovules and producing highly uniform progeny (Acuña et al. 2007). Apomictic bahiagrass plants set similar amounts of seed when self- or cross-pollinated (Quarin 1999; Acuña et al. 2007), which is an evolutionary characteristic of most apomictic species.

Aposporous apomixis in bahiagrass was determined to be inherited as a single Mendelian factor present in the simplex condition in the apomictic parents (Martínez et al. 2001). The excess of sexual plants in the progeny obtained by hybridizing artificially induced sexual and apomictic tetraploid plants was related to a possible partial lethal effect of the apomixis gene. The distorted segregation ratio and the lack of recombination observed around the apospory-controlling locus (Martínez et al. 2003) seem to be the result of an inverted chromosomal segment (Stein et al. 2004).

The nontransmission of the apospory-controlling locus through monoploid ($x = 10$) gametes has been suggested as the reason for the absence of apomixis in diploid plants (Martínez et al. 2007). In contrast, other authors have suggested that the apomixis genes are present in diploid plants with low or no expression due to a dosage requirement (Quarin et al. 2001).

Apomixis is considered one of the keys for the evolutionary success of bahiagrass in colonizing a vast area of the New World. Rare apomictic triploids growing near the diploid populations would be the source of novel apomictic tetraploids (Quarin, Norrmann, and Urbani 1989). This hypothesis is based on the confirmed ability of these triploids to produce B_{III} hybrids in combination with pollen from diploids growing in the same area (Burton and Hanna 1986). The few novel tetraploid hybrids that are successful in an environment would be perpetuated by apomixis through successive generations and would be able to colonize the area.

4.6 GERMPLASM ENHANCEMENT

4.6.1 Diploid Bahiagrass

4.6.1.1 Diploid Breeding

Sexual diploid germplasm has been improved using two approaches. The discovery of self-incompatibility in most Pensacola plants suggested that commercial F_1 hybrid seed could be produced by harvesting year after year all seed produced in a field planted to alternate strips of two self-sterile cross-fertile clones (Burton 1984). Two such clones, whose F_1 hybrids yielded 17% more dry matter than a Pensacola check in clipping tests, were sought and found by screening several dialleles. A pilot seed production field planted vegetatively produced seed to plant in pasture that gave 17% more live weight gain per acre than Pensacola bahiagrass when grazed for 4 years. Two of these hybrids, Tifhi 1 (Hein 1958) and Tifhi 2, were released; however, the hand labor required to establish the seed fields and the cold injury to one parent clone in a severe winter kept them from becoming important on the farm (Burton 1984).

In 1961, G. W. Burton collected a mixture of Pensacola bahiagrass seed from 39 farms and from 75 clones of Tifhi 2 to initiate what is considered the longest effort to improve this forage crop (Burton 1984; Gates et al. 2004). Using an innovative approach specially adjusted to the characteristics of this species called restricted recurrent phenotypic selection (RRPS), he completed 23 cycles with the specific objective of increasing seasonal forage production. Tifton 9 was released as a cultivar from the ninth selection cycle (Burton 1989). Because the selection was based on aboveground herbage accumulation of individual plants growing in a noncompetitive environment, more advanced selection cycles (subsequent to cycle 9) were not persistent in a grazing environment (Gates et al. 2004).

A recent cultivar, TifQuik, was developed from seed of Tifton 9 selected for increased rate and extent of emergence through reduced hard-seededness using RRPS. Commencing in 1996, seed from Tifton 9 bahiagrass was planted and the first emerging seedlings were selected and used for cross-pollination. This selection was repeated over four generations. The resulting cycle 4 population had a fourfold improvement of germination compared to Tifton 9 after 1 week in greenhouse trials. In replicated field trials, germination rates after 1 and 2 weeks were four- and twofold greater for TifQuik compared to Tifton 9. Plant height and yield were also greater. In 2006 yield trials in Florida, although during a drought year, TifQuik yielded 9,983, compared with 9,685 and 10,596 kg ha^{-1} for Tifton 9 and UF-Riata, respectively.

A physiological basis for the yield improvement in diploid bahiagrass may be associated with increased photoperiod insensitivity. Progressive increases in the forage yield of the selection cycles made by Dr. Glen Burton may have resulted specifically from the selection for greater insensitivity of plants to day length. Based on this concept, a breeding program at the University of Florida, initiated in 2001, focused on developing a diploid bahiagrass population through RRPS, selecting plants with greater insensitivity to day length. This trait, coupled with improved leaf tissue cold tolerance, resulted in the release of UF-Riata bahiagrass in 2007 (Blount et al. 2008).

4.6.1.2 Diploid Cultivars

AU Sand Mountain was released in 1999 by Alabama Agriculture Experiment Station (AES). AU Sand Mountain is the result of a natural selection in a plant introduction thought to have been planted at the Sand Mountain Substation some 30 years ago. This cultivar has narrow leaves, fine tillers, and a short inflorescence. In the northern part of Alabama, this new cultivar has yielded more than Tifton 9 (Mislevy and Quesenberry 1999).

Nangoku originated in Kagoshima, Japan, in 1983. It is a synthetic sexual diploid originating from five selected clones of Pensacola 64-P. Nangoku has a semi-erect growth habit and was selected for palatability, high yield, winter survival, and early spring vigor in Japan (Takai and Komatsu 1998). It is used for grazing and hay at altitudes below 400 m in southern Kyushu, Japan.

Nanpu (PI 420018) is a diploid bahiagrass originating from two generations of maternal selection from four diploid plant introductions. Nanpu is similar in yield and palatability to Shinmoe but is less productive than Nangoku. It has good winter survival and is used primarily for grazing. Nanpu was replaced by Shinmoe, and commercial seed is no longer available.

Pensacola (PI 422024) was first distributed to producers in 1943. Pensacola bahiagrass is believed to have originated in Argentina and was brought to Florida in the digestive tract of cattle shipped from the Sante Fe region of Argentina to the Perdido dock near Pensacola. It was found growing in 1935 near the docks in Pensacola by the county agricultural agent E. H. Finlayson (1941), who popularized the grass. It is a sexual diploid, is propagated by seed, and is more cold tolerant than the more common apomictic types. Pensacola bahiagrass has long, narrow leaves and taller seed stalks, and it flowers earlier than other cultivars. It has a fibrous root system capable of growing to a depth of 2–3 m or more. Although Pensacola bahiagrass has some cold tolerance, top growth is killed by moderate frost. Pensacola is the predominant cultivar grown in Florida pastures (Chambliss and Jones 1980; Chambliss and Sollenberger 1991). This cultivar is also popular as a hay crop (Killinger 1959) and utility turfgrass.

Shinmoe (PI 420019) is a diploid bahiagrass that was developed from five clones selected from 70 elite diploid clones. It has improved germination, good vigor, and earlier spring growth than Nanpu. It is adapted to southern Kyushu, Japan. Shinmoe has been replaced by Nangoku, and commercial seed is no longer available.

Tifhi-1 (NSL 4715) was released by Dr. Glen Burton at the USDA-ARS Coastal Plain Experiment Station (CPES), Georgia, and the Georgia AES in 1958. *Tifhi-2* (NSL 20064) was released in 1961 (same source). Both are sexual diploid, F_1 hybrids developed from Pensacola using two self-sterile, cross-fertile clones and selecting a superior F_1 hybrid (Hein 1958). Tifhi-1 and Tifhi-2 are leafier, have reduced seed shattering, and are higher yielding than Pensacola. These F_1 hybrids yielded 25% more forage when clipped and 17% more live weight gain when grazed by livestock compared with Pensacola (Burton 1982). These cultivars were not popularized because of the difficulty and cost related to seed production.

T 18 (PI 648217) and *T 23* (PI 648218) are diploid bahiagrass germplasm developed by Dr. Glen Burton at the USDA-ARS, CPES, Tifton, Georgia. These germplasm resulted from 18 and 23 cycles of RRPS, respectively. T 18 and T 23 are higher yielding than both Pensacola and Tifton 9; however, they were not released as cultivars because they lacked grazing persistence (Mislevy and Quesenberry 1999). They were described and entered into the USDA ARS National Genetic Resource Program (2007 and 2008).

TifQuik (PI 648303) is a new cultivar released in 2007 by the USDA-ARS. This novel cultivar resulted from selection for rapid germination from four selection cycles of RRPS from seed of Tifton 9. The resulting population exhibits a fourfold improvement of germination compared to Tifton 9 after 1 week in greenhouse trials and in field trials. When compared in field trials in Georgia (Anderson et al. 2005) and Florida (Blount et al. 2008), plant height, germination, and forage yield were greater for TifQuik than for Tifton 9.

Tifton 9 (PI 531086) was released in 1987 by Dr. Glen Burton at the USDA-ARS, CPES, Tifton, Georgia. Starting in 1980, Dr. Burton used RRPS to develop improved bahiagrass populations from the diploid cultivar Pensacola. Tifton 9 resulted from nine cycles of RRPS; it is higher yielding than Pensacola and was released in 1987 (Burton 1989). Tifton 9 has greater seedling vigor, develops longer leaves, and is 30% higher yielding than Pensacola.

UF-Riata (PI pending) is a novel population developed for fall and early spring forage production for the southeastern United States. It has improved forage growth under short day lengths and during the cool season. The population exhibits lower photoperiod sensitivity, improved leaf tissue cold tolerance, and increased forage production during the cool season compared to the standard bahiagrass cultivars Argentine and Pensacola, and Tifton 9 and TifQuik to a lesser extent. UF-Riata was released by the University of Florida, Florida Agricultural Experiment Station (UF-FAES) in 2007 (Blount, Quesenberry, et al. 2008).

4.6.2 Tetraploid Bahiagrass

4.6.2.1 Tetraploid Breeding

4.6.2.1.1 Ecotype Selection

More than 80 tetraploid ecotypes collected in Brazil, Uruguay, and Northern Argentina have been introduced into the United States during the last 60 years (Burton 1992). Although modern breeding procedures have been developed for apomictic species, this process of collection, evaluation, and selection, adapted from vegetatively propagated species, has proved to be an important breeding approach, resulting in the identification of several superior apomictic ecotypes cultivated in an extensive area that have also been the bases for hybridization and molecular transformation. Among those identified as well adapted, Argentine and Paraguay 22—formerly PI 14896 and PI 158822, respectively—have been the most accepted by local beef-cattle producers because of their superior forage yields and cold tolerance.

4.6.2.1.2 Hybridization

Although meiosis is bypassed in the female organs of tetraploid bahiagrass ecotypes as part of the apomictic process, the male gametes are meiotically derived ($n = 20$). Because of this characteristic, tetraploid ecotypes are restricted to be used as male parents for hybridization.

Crosses between sexual diploid plants as female parents and tetraploid plants as male parents result in a reduced number of mainly triploid progeny (Burton and Hanna 1986). These unique triploid plants that usually carry the genes for apomixis can be crossed as female parents with diploid plants as male parents, resulting in triploid progeny, genetic copies of the mother plant, and tetraploid progeny when the unreduced egg cell ($2n = 3x = 30$) is fertilized by a reduced sperm ($n = 10$) coming from the diploid plant. Although this approach should be considered as viable, it has several disadvantages. One of them relates to the difficulty involved in obtaining apomictic triploid progeny from diploid × tetraploid crosses that results from the variable ability of diploid plants to set seed when self-pollinated (Acuña et al. 2007) and from the expected reduced number of triploid progeny (Burton and Hanna 1986). The screening of a large progeny resulting from triploid × diploid crosses for ploidy level is another disadvantage.

The generation of artificially induced tetraploid plants from diploid plants by using chromosome doubling agents and the use of these usually sexually induced tetraploid plants as female parents for crosses with the apomictic tetraploid plants as male parents are considered the most feasible hybridization procedure. A large number of sexual autotetraploid plants was obtained by Forbes and Burton (1957) by treating seeds from selected Pensacola bahiagrass clones with colchicine, and also by Quesenberry and Smith (2003) by treating callus cultures from Tifton 9 with different chromosome doubling agents. However, breeders should be cautious, because chromosome doubling of diploid plants can result in facultative apomictic tetraploid plants (Quarin et al. 2001).

Although the fertility of induced sexual tetraploid plants is significantly reduced compared with the original diploid plants (Acuña et al. 2007), the self-incompatibility system usually present in diploid plants remains unchanged after chromosome doubling (Burton et al. 1970; Acuña et al. 2007). This characteristic reduces to a minimum the number of progeny resulting from self-fertilization of the induced tetraploid plants when they are used as female parents.

Beginning in 1956 and continuing for more than 30 years, G. W. Burton developed a breeding program based on the hybridization of artificially induced sexual tetraploid and apomictic tetraploid plants (Burton 1992). Tifton 7 and Tifton 54, among others, were apomictic clones that resulted from this program. Although they produce higher forage yields than Argentine and are more tolerant to

Figure 4.3 Bahiagrass tetraploid ($2n = 4x = 40$) progeny generated by crossing sexual and apomictic genotypes.

ergot, they have remained as experimental clones. A continuation of this program was initiated at the University of Florida in 2004. The information about the genetic control of apomixis and ecological factors restricting the growing season of bahiagrass that has accumulated during the last 20 years is expected to enhance the productivity of this new breeding program.

The new set of hybridizations involved a few of the induced tetraploids generated from Tifton 9 by Quesenberry and Smith (2003), determined to be sexual and cross-pollinated (Acuña et al. 2007), and the cultivars Argentine, Tifton 7, and a few other superior apomictic clones. The progeny generated from these crosses (Figure 4.3) was highly variable because of the heterozygous condition present in both diploid and tetraploid populations. An excess of sexual progeny resulted from the suspected partial pleiotropic lethal effect of the apomictic locus (see Section 4.5.2). A ratio of three sexual to one apomictic with different degrees of expression was found among the progeny. However, the ratio between obligate apomictic progeny (potential new cultivars) and sexual progeny was 1:8. This high proportion of sexual and intermediate plants indicates that a large number of progeny should be generated and screened to find a superior, highly apomictic clone among the progeny. A group of superior sexual clones and another of highly apomictic clones were selected for further analysis. Preliminary data show that most of the selected apomictic clones have the ability to extract higher amounts of nitrogen and phosphorous from the soil profile and to produce higher forage yields when compared with the cultivar Argentine, especially during the cool-season.

Although the apomictic progeny retains the levels of self- and cross-fertility observed for the apomictic parents, the sexual progeny are self-fertile. These results contrast with those observed for brachiaria grass (Ngendahayo, Coppens D'Eeckenbrugge, and Louant 1988) and guineagrass (Savidan 1980), where the sexual progeny are self-sterile.

4.6.2.2 Tetraploid Cultivars and Advanced Germplasm

Argentine (PI 148996) was introduced into Florida in 1945 by the USDA and was released to producers in 1951 (Killinger et al. 1951). This cultivar is a semierect tetraploid with wide leaves (Mislevy and Quesenberry 1999). Leaves can range from glabrous to pubescent, depending on the age of the plant, stage of growth, management, and environment. It is not as cold tolerant as other bahiagrass and does not produce as much forage in late fall or early spring. It is popular in the sod industry because it has fewer seed heads. Argentine bahiagrass seed heads are affected by ergot (*Claviceps paspali* Stevens & Hall), which is toxic to cattle, and *Fusarium* spp. often is present on the seed heads.

Batatais is a common bahiagrass from Brazil. It is short, with wide leaves and few racemes, and it forms a dense sod. It is used as a turfgrass in Brazil and may be vegetatively propagated (Loch and Ferguson 1999).

Common bahiagrass was introduced into Florida in 1913 by the Florida Agricultural Experiment Station. It was thought to be a potentially improved pasture plant in Florida because it could be easily established on sandy soils. Common bahiagrass plants consist of short, broad, pubescent leaves. However, it was slow to establish, low in productivity, and sensitive to the cold temperatures that sporadically occur throughout the state during the winter period. Therefore, it was never adopted by Florida ranchers (Mislevy and Quesenberry 1999).

Competidor (Kf.88, later P.1308) is an Australian cultivar released in 1986 in New South Wales, Australia (Wilson 1987). It was introduced to Grafton Research Station from the United States in 1953 as an Argentine-type bahiagrass. It performed well as a pasture grass on soils of moderate to low fertility. Competidor produces fewer seed heads and is more palatable and more shade tolerant than Pensacola.

Nan-ou (PI 556953) originated in Kagoshima, Japan, in 1991 (Takai and Komatsu 1998). Nan-ou is an apomictic tetraploid cultivar derived from mass selection from PI 284172 (CPI 21339), a plant introduction collected in Costa Rica in 1962. It has wider leaves, shorter plant height, and fewer numbers of seed heads, and it is more palatable than diploid cultivars such as Nangoku. Nan-ou produces more forage than Nangoku in summer and early autumn and is used for grazing and hay in Japan.

Paraguay was introduced to the United States from Paraguay in 1938. Its leaf blades are pubescent and narrower than common bahiagrass, and it is a heavy seed producer. Paraguay is similar to Argentine in appearance and is less cold tolerant and less winter productive than Pensacola bahiagrass. Pure stands in Florida are rare. It was used predominantly as a pasture and turf.

Paraguay 22 (PI 158822) was released in 1942 (McCloud 1953). It has short, narrow, hairy leaves and is a much greater seed producer than common bahiagrass. As a forage, Paraguay 22 is more productive than Paraguay. It is similar to Argentine in growth habit and cold tolerance; however, unlike Argentine, it is reported as resistant to ergot (McCloud 1953), although the cultivar occasionally does succumb to the seed head fungus.

Q4188 (PI 619631) and *Q4205* (PI 619632) are sexual, tetraploid bahiagrass germplasm (Quarin et al. 2003) used in the development of novel apomictic and sexual bahiagrass lines. Q4188 has short, stout, ascending rhizomes; an erect growth habit; red-purple basal leaf sheaths; purple anthers; and purple stigmas. It was derived from a cross between Q3664 and Q3853. Parent Q3664 originated from a cross between a sexual tetraploid plant (PT-2), induced by colchicine treatment of the sexual diploid Pensacola bahiagrass biotype (*P. notatum* var. *saurae*), and a white stigma bahiagrass strain (WSB). The original cross was made by Burton and Forbes (1960). In 1979, the Q3664 plant was introduced to FCA-UNNE, Argentina. Parent Q3853 was introduced to FCA-UNNE by pieces of rhizomes from Brazil collected by J. F. Valls and coworkers (accession no. 4751, found near Osorio and Capivari, state of Rio Grande do Sul). Embryological analyses showed that this plant is highly apomictic. Genetic fingerprinting using restriction fragment length polymorphism (RFLP) and random amplified polymorphic DNA (RAPD) indicated that self-pollinated progeny of plant Q4188 (experimental number F131) originated exclusively by sexual means (Ortiz et al. 1997). Q4205 has short rhizomes, an upright growth habit, red-purple leaf sheaths, and white stigmas and is highly sexual. Plant Q4205 is a selected, selfed progeny of plant Q3664. Both lines are maintained as a clone by rhizome propagation.

Riba (CPI 23944) was released in Australia in 1994 (Loch and Ferguson 1999). Riba originated in Uruguay. It is considered to be a turf type, with shorter, stiffer leaves and shorter racemes than Pensacola and Competidor. Riba was selected for its low growth habit, dark-green leaf color, and ergot resistance. It is well adapted over a wide range of soil types and maintains a better turf appearance under low fertility and reduced mowing than other varieties. In Florida trials, Riba has been slow to establish, eventually forming a dense sod.

Tifton 54 and *Tifton 7* are apomictic hybrid bahiagrass germplasm developed by Dr. Glen Burton at the CPES in Tifton, Georgia. He considered them to be superior to apomictic introductions in yield and winter hardiness. Sexual tetraploids were created by doubling the chromosomes in the sexual diploid 'Pensacola' bahiagrass (Burton and Forbes 1960). These sexual tetraploids were used as female parents and, when crossed with apomictic males, produced highly variable progeny that contained both sexual and apomictic plants. Although not released cultivars, Tifton 54 and Tifton 7 were developed through that hybridization scheme and performed well in bahiagrass yield trials in Florida.

Wilmington (PI 434189) is a tetraploid bahiagrass with characteristic narrow leaves and is purported to have improved winter survival. Although it is less productive than 'Pensacola' bahiagrass and is a poor seed producer, it is considered to be the most cold hardy of the U.S. cultivars. Few stands of Wilmington are found in the United States today.

4.6.3 Molecular Transformation

The generation of transgenic plants allows breeders to clone genes from virtually any living organism and insert the cloned gene into another organism, including turf and forage crops. An effective transformation system requires a gene delivery system, an appropriate tissue and regeneration system, desirable genes, promoter genes, and selectable markers (Vogel and Burson 2004). Bahiagrass regeneration from *in vitro* cultures has been demonstrated via organogenesis from inflorescence-derived callus (Bovo and Mroginski 1986), via somatic embryogenesis from seed-derived callus (Marousky and West 1990; Akashi, Hashimoto, and Adachi 1993; Gondo et al. 2005), and from callus derived from germinating seedlings (Grando, Franklin, and Shatters 2002; Altpeter and James 2005).

Biolistics or microprojectile bombardment has been used for transient gene studies and for the recovery of transgenic plants in diploid and tetraploid bahiagrass. Transgenic bahiagrass plants have been obtained by microprojectile bombardment of embryogenic cells (Smith et al. 2002; Gondo et al. 2005; Agharkar et al. 2007; James, Neibaur, and Altpeter 2008; Luciani et al. 2007; Sandhu, Altpeter, and Blount 2007; Zhang, Lomba, and Altpeter 2007). In these studies, phosphinothricin acetyltransferace (*bar* and *pat*) and neomycin phosphotransferace II (*npt2*) were used as selectable markers. These genes were driven by the cauliflower mosaic virus (CaMV) 35S promoter or by the rice actin promoter. Transgenic diploid and tetraploid bahiagrass plants containing the *bar* gene that confers resistance to phosphinothricin (glufosinate) were obtained (Smith et al. 2002; Gondo et al. 2005; Sandhu et al. 2007). The generated transgenic bahiagrass plants of the cultivar Argentine were resistant to greenhouse and field applications of high doses of glufosinate ammonium (Sandhu et al. 2007). Although there are no data showing the absence of resistance in diploid bahiagrass plants for fall armyworm (*Spodoptera frugiperda* Smith), a few transgenic Tifton 9 plants, containing the *cry1Fa* gene encoding a δ-endotoxin from *Basilus thurigiensis,* were also generated (Luciani et al. 2007).

The marked adaptability of bahiagrass to the variable Florida environment, scarce management, and tolerance of intense defoliation make it the dominant utility turf in the region. However, it has poor turf quality resulting from its tall inflorescences, which is present during most of the growing season, and its open canopy. Transgenic 'Argentine' bahiagrass plants with the *Arabidopsis ATHB16* transcription factor were obtained (Zhang et al. 2007). The evaluation of these transgenic plants overexpressing the *ATHB16* growing in hydroponics and greenhouse showed some promising morphological changes, such as a reduction in the number of inflorescences and a denser canopy. The endogenous giberellin catabolizing *AtGA2ox1* was also inserted into 'Argentine' bahiagrass plants in an attempt to improve the turf quality of this ecotype (Agharkar et al. 2007). These transgenic plants evaluated under field conditions showed an increased number of vegetative tillers increasing the turf density, shorter tillers, and delayed flowering.

The transcription factor *DREB1A* from xeric *Hordeum spontaneum* L. was also introduced into Argentine bahiagrass with the objective of increasing water-stress tolerance (James et al. 2007). The results from plants grown in hydroponics showed the expression of this transgene under severe water-stress and high-salinity conditions. However, the utility of these transgenic plants needs to be tested under field conditions, especially because the usefulness of genes related to plant survival under severe stress has been questioned (Sinclair, Purcell, and Sneller 2004).

The natural variability of bahiagrass is scarcely represented by the germplasm contained in the United States. The necessary research to increase the knowledge about the rich diversity of this species may be bypassed by molecular transformation and the urgent need of superior turf and forage types.

4.7 DISEASES, INSECTS, AND NEMATODES

Many insect pests, nematodes, and disease organisms have been reported associated with bahiagrass; however, few, if any, have any significant impact on the long-term performance and viability of this grass species. Because bahiagrass has few pest-related problems, it is often used in crop rotation to reduce pest pressure prior to planting subsequent crops (Brenneman et al. 2003; Norden et al. 1977; White et al. 1962; Katsvairo et al. 2006; Rodriguez-Kabana et al. 1989a, 1989b).

4.7.1 Diseases

Although disease occurrence in bahiagrass may not be considered a serious threat to its persistence as a forage or turf plant, there are some economic losses due to plant pathogens (Blount et al. 2002). Some diseases often are undetected and, as in the case of some fungal diseases like ergot (*Claviceps* spp.) or dollar spot (*Sclerotinia homoeocarpa* F. T. Benn), may play a more prominent role in loss of yield than previously thought (Blount et al. 2002).

Ergot infects the florets of bahiagrass, especially the cultivar Argentine. Typically, this prevents the formation of viable seed and reduces potential seed yield. In the southern United States, where many bahiagrass seed production fields are also grazed by livestock or harvested for hay, few legal fungicides are available to reduce ergot. Losses from potential seed production on 'Argentine' bahiagrass in the southeastern United States have varied from fewer than 5% to 30% in years of severe ergot infection. Infected florets are typically collected from 'Argentine' bahiagrass, but infected florets from 'Pensacola' and Tifton 9 have also been identified (J. W. Kimbrough, fungal mycologist, University of Florida, personal communication).

Dollar spot is a leaf spot disorder associated with bahiagrass. In the southeastern United States, warm, humid conditions promote the occurrence of the dollar spot lesions, tip dieback, and subsequent leaf tissue burn. Cultivar susceptibility to dollar spot may have contributed to the pronounced outbreak in bahiagrass hay production fields in 2001 and 2002 in north Florida. Several Tifton 9 pastures in Jackson and Walton counties in the Florida Panhandle were devastated by dollar spot. A recent study in Florida evaluated diploid and tetraploid bahiagrass cultivars under natural field infection for reaction to the fungus. Results from multilocation and multiyear evaluations showed little leaf tissue damage to 'Argentine' (average 2–5% leaf area) compared to Tifton 9 or 'Pensacola' (average 52–54% leaf area) bahiagrass. Although infected, other tetraploid cultivars and experimental lines also had few leaf lesions. The disease severity rankings among the diploid cultivars varied in each year. Soil pH and fertility of the pastures appear to play a role in disease severity when environmental conditions are optimum for infection. Timely fungicide applications reduce leaf losses; however, few legal fungicides can be applied to bahiagrass pastures that are grazed by livestock or harvested for hay.

Bahiagrass is relatively free of virus diseases (Burson and Watson 1995); however, barley yellow dwarf virus (BYDV) infection has been recently confirmed in bahiagrass (Blount, Sprenkel, et al. 2008). Preliminary results from studies in Florida implicate bahiagrass as a host of BYDV. Both tetraploid and diploid cultivars may be infected, although the virus is more commonly found in tetraploid cultivars.

Although other minor fungal diseases affect bahiagrass, they are not considered to be serious in nature, except in rare circumstances. These include, but are not limited to, *Helminthosporium* spp. (*Bipolaris* spp.), *Claviceps* spp., *Colletotrichum* spp., *Fusarium* spp., and *Puccinia* spp.

4.7.2 Insects

Generally, bahiagrass has few insect pest problems that pose a serious threat to the viability of the plant. In the southeastern United States, particularly in central Florida, non-native mole crickets (*Scapteriscus* spp.) have been the cause of bahiagrass pasture decline (Adjei et al. 2001). The tawny mole cricket (*Scapteriscus vicimus* Scudder) has been implicated as the most economically damaging of the species, devastating more than 150,000 ha of bahiagrass pasture in Florida (Adjei et al. 2001). In pastures and turf plantings in Florida, estimates of nearly $50 million (U.S. currency) can be attributed to damage, replanting, and chemical control of that mole cricket species (Adjei, Frank, and Gardner 2003). Damage to bahiagrass pastures has also been reported caused by other non-native mole cricket species, including the short-winged mole cricket (*Scapteriscus abbreviatus* Scudder) and the southern mole cricket (*Scapteriscus borelli* Giglio-Tos); however, they are not considered as serious an insect pest as the tawny mole cricket. Sod harvested in central Florida and ornamental plants transported to more northern locations in the state may have led to the spread of the tawny mole cricket into the Florida Panhandle.

Genetic resistance in bahiagrass for mole cricket susceptibility among common bahiagrass cultivars does not appear to be promising. Preliminary greenhouse and field studies (R. K. Sprenkel, extension entomologist, and A. R. Blount, forage breeder, North Florida Research and Education Center, University of Florida, unpublished results) suggested that tetraploid bahiagrass cultivars Argentine and Paraguay 22 were slightly more tolerant of tawny mole cricket feeding than diploid cultivars (Pensacola, Tifton 9, and Sand Mountain). However, all bahiagrass cultivars eventually succumbed to root damage by the tawny mole cricket when populations exceeded threshold limits.

In turfgrass situations where the mole cricket has presented a problem on bahiagrass, chemical control has been used successfully. Chemical control products include liquid, granular, and bait formulations containing a number of active ingredients, including bifenthrin, cyfluthrin, fipronil, imidacloprid, or permethrin. Combinations of chemical and biological control measures have also been used successfully in turfgrass and in pasture situations. Much emphasis has been placed more recently on biological control mechanisms for control of mole crickets on bahiagrass in the United States because of the economics of biological versus chemical control, pasture use restrictions, and general environmental considerations.

In Florida, the introduction of biological control methods to reduce mole cricket populations has been a statewide effort (Adjei, Smart, and Adams 2002; Buss, Capinera, and Leppla 2006). The nematode *(Steinernema scapterisci* Nguyen & Smart) that infects the mole cricket has been introduced throughout Florida from its native home in South America (Buss et al. 2006). Other biological control agents that are natural enemies of the mole cricket have also been introduced from South America (Frank et al. 1995). These include the parasitoid fly (*Ormia depleta* Wiedemann) and parasitoid wasp (*Larra bicolor* Fabricius). Both parasitoid insects have been shown to suppress mole crickets and have begun to establish themselves in Florida.

On occasion, the fall armyworm (*Spodoptera* spp.) may defoliate bahiagrass pastures when they are well fertilized and when other preferred grass species are unavailable. Generally, bahiagrass withstands the defoliation with no apparent stand loss.

4.7.3 Nematodes

Bahiagrass is considered a nonhost of most plant parasitic nematode species, particularly root knot nematodes (*Meloidogyne* spp.), and is used in rotations with crops susceptible to nematodes to reduce populations (Katsvairo et al. 2006). When bahiagrass is used in rotation with peanut or cotton, it reduces nematode populations below the threshold levels where economic damage to the row crop is compromised. Use of bahiagrass in crop rotations as a nematode control method is practical in field crops because the value of the row crop cannot justify the cost of using chemical nematicides. *Belonolaimus longicaudatus* Rau, a sting nematode, can attack bahiagrass roots; damage varies among the different cultivars (Baki, Ipon, and Chen 1992). Paraguay 22 is considered resistant to the sting nematode that affects 'Pensacola.'

4.8 FUTURE BREEDING AND DEVELOPMENT OF NEW CULTIVARS

Research on ecophysiological factors affecting forage yield, nutritive value and turf quality is needed to continue the genetic improvement of the diploid germplasm. One example is the promising high variability observed for the extension of the reproductive phase resulting from variable photoperiod sensitivities. We believe that plants with shorter reproductive phases would have better nutritive value, resulting in better forages, and would require a lower mowing frequency, resulting in better turf.

We believe that more emphasis should be given to the genetic improvement of the bahiagrass polyploid germplasm. The tetraploid germplasm of this species offers an excellent opportunity for manipulating apomixis to generate highly productive hybrids that may result in new cultivars. However, a more efficient screening for apomictic progeny is needed to overcome the low number of highly apomictic progeny resulting from crosses between sexual and apomictic genotypes. More research is also needed for a better understanding of the variable degrees of apomixis expression among apomictic plants. The generation and use of apomictic triploids to generate apomictic tetraploids should also be further analyzed.

4.9 FUTURE OUTLOOK

At present, there is a need for new bahiagrass cultivars that can produce more foliage for a longer period of the year and increase productivity of the cattle production systems in subtropical regions of the world. Also, as fertilizer costs increase on a worldwide basis, the demand for bahiagrass could increase as a low-fertilizer input utility turf, for soil stabilization, and, possibly as a cellulose crop for biofuel production. Thus, generation of new genetic materials should focus not only on increasing the forage productivity of the species but also on multiple uses in low-input grass systems. The creation of novel bahiagrass germplasm and breeding of new cultivars should provide new opportunities for the use of this unique plant species.

REFERENCES

Acuña, C. A., A. S. Blount, K. H. Quesenberry, W. W. Hanna, and K. E. Kenworthy. 2007. Reproductive characterization of bahiagrass germplasm. *Crop Science* 47:1711–1717.

Adjei, M. B., J. H. Frank, and C. S. Gardner. 2003. Survey of pest mole cricket (Orthoptera: Gryllotalpidae) activity on pasture in south-central Florida. *Florida Entomologist* 86:199–205.

Adjei, M. B., W. T. Crow, G. C. Smart, Jr., J. H. Frank, and N. C. Leppla. 2001. Biological control of pasture mole crickets with nematodes. ENY-9, Florida Coop. Ext. Ser., Univ. of Florida, 3 pp.

Adjei, M. B., P. Mislevy, and W. Chason. 1992. Seed yield of bahiagrass in response to sward management by phenology. *Agronomy Journal* 84:599–603.

———. 2000. Timing, defoliation management, and nitrogen effects on seed yield of 'Argentine' bahiagrass. *Agronomy Journal* 92:36–41.

Adjei, M. B., G. C. Smart, Jr., and B. J. Adams. 2002. Infectivity and persistence of *Ss* nematodes on pasture mole crickets in south-central Florida. [Online]. [2 p.] Available at http://edis.ifas.ufl.edu/IN416 (accessed January 15, 2003). Florida Coop. Ext. Serv., Inst. Food Agric. Sci., Univ. of Florida, Gainesville.

Agharkar M., P. Lomba, F. Altpeter, H. Zhang, K. Kenworthy, and T. Lange. 2007. Stable expression of *AtGA2ox1* in a low-input turfgrass (*Paspalum notatum* Flugge) reduces bioactive gibberellin levels and improves turf quality under field conditions. *Plant Biotechnology Journal* 5:791–801.

Akashi R., A. Hashimoto, and T. Adachi. 1993. Plant regeneration from seed-derived embryogenic callus and cell suspension cultures of bahiagrass (*Paspalum notatum*). *Plant Science* 90:73–80.

Altpeter, F., and V. A. James. 2005. Genetic transformation of turftype bahiagrass (*Paspalum notatum* Flügge) by biolostic gene transfer. *International Turfgrass Society Research Journal* 10:485–489.

Anderson, W. F., R. N. Gates, W. W. Hanna, and A. Blount. 2005. Rapid germinating forage bahiagrass. ASA-CSSA-SSSA International Annual Meetings, abstract.

Baki, B. B., I. B. Ipon, and C. P. Chen. 1992. *Paspalum notatum* Fluegge. In *Plant resources of South-east Asia. No 4. Forages,* ed. L. 't Mannetje and R. M. Jones, 181–183. Wageningen, the Netherlands: Pudoc-DLO.

Beaty, E. R., Brown, R. H., and J. B. Morris. 1970. Response of Pensacola bahiagrass to intense clipping. In: Norman, M. J. T., ed. *Proceedings International Grasslands Congresses,* 11th, Surfer's Paradise, QLD, Australia. 13–23 Apr. St. Lucia: University of Queensland Press, 1970:538–542.

Blount, A. R., H. Dankers, M. Timur Momol, and T. A. Kucharek. 2002. Severe dollar spot fungus on bahiagrass in Florida. [Online]. [3 p.] Available at http://www.plantmanagementnetwork.org/pub/cm/research/dollarspot/ (accessed January 15, 2003). Crop Management: 2002 plant management network.

Blount, A. R., K. H. Quesenberry, T. R. Sinclair, and P. Mislevy. 2008. *Journal of Plant Registrations* (submitted for publication).

Blount, A. R., R. K. Sprenkel, S. M. Gray, E. C. Benson, C. Malmstom, J. M. Anderson, B. A. R. Hadi, K. L. Flanders, and P. Mislevy. 2008. Preliminary report of barley yellow dwarf virus infection in bahiagrass (*Paspalum notatum* Flugge). *Plant Management* (submitted for publication).

Bovo, O. A., and L. A. Mroginski. 1986. Tissue culture in *Paspalum* (Gramineae): Plant regeneration from cultured inflorescences. *Journal of Plant Physiology* 124:481–492.

Brenneman, T. B., P. Timper, N. A. Minton, and A. W. Johnson. 2003. Comparison of bahiagrass, corn, and cotton as rotational crops for peanut. *Proceedings of the Sod Based Cropping Systems Conference.* Available: http://www.nfrec.ifas.ufl.edu/sodrotation/postconference.htm.

Burson, B. L., and V. H. Watson. 1995. Bahiagrass, dallisgrass, and other Paspalum species. In *Forages—An introduction to grassland agriculture,* vol. 1, 5th ed., ed. R. F. Barnes et al., 431–440. Ames: Iowa State Univ. Press.

Burton, G. W. 1940. A cytological study of some species in the genus *Paspalum. Journal of Agricultural Research* 60:193–197.

———. 1946. Bahia grass types. *Journal of the American Society of Agronomy* 38:273–281.

———. 1948. The method of reproduction in common Bahia grass, *Paspalum notatum. Journal of the American Society of Agronomy* 40:443–452.

———. 1955. Breeding Pensacola bahiagrass, *Paspalum notatum:* I. Method of reproduction. *Agronomy Journal* 47:311–314.

———. 1967. A search for the origin of Pensacola bahiagrass. *Economic Botany* 21:319–382.

———. 1982. Improved recurrent restricted phenotypic selection increases bahiagrass forage yields. *Crop Science* 22:1058–1061.

———. 1984. Plant breeding 1910–1984. In *Gene manipulation in plant improvement,* ed. J. P. Gustafson. New York: Plenum Publishing Corporation.

———. 1989. Registration of Tifton 9 Pensacola bahiagrass. *Crop Science* 29:1326.

Burton, G. W., and I. Forbes, Jr. 1960. The genetics and manipulation of obligate apomixis in common bahiagrass (Paspalum *notatum* Flüggé). 66–71. In C. L. Skilmore (ed.) *Proceedings of the International Grassland Congress*, 8th, Reading, England. 11–21 July 1960. Alden Press, Oxford, UK.

———. 1992. Manipulating apomixis in *Paspalum*. In *Proceedings of Apomixis Workshop,* ed. J. H. Elgin and J. P. Miksche, 16–19. Atlanta, Georgia. February 11–12, 1992. USDA-ARS, ARS-104. Washington, D.C.: U.S. Gov. Print. Office.

Burton, G. W., I. Forbes, Jr., and J. Jackson. 1970. Effect of ploidy on fertility and heterosis in Pensacola bahiagrass. *Crop Science* 10:63–66.

Burton, G. W., and W. W. Hanna. 1986. Bahiagrass tetraploids produced by making (apomictic tetraploid × diploid) × diploid hybrids. *Crop Science* 26:1254–1256.

Busey, P. 1989. Progress and benefits to humanity from breeding warm-season grasses for turf. In *Contributions from breeding forage and turf grasses,* ed. D. A. Sleper et al., 49–70. Madison, WI: CSSA Spec. Publ. 15. CSSA.

Buss, E. A., J. L. Capinera, and N. C. Leppla. 2006. EDIS, University of Florida, Institute of Food and Agricultural Sciences online publication. [Online]. [9 p.] Available at http://edis.ifas.ufl.edu/LH039 (accessed March 14, 2008). Florida Coop. Ext. Serv., Inst. Food Agric. Sci., Univ. of Florida, Gainesville.

Chambliss, C. G., and M. B. Adjei. 2006. Bahiagrass. SS-AGR-36. Agron. Dept., Univ. of Florida, Gainesville.

Chambliss, C. G., and D. W. Jones. 1980. Bahiagrass. Ext. Cir. 321B. Florida Coop. Ext. Serv., Univ. of Florida, Gainesville.

Chambliss, C. G., and L. E. Sollenberger. 1991. Bahiagrass: The foundation of cow–calf nutrition in Florida. *Proceedings of Beef Cattle Short Course* 40: 74–80. IFAS/Univ. of Florida, Gainesville.

Chase, A. 1929. The North American species of *Paspalum,* 28:1–310. Contr. U.S. Natl. Herb. Washington, D.C.: U.S. Gov. Print. Office.

Cuomo, G. J., D. C. Blouin, D. L. Corkern, J. E. McCoy, and R. Walz. 1996. Plant morphology and forage nutritive value of three bahiagrasses as affected by harvest frequency. *Agronomy Journal* 88:85–89.

Daurelio, L. D., F. Espinoza, C. L. Quarin, and S. C. Pessino. 2004. Genetic diversity in sexual diploid and apomictic tetraploid populations of *Paspalum notatum* situated in sympatry and allopatry. *Plant Systematics Evolution* 244:189–199.

Duke, J. A. 1981. The gene revolution. Paper 1, 89–150. Office of Technology Assessment, Background papers for innovative biological technologies for lesser developed countries. USGPO. Washington, D.C.

Finlayson, E. H. 1941. Pensacola—A new fine leaved bahia. *Southern Seedsman* 4 (12): 9.

Forbes, I., Jr., and G. W. Burton. 1957. The induction and some effects of autotetraploidy in Pensacola bahiagrass, *Paspalum notatum* var. *Saurae Parodi.* Agronomy Abstracts, ASA Annual Meetings, p. 53.

———. 1961. Cytology of diploids, natural and induced tetraploids, and intra-specific hybrids of bahiagrass, *Paspalum notatum* Flügge. *Crop Science* 1:402–406.

Frank, J. H., T. R. Fasulo, D. E. Short, and A. S. Weed. 1995. Alternative methods of mole cricket control. University of Florida. Online publication: http://molecrickets.ifas.ufl.edu/mcri0002.htm

Gates, R. N., and G. W. Burton. 1990. Sampling procedures allowing selection of bahiagrass for improved digestibility. In *Agronomy Abstracts*, 12, Appdx. 1. Madison, WI: ASA.

Gates, R. N., G. M. Hill, and G. W. Burton. 1999. Response of selected and unselected bahiagrass populations to defoliation. *Agronomy Journal* 91:787–795.

Gates R. N., C. L. Quarin, and C. G. S. Pedreira. 2004. Bahiagrass. In *Warm-season (C4) grasses,* ed. L. E. Moser, B. L. Burson, and L. E. Sollenberger. Madison, WI: ASA, CSSA, SSSA.

Gondo, T., S. Tsuruta, R. Akashi, O. Kawamura, and F. Hoffmann. 2005. Green, herbicide-resistant plants by particle inflow gun-mediated gene transfer to diploid bahiagrass (*Paspalum notatum*). *Journal of Plant Physiology* 162:1367–1375.

Grando, M. F., C. I. Franklin, and R. G. Shatters, Jr. 2002. Optimizing embryogenic callus production and plant regeneration from 'Tifton 9' bahiagrass seed explants for genetic manipulation. *Plant Cell, Tissue and Organ Culture* 71:213–22.

Hein, M. A. 1958. Registration of varieties and strains of grasses. *Agronomy Journal* 50:399–401.

Hitchcock, A. S. 1971. *Manual of the grasses of the United States,* vol. 2, 605–606. New York: Dover Publications, Inc.

James, V. A., I. Neibaur, and F. Altpeter. 2008. Stress inducible expression of the *DREB1A* transcription factor from xeric, *Hordeum spontaneum* L. in turf and forage grass (*Paspalum notatum* Flugge) enhances abiotic stress tolerance. *Transgenic Research* 17:93–104.

Jean, S.-Y., and Tzo-chuan Juang. 1979. Effect of bahia grass mulching and covering on soil physical properties and losses of water and soil of slopeland (first report). *Journal of the Agricultural Association of China (Taipei)* 105:57–66.

Katsvairo, T. W., D. L. Wright, J. J. Marois, D. Hartzog, J. R. Rich, and P. J. Wiatrak. 2006. Sod–livestock integration into the peanut–cotton rotation, a systems farming approach. *Journal of Agronomy* 98:1156–1171.

Killinger, G. B. 1959. Pasture herbage changes in Florida during the past two decades (1939–1959). *Soil Crop Science Society of Florida Proceedings* 19:162–166.

Killinger, G. B., G. E. Ritchey, C. B. Blickensderfer, and W. Jackson. 1951. *Argentine bahia grass.* Agricultural Experiment Station Annual Report. University of Florida, Gainesville.

Kretschmer, A. E., Jr., and N. C. Hood. 1999. *Paspalum notatum* in Florida, USA. In *Forage seed production.* Vol. 2. *Tropical and subtropical species,* ed., D. S. Loch and J. E. Ferguson, 335. Wallingford, U.K.: EDS., CABI Publishing.

Loch, D. S., and J. E. Ferguson. 1999. Tropical and subtropical forage seed production: An overview. In *Forage seed production.* Vol. 2. *Tropical and subtropical species,* ed., D. S. Loch and J. E. Ferguson, 1–40. Wallingford, U.K.: EDS., CABI Publishing.

Luciani, G., F. Altpeter, J. Yactayo-Chang, H. Zhang, M. Gallo, R. L. Meagher, and D. Wofford. 2007. Expression of cry1Fa in bahiagrass enhances resistance to fall armyworm. *Crop Science* 47:2430–2436.

Marousky, F. J., and S. H. West. 1990. Somatic embryogenesis and pant regeneration from cultured mature caryopses of bahiagrass (*Paspalum notatum* Flugge). *Plant Cell, Tissue and Organ Culture* 20:125–129.

Martínez, E. J., C. A. Acuña, D. H. Hojsgaard, M. A. Tcach, and C.L. Quarin. 2007. Segregation for sexual seed production in *Paspalum* as directed by male gametes of apomictic triploid plants. *Annals of Botany* 100:1239–1247.

Martínez E. J., J. Stein, H. E. Hopp, and C. L. Quarin. 2003. Genetic characterization of apospory in tetraploid *Paspalum notatum* based on the identification of linked molecular markers. *Molecular Breeding* 12:319–327.

Martínez, E. J., M. H. Urbani, C. L. Quarin, and J. P. A. Ortiz. 2001. Inheritance of apospory in bahiagrass, *Paspalum notatum. Hereditas* 135:19–25.

McCloud, D. E. 1953. Forage and cover plant introduction by the Florida Agricultural Experiment Station. *Soil Crop Science Society of Florida Proceedings* 13:32–38.

Mills, P. F. L. and J. N. Boultwood. 1978. A comparison of *Paspalum notatum* accessions for yield and palatability. *Zimbabwe Agricultural Journal* 75:71–74.

Mislevy, P., and K. H. Quesenberry. 1999. Development and short description of grass cultivars released by the University of Florida (1892–1995). *Soil Crop Science Society of Florida Proceedings* 58:12–19.

Mislevy, P., T. R. Sinclair, and J. D. Ray. 2001. Extended daylength to increase fall/winter yields of warm-season perennial grasses. In *Proceedings of the 19th International Grassland Congress,* ed. J. A. Gomide et al., 256–257. Sao Pedro, SP, Brazil, February 11–21, 2001. FEALQ, Piracicaba, SP, Brazil.

Moore, J. E., O. C. Ruelke, C. E. Rios, and D. E. Franke. 1971. Nutritive evaluation of Pensacola bahiagrass hays. *Soil Crop Science Society of Florida Proceedings* 30:211–221.

Ngendahayo, M., G. Coppens D'Eeckenbrugge, and B. P. Louant. 1988. Self-incompatibility studies in *Brachiaria ruziziensis* Germain et Evrard, *Brachiaria decumbens* Stapf and *Brachiaria brizantha* (Hochst) Stapf and their interspecific hybrids. *Phytomorphology* 38:47–51.

Norden, A. J., V. G. Perry, F. G. Martin, and J. NeSmith. 1977. Effect of age of bahiagrass sod on succeeding peanut crops. *Peanut Science* 4:71–74.

Ortiz, J. P. A., S. C. Pessino, O. Leblanc, M. D. Hayward, and C. L. Quarin. 1997. Genetic fingerprinting for determining the mode of reproduction in *Paspalum notatum,* a subtropical apomictic forage grass. *Theoretical and Applied Genetics* 95:850–856.

Parodi, L. R. 1948. Gramíneas Argentinas nuevas o críticas. I. La variacíon en Paspalum notatum Fluegge. *Revista Argentina de Agronomia* 15:53-57.

Pozzobon, M. T., and J. F. M. Valls. 1997. Chromosome number in germplasm accessions of *Paspalum notatum* (Gramineae). *Brazilian Journal of Genetics* 20:29–34.

Quarin, C. L. 1992. The nature of apomixis and its origin in Panicoid grasses. *Apomixis Newsletter* 5:8–15.

———. 1999. Effect of pollen source and pollen ploidy on endosperm formation and seed set in pseudogamous apomictic *Paspalum notatum. Sexual Plant Reproduction* 11:331–335.

Quarin, C. L., B. L. Burson, and G. W. Burton. 1984. Cytology of intra- and interspecific hybrids between two cytotypes of *Paspalum notatum* and *P. cromyorrhizon. Botanical Gazette* 145:420–426.

Quarin, C. L., F. Espinoza, E. J. Martínez, S. C. Pessino, and O. A. Bovo. 2001. A rise of ploidy level induces the expression of apomixis in *Paspalum notatum. Sexual Plant Reproduction* 13:243–249.

Quarin C. L., M. H. Urbani, A. R. Blount, E. J. Martínez, C. M. Hack, G. W. Burton, and K. H. Quesenberry. 2003. Registration of Q4188 and Q4205, sexual tetraploid germplasm lines of bahiagrass. *Crop Science* 43:745–746.

Quarin, C. L., G. A. Norrmann, and M. H. Urbani. 1989. Polyploidization in aposporous *Paspalum. Apomixis Newsletter* 1:28–29.

Quesenberry, K. H., and R. Smith. 2003. Production of sexual tetraploid bahiagrass using in vitro chromosome doubling agents. In *Molecular breeding of forage and turf,* ed. A. Hopkins and R. Barker, 145. Abstracts, Third International Symposium, Dallas, TX, May 18–22, 2003.

Rodriguez-Kabana, R., C. F. Weaver, R. Garcia, D. G. Robertson, and E. L. Carden. 1989a. Additional studies on the use of bahiagrass for the management of root-knot and cyst nematodes in soybean. *Nematropica* 21:203–210.

———. 1989b. Bahiagrass for the management of root-knot and cyst nematodes in soybean. *Nematropica* 19:185–193.

Rollins, G. H., and C. S. Hoveland. 1960. Wanted: Good summer perennial grasses for dairy cows. *Alabama Agric. Exp. Stn. Highlights Agricultural Research* 7 (2).

Sandhu S., F. Altpeter, and A. R. Blount. 2007. Apomictic bahiagrass expressing the bar gene is highly resistant to glufosinate under field conditions. *Crop Science* 47:1691–1697.

Savidan, Y. 1980. Chromosomal and embryological analyses in sexual × apomicitc hybrids of *Panicum maximum* Jacq. *Theoretical and Applied Genetics* 57:153–156.

Scott, J. M. 1920. Bahia grass. *Journal of the American Society of Agronomy* 12:112–113.

Sinclair, T. R., L. C. Purcell, and C. H. Sneller. 2004. Crop transformation and the challenge to increase yield potential. *Trends in Plant Science* 9:70–75.

Smith, R. L., M. F. Grando, Y. Y. Li, J. C. Seib, and R. G. Shatters. 2002. Transformation of bahiagrass (*Paspalum notatum* Flugge). *Plant Cell Reports* 20:1017–1021.

Sollenberger, L. E., W. R. Ocumpaugh, V. P. B. Euclides, J. E. Moore, K. H. Quesenberry, and C. S. Jones, Jr. 1988. Animal performance on continuously stocked 'Pensacola' bahiagrass and Floralta limpograss pastures. *Journal of Production Agriculture* 1:216–220.

Stein J., C. L. Quarin, E. J. Martínez, S. C. Pessino, and J. P. A. Ortiz. 2004. Tetraploid races of *Paspalum notatum* show polysomic inheritance and preferential chromosome pairing around the apospory-controlling locus. *Theoretical and Applied Genetics* 109:186–191.

Sugimoto, Y., M. Hirata, and M. Ueno.1985. Fate of 15N-labeled fertilizer nitrogen applied at different times of the year on bahiagrass pasture. In *Proceedings of 15th International Grasslands Congress,* 598–600. Kyoto, Japan, August 24–31, 1985. Nishi–nasuno, Japan: Science Council of Japan and the Japanese Society of Grassland Science.

Takai, T., and T. Komatsu. 1998. Comparison on physical strength and structure of the leaf blade between Nan-ou and Nangoku varieties of Bahiagrass (*Paspalum notatum* Flugge). *Sochi Shikenjo Kenkyu Hokoku* (56): 13–20. {a} Hokkaido Natl. Agric. Exp. Stn., Hitsujigaoka, Toyohira-ku, Sapporo, Hokkaido 062-8555, Japan.

Tischler C. R., and B. L. Burson. 1995. Evaluating different bahiagrass cytotypes for heat tolerance and leaf epicuticular wax content. *Euphytica* 84:229–235.

USDA, ARS, National Genetic Resources Program. Germplasm Resources Information Network (GRIN). 2001. [Online database] National Germplasm Resources Laboratory, Beltsville, MD. Available: http://www.ars-grin.gov/cgi-bin/npgs/acc/display.pl?1624948

———. 2001. Germplasm Resources Information Network (GRIN). [Online database] National Germplasm Resources Laboratory, Beltsville, MD. Available: http://www.ars-grin.gov/cgi-bin/npgs/acc/display.pl?1624956

———. 2006. Germplasm Resources Information Network (GRIN). [Online database] National Germplasm Resources Laboratory, Beltsville, MD. Available: http://www.ars-grin.gov/cgi-bin/npgs/acc/display.pl?1728159 T 18

———. 2006. Germplasm Resources Information Network (GRIN). [Online database] National Germplasm Resources Laboratory, Beltsville, MD. Available: http://www.ars-grin.gov/cgi-bin/npgs/acc/display.pl?1728160 T 23

Vogel, K. P., and B. L. Burson. 2004. Breeding and genetics. In *Warm-season (C4) grasses,* ed. L. E. Moser, B. L. Burson, and L. E. Sollenberger, 51–94. Agronomy Monograph 45. Madison, WI: ASA, CSSA, SSSA.

Watson, V. H., and B. L. Burson. 1985. Bahiagrass, carpetgrass, and dallisgrass. In *Forages, the science of grassland agriculture,* 4th ed., ed. M. E. Heath et al., 255–262. Ames: Iowa State Univ. Press.

White, A. W., Jr., G. N. Sparrow, and R. L. Carter. 1962. Peanuts and corn in sod-based rotations. *Georgia Agricultural Research* 4 (2): 5–6.

Wilson, G. P. M. 1987. *Paspalum notatum* Flügge (Bahia grass). cv. Competidor (Reg. no. A-7c-1). *Tropical Grasslands* 21:93–94.

Zhang H., P. Lomba, and F. Altpeter. 2007. Improved turf quality of transgenic bahiagrass (*Paspalum notatum* Flugge) constitutively expressing the ATHB16 gene, a repressor of cell expansion. *Molecular Breeding* 20:415–423.

Zuloaga, F. O., and O. Morrone. 2005. Revision de las especies de *Paspalum* para America del Sur Austral (Argentina, Bolivia, Sur del Brasil, Chile, Paraguay y Uruguay). *Monograph System Botany Missouri Botanical Garden* 102:1–297.

CHAPTER 5

Biology, Cytogenetics, and Breeding of *Brachiaria*

Cacilda Borges do Valle and Maria Suely Pagliarini

CONTENTS

5.1 INTRODUCTION

Tropical grasses for pastures represent the single most valuable resource in animal production worldwide. Besides providing means of transforming roughage grown most commonly on soils of low fertility levels into high-quality protein for human consumption, they convey an ecological and sustainable approach for doing so. Native and cultivated pastures cover wide extensions of land in the tropics, and these natural environments encompass a wide array of genera and species of forage plants. Cultivated pastures in the tropics, however, are dangerously composed of a few varieties and—in the case of grasses, such as the ones from the genus *Brachiaria* (Poaceae: Panicoideae: Paniceae)—of apomictic ecotypes that, for all practical purposes, are "clones" through seeds, therefore creating monocrops. This lack of biodiversity exposes the ecosystems by exerting tremendous selection pressure on pests and/or diseases and justifies the urgency in developing and releasing new cultivars by breeding and/or selection.

Brachiaria is the single most important genus of forage grasses for pastures in the tropics. Species of this genus are indigenous to African savannas as most of the important grasses currently used as pastures in the tropics (Figure 5.1). The forage potential of the genus *Brachiaria* was recognized in the 1960s, initially in Australia and a decade later in South America. A handful of *Brachiaria* cultivars, derived directly from naturally occurring germplasm, have greatly impacted animal production systems throughout the tropics in the past 25–30 years by becoming the major component of sown pastures in the lowlands and savannas of tropical America, Asia, the South Pacific, and Australia (Miles et al. 1996; 2004). These apomictic cultivars respond to over 65% of seed commercialized in Brazil alone which is by far the largest user and producer of *Brachiaria* seed for internal use and for export. The employed cultivars belong to four African species: *Brachiaria brizantha* (A. Rich.) Stapf; *B. decumbens* Stapf; *B. humidicola* (Rendle) Schweickert.; and *B. ruziziensis* Germain & Evrard. *B. dictyoneura* (Fig. & De Not.). Stapf, often included in the above list, refers to *B. humidicola* wrongly classified (Maass, 1996).

Cultivars of *Brachiaria* cover large expanses of pasture in major ecosystems of tropical America. The humid lowlands and the savannas, also called "cerrados," occupy about 50% of Brazil; 60% of the area encompassed by Bolivia, Peru, and Ecuador; 14% of Mexico; and significant areas in other countries of Latin America. The cerrados cover about 250 million ha (mostly in Brazil). These areas are characterized by a well-defined dry season and acid soils. Cattle farming in these regions is based almost solely on cultivated pastures of *Brachiaria* cultivars. These cultivated pastures are often established on low- to medium-fertility soils. Some cultivars have adapted well to this ecosystem, providing good forage and animal production.

A number of cultivars have been released in the last 30 years (Miles and Lapointe 1992; Argel and Keller-Grein 1996). In Brazil, there are 10 registered cultivars listed on the National Service for Cultivar Protection website (http://www.agricultura.gov.br). *Brachiaria decumbens* cv. Basilisk is the most widely used worldwide because it adapts to a wide range of soils, is easy to manage, and readily establishes from seed. It is, however, highly susceptible to a group of sucking insects called "spittlebugs" (Homoptera: Cercopidae), and it has been associated with photosensitization in cattle in Brazil (Dobereiner et al. 1976; Schenk, Nunes, and Silva 1991; Driemeier et al. 1999; Barbosa et al. 2006). Another widely used cultivar since the late 1980s, *Brachiaria brizantha* cv. Marandu, is resistant to spittlebugs but requires soils of medium to high fertility and does not tolerate waterlogged sites (Dias Filho 2002). Common *B. humidicola* and cultivar Llanero are better adapted to poorly drained soils but have lower production and only medium to low nutritional quality (Argel and Keller-Grein 1996).

Brachiaria is and will continue to be the foremost genus for the acid and low-fertility soils of tropical America (Pizarro et al. 1996). Ironically, the commercial use of *Brachiaria* species is minimal in its native Africa. The limited use of cultivated pastures is not based on lack of adaptation to the physical or biotic environment, but their attributes make them less appropriate than other

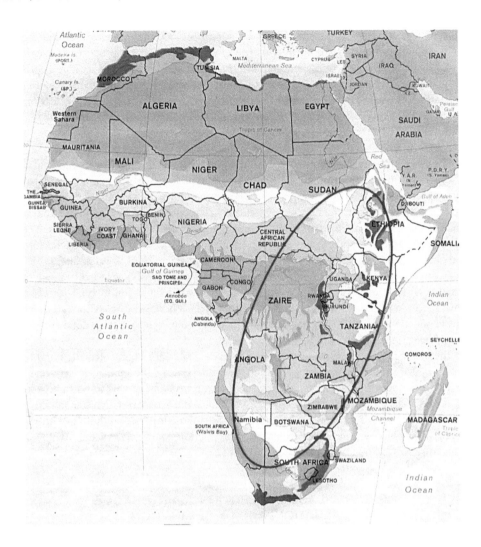

Figure 5.1   Map of Africa showing distribution of *Brachiaria*.

forages for sown pastures in the livestock production systems prevailing in Africa (Ndikumana and Leeuw 1996). The most common *Brachiaria* species cultivated as pastures in tropical Africa are *B. brizantha, B. ruziziensis, B. decumbens,* and *B. mutica.*

In Asia, the South Pacific, and Australia, *Brachiaria* species occupy about 300,000 ha. In the first two regions, they are the most widely grown pasture grasses in the humid and subhumid tropics. In Australia, the area to which they are adapted is relatively small, but within it, *Brachiaria* occupies more than half of the improved pastures. *Brachiaria mutica* was the first species introduced into the region in the late 1800s. However, in recent years, *B. decumbens, B. humidicola,* and *B. ruziziensis* have become the most extensively cultivated forage grasses. The success of *Brachiaria* cultivars in these regions can be attributed to their broad adaptation, aggressiveness, and resilience, which enable them to persist even under unfavorable conditions (Stür, Hopkinson, and Chen 1996).

Attributes such as adaptation to shade, drought, water logging, low fertility, and high aluminum soils; tolerance of heavy defoliation; good seed production; apomixis; and rendering uniform pastures are associated with the widespread use of *Brachiaria* species in production systems worldwide. As C4 grasses, *Brachiaria* species generally display strong regrowth under frequent defoliation. This leads to persistence but generally impairs association with legumes due to competition for light and nutrients. Also, the adoption of grazing management practices is usually unfavorable to most legumes sown in association with *Brachiaria*. Fischer and Kerridge (1996) stated that despite the negative aspects of pasture degradation due to overgrazing, the much more positive effects of carbon sequestration measured in *Brachiaria* pastures should stimulate selection of new *Brachiaria* accessions for the physiological and agronomic characteristics that will make the genus even more useful.

The availability of abundant high-quality seed is vital to widespread adoption of current and future *Brachiaria* cultivars. *Brachiaria* seed production varies according to geographical distribution, photoperiod, management practices, and edaphic conditions. Within suitable environments, *Brachiaria* species are remarkable for their success as seed crops, considering these are not domesticated species. Low seed production may be due to cytological and embryological causes, predominance of polyploidy, and apomixis with pseudogamy in the commercial cultivars. These will be discussed later in the chapter. Seed of commercial species is extensively produced for pasture use. Seed quality is heavily influenced by vitality and dormancy (Hopkinson et al. 1996). Vitality depends mostly on maturity of seed at harvest. Dormancy is strongly developed in the genus after harvest and persists in most taxa into the season. Currently available commercial *Brachiaria* cultivars have merits which have stimulated their rapid expansion worldwide. Although they display enormous economic and social impact, they also display some compromising deficiencies (Miles, Maass, and Valle 1996). These will be presented in Section 5.7.

Over the past decade, extensive efforts have been dedicated for characterization and breeding of *Brachiaria*. The studies on cytology, mode of reproduction of different accessions, and hybrids have created new opportunities and challenges in the improvement of this important genus. New cultivars have recently been released. New ecotypes and many superior hybrids should promote pasture diversification. A thorough evaluation of these new materials is required, with particular emphasis on persistence, animal productivity, seed production, tolerance to water logging, reaction to pests and diseases, and compatibility with legumes.

5.2 GERMPLASM COLLECTION, MAINTENANCE, EVALUATION, AND DISSEMINATION

5.2.1 Germplasm Collection

Germplasm of *Brachiaria* has been collected since the 1950s, but the first comprehensive exploration for this genus was carried out in the mid-1980s in eastern Africa. This exploration led to the acquisition of a very large and diverse collection.

As early as 1930, some ecotypes of *Brachiaria* reached Australia, including the one that became the most widespread of all: *B. decumbens* cv. Basilisk (Oram 1990). In the 1950s, the National Agricultural Research Station at Kitale, Kenya, collected 154 accessions emphasizing *B. brizantha* and *B. ruziziensis*. Some accessions were exchanged with other African institutions, and a few others were incorporated into the U.S. Department of Agriculture (USDA) collection or into the Commonwealth Scientific and Industrial Research Organization (CSIRO) collection in Australia. In the 1960s and mid-1970s, some institutions in Brazil, such as the International Research Institute (IRI), in Matão, SP; the Instituto de Pesquisas Agropecuárias do Norte (IPEAN) (later Centro de Pesquisa de Pecuária do Trópico Úmido [CPATU]); and the Empresa de Pesquisa Agropecuária de Minas Gerais (EPAMIG), shared small working collections of 10–30 accessions each, which they exchanged and evaluated for forage potential (Keller-Grein, Maass, and Hanson 1996).

Germplasm collection increased significantly between 1974 and 1984, when about 220 *Brachiaria* accessions, particularly of *B. brizantha, B. decumbens,* and *B. jubata,* were collected by the former

Trust Fund Forage Collection and Evaluation Project of the Food and Agriculture Organization of the United Nations (FAO), based at the National Agriculture Research Station at Kitale (Ibrahim 1984). Part of this collection was sent to the International Center for Tropical Agriculture (CIAT) in 1981. Seed viability was so low that most accessions were lost. Until the mid-1980s, no more than 150 accessions of the genus *Brachiaria* existed in the major collections of tropical forage germplasm, some of which were duplicates due to re-introductions from various countries under different codes. For lack of protocols to distinguish among duplicates, some of these accessions are kept as individuals to this day.

The economic impact and quick expansion of some varieties contrasting with the narrow germplasm base motivated CIAT to broaden the range of its *Brachiaria* holdings through direct collection in the sites of origin. A great expedition was carried out in six countries in eastern Africa in 1984 and 1985. It was supported by the International Board for Plant Genetic Resources (IBPGR) and with the collaboration of the former International Livestock Center for Africa (ILCA) (today, the International Livestock Research Institute, ILRI) and several African institutions. About 800 accessions covering at least 23 known species were collected, thus increasing the *Brachiaria* germplasm collection held at CIAT. Further collection was done by ILRI, IBPGR, and Roodeplaat Grassland Institute in southern and South Africa. Seed samples of some accessions are stored in the gene bank of the Royal Botanic Gardens, Kew, United Kingdom (Keller-Grein, Maass, and Hanson 1996). As stated by Miles et al. (2004), additional germplasm acquisition is still warranted because several regions indigenous to *Brachiaria,* such as Uganda and Mozambique, have not been visited. Moreover, further exploration of sites where sexual ecotypes have been identified should produce important novel variations for breeding purposes, in both sexual and apomictic forms.

5.2.2 Existing Collections

Four major and three minor collections of *Brachiaria* exist *ex situ.* According to Keller-Grein, Maass, and Hanson (1996), in 1994, these collections held a total of 987 distinct accessions of 33 known species. About 40% of existing accessions are of *B. brizantha* and another 39% are of *B. humidicola, B. decumbens, B. nigropedata, B. jubata,* and *B. ruziziensis.*

CGIAR (Consultative Group on International Agricultural Research; http://singer.grinfo.net) reports the existence of 1,805 accessions in the *Brachiaria* world collection, of which 63% are in CIAT (Colombia: 1,133 accessions) and 37% in ILRI (Kenya/Ethiopia: 672 accessions). *Brachiaria brizantha* alone constitutes 45% of all accessions in the bank, reflecting its wide distribution and the focus on this species as a major pasture grass. Accessions of *B. decumbens* and *B. ruziziensis* account for only 17% of the collection, illustrating their narrower natural distribution (Keller-Grein, Maass, and Hanson 1996). Together with *B. humidicola,* these four species add up to 72% of the total collection. Embrapa (Brazil) has 455 accessions of 13 different species in the germplasm bank, maintained live in the field. The Australian Tropical Forage Genetic Resources Center (ATFGRC) of CSIRO maintains 169 accessions of *Brachiaria* under the denomination of *Urochloa.* There are three minor collections: (1) USDA, also as *Urochloa* (53 accessions), (2) Gene bank of Kenya (GBK; 51 accessions), and (3) Roodeplaat Grassland Institute of the African Research Council (RGI/ARC, South Africa, 39 accessions).

5.2.3 Germplasm Evaluation

The collections at CIAT and Embrapa are used to develop cultivars through selection and/ or breeding. CIAT established a large international network (Red Internacional de Ensayos de Pasturas Tropicales [RIEPT]) to evaluate this valuable collection in several major screening sites in Colombia, Brazil, Peru, and Costa Rica. Promising accessions not only of *Brachiaria* but also of

Table 5.1   Germplasm of *Brachiaria* in Brazil, Analyzed for Ploidy Level and Mode of Reproduction

Species	Accessions Analyzed	Ploidy Levels					Accessions Analyzed	Mode of Reproduction	
		2n	4n	5n	6n	7n		Sexual[a]	Apomictic[b]
B. adspersa	—	—	—	—	—	—	1	1	0
B. arrecta	5	—	5	—	—	—	4	4	0
B. bovonei	—	—	—	—	—	—	2	0	2 (11–14%)
B. brizantha	222	2	157	41	22	—	275	1	274 (0–63%)
B. comata	—	—	—	—	—	—	1	1	0
B. decumbens	51	23	23	5	—	—	65	24	41 (0–56%)
B. deflexa	—	—	—	—	—	—	1	1	0
B. dictyoneura	8	—	6	—	2	—	8	1	7 (0–43%)
B. dura	2	—	—	—	2	—	3	3	0
B. humidicola	60	—	22	18	19	1	64	3	61 (0–66%)
B. jubata	30	4	12	13	—	1	43	8	35 (0–47%)
B. nigropedata	21	—	19	—	2	—	21	0	21 (5–20%)
B. platynota	2	2	—	—	—	—	4	3	1 (61%)
B. plantaginea	—	—	—	—	—	—	1	1	0
B. ruziziensis	29	24	5[c]	—	—	—	36	36	0
B. serrata	—	—	—	—	—	—	2	2	0
B. subquadripara	—	—	—	—	—	—	2	2	0
B. subulifolia	—	—	—	—	—	—	5	0	5 (7–38%)
Total	430	55	249	77	47	2	538	91	447

Sources: Penteado, M. I. O. et al. 2000. *Boletim de Pesquisa*, 11, Campo Grande: Embrapa Gado de Corte, 32 pp.; Valle, C. B. do, and Y. H. Savidan. 1996. In *Brachiaria: Biology, agronomy, and improvement*, ed. J. W. Miles, B. L. Maass, and C. B. do Valle, 147–163. Cali: CIAT/Brasília:EMBRAPA-CNPGC.
[a] Aposporous sacs were not observed.
[b] Variation in potential sexuality (percent of meiotic sacs) observed in apomictic accessions.
[c] Colchicine-induced tetraploid acessions.

several forages (grasses and legumes) were distributed to national programs during the 1980s and 1990s and evaluated under a standardized methodology. Results from these evaluations were pooled into a database available to users at the site www.tropicalforages.info, where taxonomic descriptions, production characteristics, and recommended use can be easily accessed.

A large part of this collection transferred to Brazil was evaluated by Embrapa at various sites. Ploidy level was determined by flow cytometry (Penteado et al. 2000). Mode of reproduction was established (Table 5.1) (Valle 1990; Valle and Savidan 1996). Agronomic performance (Grof et al. 1989; Valle, Calixto, and Amézquita 1993; Valle et al. 2001 [Table 5.2], Pizarro et al. 1996; Macedo, Machado, and Valle 2004), morphological characterization (Valle, Maass, et al. 1993; Assis et al. 2002, 2003), and anatomical approach for forage quality (Lempp et al. 2005) helped select promising genotypes. Resistance to spittlebugs—the most important pest in *Brachiaria* pastures—was also evaluated (Valério et al. 1996). The cultivar development scheme presented in Figure 5.2 is a long-term investment of a large multidisciplinary team and involves evaluation in small plots with large numbers of accessions and species (Valle, Calixto, et al. 1993).

This evaluation was followed by national agronomic trials (20–25 accessions) (Valle et al. 2001). Subsequent preliminary grazing trials with nine accessions (Euclides et al. 2001) were followed by animal performance trials on pastures with four accessions (Pereira et al. 2004; Euclides et al. 2005). Two of those, cv. Xaraés and cv. BRS Piatã, were recently released as new cultivars (Valle et al. 2004; Embrapa Folder 2007, www.unipasto.com.br). Other promising

Table 5.2 Average Production[a] and Agronomic Characteristics of 24 *Brachiaria* Ecotypes Evaluated in Two Sites in Brazil

Codes[b]	TDM[c]	LDM[d]	GDM[e]	%L[f]	L:S[g]	REG[h]
BRA004308	1506	987	1368	66.8	3.8:1	4.0
BRA003361	1477	909	1356	63.5	3.6:1	3.3
BRA002844	1399	821	1274	59.8	2.8:1	3.1
BRA003204	1370	850	1294	63.1	2.4:1	3.1
BRA003450	1356	792	1230	58.4	2.6:1	3.0
BRA003441	1330	968	1205	72.3	7.7:1	4.6
BRA003824	1325	724	1138	55.7	2.7:1	3.1
BRA001068[i]	1321	558	1086	45.5	2.0:1	2.6
BRA003891	1312	689	1122	54.5	2.8:1	2.8
BRA003484	1295	744	1215	59.3	2.1:1	3.4
BRA005118	1248	487	1009	39.5	2.1:1	2.9
BRA002801	1239	623	1092	53.7	2.7:1	2.9
BRA003387	1223	744	1118	62.0	3.0:1	3.2
BRA000591[i]	1199	608	987	50.5	2.9:1	2.9
BRA003948	1166	634	975	55.2	2.9:1	2.9
BRA002739	1160	620	1035	55.1	2.3:1	3.2
BRA003719	1154	459	955	42.0	3.3:1	3.0
BRA003395	1136	745	1040	66.6	3.3:1	3.5
BRA003000	1129	488	939	46.1	2.2:1	2.7
BRA004391	1126	631	949	56.8	3.4:1	2.9
BRA004499	996	441	819	44.9	1.7:1	2.3
BRA003247	981	661	904	66.3	3.7:1	3.9
BRA005011	944	621	757	59.5	8.3:1	3.4
BRA002208[i]	881	328	645	38.4	3.6:1	2.6
Average	1220	672	1063	55.7	3.7:1	3.1
LSD	176	99	153	1.1	0.7	0.2
Top 10	1369	828	1232	63.9	4.4:1	3.6

Source: Valle, C. B. do et al. 2001. *Proceedings of the International Grassland Congress*, 19, Brazil: FEALQ. ID#13–14.

[a] Kilograms per cut.
[b] Brazilian germplasm registration number.
[c] Total dry matter.
[d] Leaf dry matter.
[e] Green dry matter (leaf + stem).
[f] Percentage of leaves in the dry matter.
[g] Leaf-to-stem ratio.
[h] Regrowth after 7 days (minimum = 0 and maximum = 6).
[i] Control cultivars: BRA001068 = *B. decumbens* cv. Basilisk; BRA000591 = *B. brizantha* cv. Marandu; BRA002208 = *B. humidicola* common.

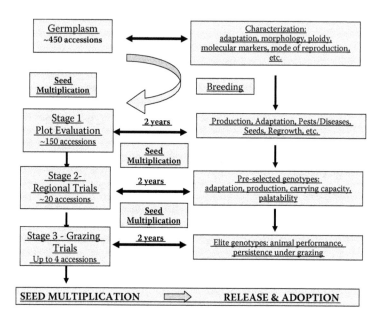

Figure 5.2 Stages of germplasm evaluation in Brazil, leading to cultivar development of *Brachiaria*. (From Valle 2001.)

genotypes are candidates for release and are awaiting final animal performance information to determine advantage over already released cultivars. Breeding was started in 1988 to determine some basic biological attributes. Breeding involved interspecific hybridization (Valle, Glienke, and Leguizamon 1993). Progress obtained will be presented in a later section. Morphological classification and agronomic evaluation were also performed on the ILCA collection (Heering, 1987).

5.2.4 Germplasm Maintenance

Proper maintenance of germplasm is essential to minimize genetic drift, shift, or erosion. Contamination due to outcrossing of accessions maintained in plots in the field is minimal due to widespread apomixis in *Brachiaria*. Sexual accessions of outcrossing diploid *B. ruziziensis* or other sexually reproducing biotypes should be multiplied in isolation or conserved *in vitro*. Protocol for *in vitro* culture of *Brachiaria* exists (Tohme et al. 1996; Rodrigues-Otubo, Penteado, and Valle 2000). However, no cost-effective conservation method has yet been developed for this purpose. Current practices employed to avoid mechanical mixtures during seed processing are adequate.

Both CIAT and Embrapa centers maintain live field collections of *Brachiaria* because seed storage procedures have not been adequate to regenerate weaker genotypes. Because *Brachiaria* seeds are orthodox, their rates of deterioration increase with rising temperature and moisture content (Hopkinson et al. 1996), which is typical of natural conditions in the tropics. Furthermore, physiology and seed production practices have not been determined for several species in this genus; therefore, the quality of the seed produced in germplasm banks is poor. In the collections of *Brachiaria* at CIAT and ILCA, seeds are stored in active collections at 5–10°C and in the base collections at CIAT at –20°C in the gene bank headquarters. Accessions that produce few viable seeds are being maintained live in field collections.

5.2.5 Germplasm Dissemination

The low levels of seed production in most species of *Brachiaria* are a constraint to the dissemination of germplasm. Most accessions flower profusely, but seed set is low because *Brachiaria* is not domesticated. Commercial varieties of *B. brizantha, B. decumbens, B. ruziziensis,* and *B. humidicola* are, however, considered good seeders and most pasture establishment is done by seed. Distribution of vegetative material is not practical or feasible because of phytosanitary risks. Some exotic seed-borne diseases have just recently been described in Brazil, such as the fungi *Claviceps sulcata* (Fernandes, Fernandes, and Bezerra 1995) and *Ustilago operta* (Verzignassi et al. 2001; Fernandes, de Jerba, and Verzignassi 2004) and could impair further distribution. To facilitate germplasm transfer and minimize phytosanitary risk, CIAT and ILRI developed techniques for *in vitro* culture of axillary buds (CIAT 1986). The success rate in culture varied among accessions and species, and only uncontaminated cultures are distributed for other centers. In some collections, such as ATFGRC/CSIRO and USDA, germplasm distribution is done only as seed.

5.3 TAXONOMY AND AREA OF DISTRIBUTION

Brachiaria, Urochloa, Eriochloa, and *Panicum*—a group of genera with still-undefined taxonomic boundaries—have the PEP-CK (phosphenolpyruvate carboxykinase) type of the C4 photosynthetic pathway (Clayton and Renvoize 1986). The main characteristics that identify *Brachiaria* within the Paniceae are the ovate or oblong spikelets, on one-sided racemes, with the lower glume adjacent to the rachis (i.e., in an adaxial position) (Renvoize, Clayton, and Kabuye 1996; Figure 5.3). These characteristics, however, are not consistent throughout the genus because sometimes they are difficult to distinguish.

Renvoize, Clayton, and Kabuye (1996) pointed out that the taxonomy is far from satisfactory and although various workers have recently sought to clarify the situation, none has provided sufficient solutions to the problems of generic identity and species composition across the entire taxon. These authors proposed a classification for the genus *Brachiaria* where the species were sorted into nine groups according to character associations. This left only a few species ungrouped. Deciding which characters were of the greatest significance as indicators of natural affinity proved difficult, and the choice was, to a large extent, arbitrary and based on the authors' experience. The most significant characteristics used for assessing relationships among groups were: spikelet outline shape, spikelet three-dimensional shape, relative length and shape of the lower glume, presence of internode between lower and upper glume, presence of a stipe, veins in the upper glume and lower lemma, rachis in cross-section, and upper lemma texture. Each group was then described and subgroups formed based on secondary characters for assessing relationships, such as raceme numbers and distribution on the axis, size of pedicel, and spikelet arrangement, size, and pubescence. The list of species in the subgroups was arranged by continent of origin—African, American, and Australian/Asian species—and is presented in Table 5.3.

Using *B. erucifomis* as the type species, Webster (1987) reclassified all other species of *Brachiaria* with disarticulation below the glumes as *Urochloa*. The presence of a mucro on the upper spikelet of *Urochloa*—also present in a few minor *Brachiaria* species—was also used as argument to group them together. Morrone and Zuloaga (1992) followed Webster's arguments and classified all South American species of *Brachiaria* in *Urochloa*. They considered the economic importance of *P. maximum* and the need for further research for not placing this species in *Urochloa*, but neglected this argument for *Brachiaria*—a genus of much greater expansion and global utilization. Recently, Torres-Gonzalez and Morton (2005) published a phylogenetic study based on the nucleotide base sequence polymorphism of the internal transcribed spacer (ITS) region of nuclear ribosomal DNA (rDNA) and were unable to separate *Brachiaria* from

Figure 5.3 Inflorescences of *Brachiaria* displaying typical one-sided racemes, (A) ovate spikelets with the first glume adjacent to the rachis; (B) and (C) on the *B. humidicola*.

Urochloa. The cladistic analysis suggested that these two genera form a paraphyletic complex with *Eriochloa* and *Melinis* and that species of all these genera belong to the same monophyletic group. These contradictory statements suggest that systematic and probably genomic studies need to be pursued before the limits between these genera can be clearly established. The present taxonomy definitely does not offer a satisfactory solution for the genera identity problem or the composition of species within the genus.

Veldkamp (1996), S. Renvoize (personal communication 2006), and J. F. M. Valls (personal communication 2006) recommend the current use of *Brachiaria* until further convincing evidence can be produced to justify transference of species between these genera. In the interest of the producers and seed companies and considering the implication of all legislation concerning registration and protection of cultivars, there is consensus in maintaining the present taxonomy for *Brachiaria*.

Besides the taxonomic confusion, Maass (1996) reported that germplasm exchange during the dissemination process contributed to increasing the confusion about the identity of accessions because some genotypes were widely distributed under several different names or erroneously under the species name, such as occurred with the cultivar Humidicola. This continues to cause confusion in the published literature. *Brachiaria dictyoneura* cv. Llanero, for example, is in fact *B. humidicola* (Renvoize, Clayton, and Kabuye 1996), and most *B. humidicola* commercialized and planted as "common" in the tropics refer to the cultivar Tully.

Despite the high number of species in the genus, only seven *Brachiaria* species are considered important for tropical pasture development and are represented by a large number of accessions in germplasm collections. Keller-Grein, Maass, and Hanson (1996) presented maps for these species,

Table 5.3 *Brachiaria* Species by Continent of Origin[a] and Taxonomic Grouping

Species	Group Number	Species	Group Number
African species:		B. ruziziensis Germain & Evrard	**5**
B. ambigens Chiov.	2	B. scalaris Pilg.	1
B. antsirabensis A. Camus	Ungrouped	B. schoenfelderi Hubbard & Schweick.	7
B. arrecta (Dur.& Schinz) Stent	3	B. semiundulata (A. Rich.) Stapf	Ungrouped
B. bemarivensis A. Camus	1	B. serpens (Kunth) Hubbard	1
B. bovonei (Chiov.) Robyns	6	B. serrata (Thunb.) Stapf	2
B. breviglumis Clayton	1	B. serrifolia (Hochst.) Stapf	1
B. brevispicata (Rendle) Stapf	6	B. stigmatisata (Mez) Stapf	6
B. brizantha (A. Rich.) Stapf	5	B. subquadripara (Trin.) Hitchc.	4
B. chusqueoides (Hack.) Clayton	1	B. subrostrata A. Camus	Ungrouped
B. clavipila (Chiov.) Robyns	Ungrouped	B. subulifolia (Mez.) Clayton	6
B. comata (A. Rich.) Stapf	1	B. subulifolia (Mez.) Clayton	6
B. coronifera Pilg.	1	B. turbinata Van der Veken	8
B. decumbens Stapf	5	B. umbellata (Trin.) Clayton	Ungrouped
B. deflexa (Schumach.) Hubbard	1	B. umbratilis Napper	1
B. dictyoneura (Fig. & De Not.) Stapf	6	B. villosa (Lam.) A. Camus	1
B. distachya (L.) Stapf	4	B. xantholeuca (Schinz) Stapf	2
B. distachyoides Stapf	3	**American species:**	
B. dura Stapf	5	B. adspersa (Trin.) Parodi	1
B. eminii (Mez) Robyns	5	B. arizonica (Scribn. & Merr.) S. T. Blake	1
B. epacridifolia (Stapf) A. Camus	Ungrouped	B. ciliatissima (Buckl.) Chase	9
B. eruciformis (J. E. Smith) Griseb.	7	B. echinulata (Mez) Parodi	1
B. glomerata (Stapf) A. Camus	2	B. fasciculata (Swartz) Parodi	1
B. grossa Stapf	1	B. lorentziana (Mez) Parodi	1
B. humbertiana A. Camus	1	B. megastachya (Nees ex Trin.) Zuloaga & Soderstrom	1
B. humidicola (Rendle) Schweick.	6	B. meziana Hitchc.	4
B. jubata (Fig. & De Not.) Stapf	6	B. mollis (Swartz) Parodi	1
B. lachnantha (Hochst.) Stapf	2	B. ophryodes Chase	9
B. lata (Schumach.) Hubbard	1	B. paucispicata (Morong) Clayton	Ungrouped
B. leersioides (Hochst) Stapf	1	B. platyphylla Nash	3
B. leucacrantha (K. Schum.) Stapf	Ungrouped	B. tatianae Zuloaga & Soderstrom	Ungrouped
B. lindiensis (Pilg.) Clayton	1	B. texana (Buckl.) S. T. Blake	1
B. longiflora Clayton	1	**Australian species:**	
B. malacodes (Mez & K. Schum.) Scholz	7	B. advena Vickery	Ungrouped
B. marlothii (Hack.) Stent	Ungrouped	B. argentea (R. Br.) Hughes	8
B. mesocoma (Nees) A. Camus	2	B. foliosa (R. Br.) Hughes	1
B. mutica (Forssk.) Stapf	3	B. gilesii (Benth.) Chase	9
B. nana Stapf	1	B. holosericea (R. Br.) Hughes	Ungrouped
B. nigropedata (Ficalho & Hiern) Stapf	2	B. piligera (F. Muell.) Hughes	4
B. oligobrachiata (Pilg.) Henr.	5	B. polyphylla (R. Br.) Hughes	1
B. orthostachys (Mez) W. D. Clayton	1	B. praetervisa (Domin) Hubbard	1
B. ovalis Stapf	1	B. pubigera (Roem. & Shult.) S. T. Blake	1
B. perrieri A. Camus	Ungrouped	B. whiteana (Domin) Hubbard	4
B. plantaginea (Link) Hitchc.	4	**Indian and Southeast Asian species**	
B. platynota (K. Schum.) Robyns	6	B. burmanica Bor	4

Continued

Table 5.3   *Brachiaria* Species by Continent of Origin[a] and Taxonomic Grouping (*Continued*)

Species	Group Number	Species	Group Number
B. psammophila (Welw.) Launert	2	*B. kurzii* (Hook. F.) A. Camus	1
B. pseudodichotoma Bosser	4	*B. nilagirica* Bor	1
B. pungipes Clayton	8	*B. remota* (Retz.) Haines	1
B. ramosa (L.) Stapf	1	*B. semiverticillata* (Rottl.) Alst.	1
B. reptans (L.) Gardner & Hubbard	1	*B. tanimbarensis* Ohwi	Ungrouped
B. reticulata Stapf	6		
B. rugulosa Stapf	3		

Source: Renvoize, S. A., Clayton, and Kabuye 1996. In *Brachiaria: Biology, agronomy and improvement,* ed. J. W. Miles, B. L. Maass, and C. B. do Valle, 1–15. Cali: CIAT/Brasília:EMBRAPA-CNPGC.

[a] *Brachiaria* species are arranged in alphabetical order by continent of origin. They were tentatively clustered into nine groups, with 14 species still outstanding.

showing the natural distribution and germplasm collection sites and thus highlighting geographical gaps in the ex situ collections:

Brachiaria brizantha (Figure 5.4) is widespread in tropical Africa, occurring in open and wooded grasslands, along margins of woodlands and thickets, and in upland grasslands. The collection sites of available germplasm cover the natural geographic range in eastern and southeastern Africa. However, considerable gaps in the collection exist for West Africa and southern tropical Africa, especially Zaire and Zambia, which are centers of diversity for this species.

Brachiaria decumbens (Figure 5.5) has a narrower distribution. Western Kenya, Rwanda, and Burundi are well covered by germplasm holdings. No germplasm has been collected from western Tanzania or Zaire and Uganda—ironically, the origin of the most widespread cultivar: Basilisk. It is found in deciduous bushland, in grasslands, and at forest edges.

Brachiaria ruziziensis (Figure 5.6) also has a narrow natural distribution. Germplasm collections were made from Burundi and Rwanda, but none from Zaire. The grass occurs in grasslands and disturbed places.

Figure 5.4   (See color insert following page 274.) Pasture of *B. brizantha* cv. Marandu illustrating its growth habit. Insets of the inflorescence, spikelets, and seeds.

Figure 5.5 **(See color insert following page 274.)** Pasture of *B. decumbens* cv. Basilisk illustrating its growth habit. Insets of the inflorescence, spikelets, and seeds.

Figure 5.6 **(See color insert following page 274.)** Plants of *B. ruziziensis* illustrating its growth habit. Insets of the inflorescence, spikelets, and seeds.

Brachiaria humidicola (Figure 5.7) is usually found in seasonally swampy grasslands. The existing germplasm collections represent the natural distribution in eastern and southeastern Africa, but gaps exist for Nigeria, Sudan, and southern Africa.

Brachiaria dictyoneura (Figure 5.8), considering the natural distribution of this species, has no germplasm material available from Sudan, Uganda, northern and western Tanzania, Zambia, or Mozambique.

Figure 5.7   Plants of *B. humidicola* illustrating its growth habit. Inset of the inflorescence.

Figure 5.8   Plants of *B. dictyoneura* illustrating its growth habit. Insets of the inflorescence, spikelets, and seeds.

Brachiaria jubata (Figure 5.9) is widely distributed in tropical Africa and occurs in seasonally moist grasslands, wet bushland, and at margins of swamps. No germplasm is available from western, central, eastern, and southeastern Africa.

Brachiaria nigropedata (Figure 5.10) usually occurs in dry, sandy soils in open or wooded grasslands in Zimbabwe. The germplasm collection does not adequately represent its natural distribution in southern and eastern Africa.

Figure 5.9 Plants of *B. jubata* illustrating its growth habit. Insets of the spikelets and seeds.

Figure 5.10 Plants of *B. nigropedata* illustrating its growth habit. Insets of the inflorescence and spikelets with dark pedicels.

5.4 REPRODUCTIVE BIOLOGY

In several families of angiosperms, sexual reproduction (amphimixis) is replaced by or combined with asexual reproduction or apomixis (Gr: without mixing). In this case, an embryo is produced without the fusion of male and female gametes (i.e., an asexual embryo is formed inside a gametophyte). The origin of this embryo may vary (Valle and Savidan 1996), but the offspring resulting from apomictic reproduction of a single plant is a clone. As a breeding tool, apomixis offers several advantages. It is the only mode of reproduction that associates fixation of hybrid vigor with seed propagation. Apomictic hybrids breed true, and superior genotypes can be rapidly increased by seed. Apomixis also simplifies commercial hybrid seed production because isolation is not necessary: parental lines need not be increased, mechanical mixture is less likely, and outcross contamination does not occur.

Although not common, apomixis is found in diverse and unrelated plant families of the kingdom. More than 500 species in the most important genera of the Panicoideae subfamily are estimated to exhibit apomixis (Valle and Savidan 1996). Apomictic reproduction is largely widespread in some important genera of forage grasses, such as *Paspalum*, *Panicum*, and *Brachiaria*. Apomixis is often associated with polyploidy. There are no reports of apomixis in diploid species. This mode of reproduction is frequently accompanied by meiotic anomalies leading to reduced pollen fertility. Sexuality has been generally found at the diploid level and is generally associated with regular chromosome pairing and division.

Mode of reproduction has been determined for a large germplasm collection of *Brachiaria*, including 510 accessions of 17 species (Valle 1990; Valle and Miles 1992, 1994). Embryo-sac structures were examined, using clearing with methyl salicylate and interference contrast microscopy. The results of the analysis, summarized by Valle and Savidan (1996), showed that 425 accessions reproduced by apomixis against 85 reproducing by sexual means. Among 275 accessions of *B. brizantha* of this collection, only one presented sexual reproduction. Of 65 accessions of *B. decumbens*, 42 were apomictic, and of 58 accessions of *B. humidicola*, only one was sexual and tetraploid.

On the other hand, *B. ruziziensis* (36 accessions) and all accessions of some poorly represented species in the Embrapa Beef Cattle collection, such as *B. adspersa* (1), *B. arrecta* (4), *B. comata* (1), *B. deflexa* (1), *B. dura* (1), *B. serrata* (2), and *B. subquadripara* (2), showed sexual reproduction. In the most important *Brachiaria* species for pastures—*B. brizantha*, *B. decumbens*, *B. humidicola*, and *B. jubata*, except for *B. ruziziensis*—the reproduction mode was predominantly apomictic.

In *Brachiaria*, megagametogenesis may follow two pathways, according to Valle and Savidan (1996), which are displayed in Figure 5.11. The first is sexual, where regular meiosis of the megaspore mother cell results in a tetrad of reduced cells. One of these (chalazal surviving megaspore) undergoes three mitoses, resulting in a *Polygonum*-type reduced embryo sac. Its nuclei differentiate into one egg cell, two short-lived synergids, two polar nuclei in the central cell, and three antipodal cells. The second pathway is asexual, where the aposporic embryo sacs develop from enlarged, unreduced nucellar cells after all four megaspore degenerate. Nucellar cells undergo two mitoses, producing four-nucleate (one egg cell, two short-lived synergids, and one polar nucleus) *Panicum*-type embryo sacs.

Determination of mode of reproduction in *Brachiaria* is routinely made by ovary extraction, clarification, and examination under interference contrast microscopy. It is tedious, time consuming, and delayed until flowering occurs. Plants are classified as sexual when only *Polygonum* type or sterile embryo sacs are observed. Whenever single or multiple aposporic sacs are identified, the plant is classified as apomictic. Figure 5.12 illustrates the aspect of ovaries displaying the different types of embryo sacs in *Brachiaria*. The search for a molecular marker for apomixis has been under way for years and would certainly simplify the identification and selection in the breeding population.

Although simple genetic models for the inheritance of apomixis have been proposed for several natural apomictic grasses, the evidence supporting monogenic determination of apomixis in

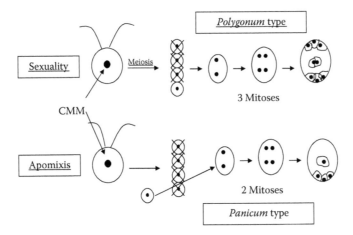

Figure 5.11 Scheme illustrating the formation of embryo sacs of the *Polygonum* type (sexual) and aposporic embryo sacs of the *Panicum* type (apomixis). CMM: megaspore mother cell.

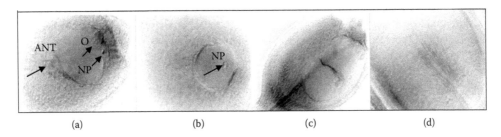

Figure 5.12 (a) Photographs of ovaries containing *Polygonum* type embryo sacs; (b) aposporous sac; (c) multiple aposporous embryo sacs; (d) sterile sac. O = egg cell; NP = polar nucleus; ANT = antipodal cells.

Brachiaria was, first, limited to the analysis by Ndikumana (1985) of a small hybrid population. Breeding work, therefore, began on an empirical basis as additional data accumulated. The current understanding of genetic control of apomixis in the *B. ruziziensis/B. decumbens/B. brizantha* complex has been expanded greatly by the work of Valle and collaborators at Embrapa Beef Cattle and CIAT (Valle and Miles 1992, 1994; Valle, Leguizamon, and Guedes 1991; Valle and Savidan 1996). Crosses were made by Valle, Glienke, and Leguizamon (1994), using sexual tetraploid clones of *B. ruziziensis* derived from the original Belgian materials and two natural tetraploid apomicts: *B. brizantha* cv. Marandu and *B. decumbens* cv. Basilisk. Sexual plants selected for good phenotypic characteristics in the field were hybridized in the greenhouse. First-generation hybrids resulted in a proportion of sexual to apomictic progeny close to 1:1 (Valle and Savidan 1996). Sexual and apomictic hybrids were selected. All selfed sexual tetraploid plants and the sexual hybrids yielded only sexual progeny. Crosses between sexual hybrids and apomictic plants gave sexual and apomictic progenies in a 1:1 proportion. Crosses between sexual plants, either at the tetraploid or diploid level, yielded only sexual progeny, indicating recessiveness of sexuality.

Segregation in the second-generation hybrids fits a single-gene model for inheritance of apomixis in *Brachiaria*. Crosses between facultative apomicts and obligate apomicts resulted in a small number of viable progenies. These did not allow for conclusive results but yielded a larger number of apomicts than sexual plants expected for a 3:1 ratio. This analysis is still in progress and is now being conducted using intraspecific hybrids of *B. humidicola* to avoid the effect of incompatibility between species. There is strong evidence that apomixis is dominant over sexuality. The proposed genotypes are *aa* for sexual diploids, *aaaa* for induced autotetraploids, and *Aaaa* for apomictic tetraploids, as presented by Savidan (1983) for *Panicum maximum*. Hybridization of obligate apomicts with sexual plants thus affords the opportunity to produce new gene combinations and permanently fix superior heterozygous progenies for immediate evaluation as potential new apomictic cultivars. Superior sexual genotypes can be integrated into the sexual genetic pool to be used in further crossings (Valle and Savidan 1996).

5.5 CYTOGENETICS

5.5.1 Chromosome Numbers

Little is known about the biology of these grasses, despite the high number of species, agronomic and economic interest, and the widespread distribution and use in the tropical world. Chromosome numbers involving some species have been occasionally reported—mainly those of agronomic importance. Table 5.4 summarizes the present knowledge on chromosome status in the genus. No more than 44 species were analyzed. Several *Brachiaria* species without agronomic interest or not present in active germplasm collections have not been cytologically studied until now. Darlington and Wylie (1955) determined $x = 7$ and $x = 9$ as the basic chromosome numbers for the genus. Although the majority of researchers agree with these numbers, Basappa, Muniyamma, and Chinnappa (1987) reported $x = 8$, and Christopher and Abraham (1976) reported $x = 5, 8, 10,$ and 12 for *Brachiaria*. More recently, Risso-Pascotto, Pagliarini, and Valle (2006c) reported $x = 6$ in *B. dictyoneura*.

Several chromosome determinations in the genus are recorded in the literature, but detailed studies on karyotype morphology are scarce. Chromosomes were studied at pachytene and an idiogram was proposed for *B. ruziziensis* (Valle, Singh, and Miller 1987; Figure 5.13). The existing data about chromosome number was determined in meiosis, where the chromosomes are larger than in mitosis. Plants with small chromosomes such as *Brachiaria* require fixation in propionic rather than acetic acid for better results. Traditional propionic or acetocarmine squashes are still the most widely used techniques for simple determination of chromosome number and meiotic analysis. More sophisticated techniques that could help identify contrasting genomes in polyploid forms, such as chromosome painting, have never been used in *Brachiaria*. A screening of ploidy levels in all the accessions maintained at Embrapa Beef Cattle (Brazil) was done using flow cytometry. This technique detected a wide variation in ploidy levels, within and between species, ranging from diploids to heptaploids (Penteado et al. 2000).

Chromosome characterization in *Brachiaria* was done by Bernini and Marin-Morales (2001) in 12 accessions of *B. brizantha, B. decumbens, B. humidicola, B. jubata,* and *B. ruziziensis*. The use of conventional Feulgen methodology for chromosome staining did not give satisfactory results. Better staining was achieved with Giemsa. The authors presented karyomorphological parameters, such as karyotypic formulae, chromosome length of the haploid set (HCL), chromosome size, arm ratio, and satellite presence and position, as well as asymmetry and symmetry indexes. The results obtained revealed that the haploid chromosome length varied among species and also among accessions within species, despite the same ploidy level. For example, in *B. decumbens,* the HCL was 25.91 µm for the tetraploid accession D58 ($2n = 4x = 36$) and 36.49 µm for D70, another tetraploid accession. On the other hand, in the tetraploid accession—H19 of *B. humidicola*—the HCL was 47.64 µm. These data reflect intra- and interspecific variability in chromosome sizes, probably due to quantitative differences in the DNA content.

Bernini (1997) utilized C-bands to differentiate accessions D58 and D70 of *B. decumbens;* the results showed similar quantities of heterochromatin, thus suggesting that the variation in the HCL is caused by small portions of heterochromatin dispersed in the euchromatin, which were not detectable by conventional chromosome banding methods. Karyotypic formulae also varied among

Table 5.4　Chromosome Number for the Genus *Brachiaria*

Species	Chromosome numbers	Ref.
B. arizonica	$2n = 4x = 36$	Reeder 1977
B. bovonei	$2n = 4x = 36$	Dujardin 1979
	$2n = 10x = 90$	Spies and Du Plessis 1986
B. brizantha	$2n = 2x = 18$	Carnahan and Hill 1961
		Mendes-Bonato 2002
	$2n = 4x = 36$	Nath and Swaminathan 1957
		De Wet 1960
		Sotomayor-Ríos et al. 1960
		Carnahan and Hill 1961
		Takeota 1965
		Sinha and Jha 1972
		Kammacher et al. 1973
		Gould and Soderstrom 1974
		Nassar 1977
		Dujardin 1978
		Basappa and Muniyamma 1981
		Mehra 1982
		Ndikumana 1985
		Valle 1986
		Basappa, Muniyamma, and Chinnappa 1987
		Sinha, Bhardwaj, and Singh 1990
		Spies et al. 1991
		Mendes-Bonato 2002b
	$2n = 5x = 45$	Risso-Pascotto et al. 2003b
	$2n = 6x = 54$	Carnahan and Hill 1961
		Moffett and Hurcombe 1949
		Sharma and Kour 1980
		Spies and Du Plessis 1987
		Basappa, Muniyamma, and Chinnappa 1987
		Mendes-Bonato 2002b
B. chennaveeraiahana	$2n = 4x = 36$	Basappa, Muniyamma, and Chinnappa 1987
B. chusqueiodes	$2n = 2x = 18$	Hoshino and Davidse 1988
B. ciliatissima	$2n = 4x = 36$	Brown 1951
B. decumbens	$2n = 2x = 18$	Valle et al. 1989
		Valle and Glienke 1991
	$2n = 4x = 36$	Zerpa 1952
		Pritchard 1967
		Nath, Swaminathan, and Mehra 1970
		Nassar 1977
		Basappa and Muniyamma 1981
		Ndikumana 1985
		Valle 1986

Continued

Table 5.4 Chromosome Number for the Genus *Brachiaria* (*Continued*)

Species	Chromosome numbers	Ref.
		Basappa, Muniyamma, and Chinnappa 1987
		do Valle et al. 1989
		Mendes-Bonato et al. 2001a; Mendes-Bonato 2002b
		Junqueira Filho et al. 2003
B. deflexa	$2n = 2x = 18$	Trouin 1972
		Basappa and Muniyamma 1981
		Basappa, Muniyamma, and Chinnappa 1987
		Hoshino and Davidse 1988
B. dictyoneura	$2n = 4x = 24$ ($n = 6$)	Risso-Pascotto et al. 2006a
	$2n = 6x = 42$ ($n = 7$)	Carnahan and Hill 1961
		Moffett and Hurcombe 1949
B. distachya	$2n = 4x = 36$	Gould and Soderstrom 1974
		Basappa and Muniyamma 1981
		Basappa, Muniyamma, and Chinnappa 1987
		Bir et al. 1987
		Sharma and Kour 1980
		Mehra and Sharma 1975
		Mehra and Chowdhury 1976
		Muniyamma 1976
		Bir and Chauhan 1990
	$2n = 8x = 72$	Takeota 1965
B. distichophylla	$2n = 4x = 36$	Olorode 1975
B. eminii	$2n = 2x = 18$	Dujardin 1979
B. eruciformis	$2n = 2x = 18$	Delay 1951
		Sarkar et al. 1973
		Basappa and Muniyamma 1981, 1983
		Basappa, Muniyamma, and Chinnappa 1987
	$2n = 4x = 36$	Hoshino and Davidse 1988
		Mehra and Sharma 1975
		Sharma and Kour 1980
		Sharma and Sharma 1979
B. extensa	$2n = 4x = 36$	Parodi 1946
		Nuñez 1952
B. fasciculata	$2n = 4x = 36$	Davidse and Pohl 1978
B. fulva	$2n = 4x = 36$	Takeota 1965
B. glomerata	$2n = 2x = 18$	Hoshino and Davidse 1988
B. humidicola	$2n = 4x = 36$	Valle and Glienke 1991
	$2n = 6x = 54$	Valle 1986
		Honfi, Quarin, and Valls 1990
	$2n = 8x = 72$	De Wet and Anderson 1956
B. hybrida	$2n = 6x = 54$	Basappa and Muniyamma 1983

Table 5.4 Chromosome Number for the Genus *Brachiaria* (*Continued*)

Species	Chromosome numbers	Ref.
B. jubata	$2n = 2x = 18$	Kammacher et al. 1973
	$2n = 4x = 36$	Kammacher et al. 1973
		Olorode 1975
		Goldblatt 1981
		Valle 1986
B. kotschyana	$2n = 2x = 18$	Dujardin 1979
B. lata	$2n = 6x = 54$	Basappa, Muniyamma, and Chinnappa 1987
B. meziana	$2n = 4x = 36$	Reeder 1967
		Reeder 1984
B. miliiformis	$2n = 6x = 54$	Larsen 1963
		Chatterjee 1975
		Basappa and Muniyamma 1981
		Basappa, Muniyamma, and Chinnappa 1987
	$2n = 8x = 72$	Takeota 1965
		Gould and Soderstrom 1974
B. munae	$2n = 4x = 36$	Basappa and Muniyamma 1981
		Basappa, Muniyamma, and Chinnappa 1987
B. mutica	$2n = 4x = 36$	Hunter 1934
		Narayan and Muniyamma 1972
		Kammacher et al. 1973
		Davidse and Pohl 1974
		Gould and Soderstrom 1974
		Muniyamma and Narayan 1975
		Christopher and Abraham 1976
		Nassar 1977
		Dujardin 1979
		Basappa and Muniyamma 1981
		Mehra 1982
		Basappa, Muniyamma, and Chinnappa 1987
		Sinha, Bhardwaj, and Singh 1990
B. nigropedata	$2n = 2x = 18$	De Wet and Anderson 1956
		Spies and Du Plessis 1986
	$2n = 4x = 36$	Moffett and Hurcombe 1949
		Hoshino and Davidse 1988
		Utsunomiya et al. 2005
B. paspaloides	$2n = 4x = 36$	Gould and Soderstrom 1974
		Sharma and Kour 1980
		Basappa and Muniyamma 1981
		Basappa, Muniyamma, and Chinnappa 1987
B. plantaginea	$2n = 8x = 72$	Reeder 1967
		Davidse and Pohl 1972

Continued

Table 5.4   Chromosome Number for the Genus *Brachiaria* (*Continued*)

Species	Chromosome numbers	Ref.
	$2n = 2x = 18$	Gould 1958
	$2n = 4x = 36$	Parodi 1946
B. ramosa	$2n = 2x = 18$	Bir and Sahni 1984
	$2n = 4x = 36$	Mulay and Leelamma 1956
		Singh and Godward 1960
		Singh 1965
		Gupta 1971
		Mehra and Sharma 1975
		Sharma and Sharma 1979
		Sharma and Kour 1980
		Basappa and Muniyamma 1981
		Sinha, Bhardwaj, and Singh 1990
B. reptans	$2n = 2x = 14$	Sharma and Jhuri 1959
		Sinha and Jha 1972
		Gould and Soderstrom 1974
		Christopher and Abraham 1976
		Sharma and Sharma 1979
		Sharma and Kour 1980
		Sarkar et al. 1973
		Basappa and Muniyamma 1981
		Mehra 1982
		Basappa, Muniyamma, and Chinnappa 1987
	$2n = 4x = 28$	Bir and Sahni 1984
B. ruziziensis	$2n = 2x = 18$	Schank and Sotomayor-Ríos 1968
		Sotomayor-Ríos et al. 1970
		Ferguson 1974
		Ferguson and Crowder 1974
		Basappa and Muniyamma 1981
		Ndikumana 1985
		Valle 1986
		Valle, Singh, and Miller 1987
		Basappa, Muniyamma, and Chinnappa 1987
B. semiundulata	$2n = 4x = 36$	Muniyamma 1976
		Nagabushana 1980
		Basappa and Muniyamma 1981
		Basappa, Muniyamma, and Chinnappa 1987
B. semiverticillata	$2n = 4x = 36$	Narayan and Muniyamma 1972
		Basappa, Muniyamma, and Chinnappa 1987
B. serrata	$2n = 4x = 36$	De Wet 1958
		Moffett and Hurcombe 1949
		Du Plessis and Spies 1988
B. setigera	$2n = 4x = 36$	Narayan and Muniyamma 1972
		Gould and Soderstrom 1974

Table 5.4   Chromosome Number for the Genus *Brachiaria* (Continued)

Species	Chromosome numbers	Ref.
	2n = 4x = 28	Basappa and Muniyamma 1981
		Basappa, Muniyamma, and Chinnappa 1987
B. stapfiana	2n = 4x = 36	Basappa and Muniyamma 1981
B. stigmatisata	2n = 4x = 36	Basappa, Muniyamma, and Chinnappa 1987
B. subquadripara	2n = 8x = 72	Basappa, Muniyamma, and Chinnappa 1987
		Trouin 1972
		Larsen 1963
		Chen and Hsu 1964
		Davidse and Pohl 1972
		Hsu 1972
		Gould and Soderstrom 1974
		Basappa and Muniyamma 1981
		Mehra 1982
		Basappa and Muniyamma 1983
B. subulifolia	2n = 4x = 36	Spies and Du Plessis 1987
B. sylvaticum	2n = 4x = 28	Mehra and Sood 1974
B. villosa	2n = 4x = 36	Chen and Hsu 1961
		Mehra and Sharma 1975
		Basappa and Muniyamma 1981
		Mehra 1982
		Basappa, Muniyamma, and Chinnappa 1987
B. Viridula	2n = 4x = 36	Moffett and Hurcombe 1949

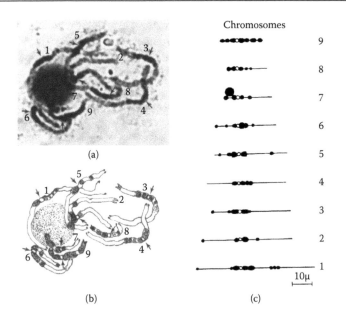

Figure 5.13   Chromosome complement of *B. ruziziensis* ($n = 9$): (a) actual chromosomes in pachytene and (b) interpretive drawing (×880); arrows indicate centromere. (c) Idiogram. Centromere location is indicated by an open circle. (From Valle, Singh, and Miller 1997.)

species and accessions with the same chromosome number. Analysis of karyotype symmetry indicated a slight increase in asymmetry in the direction of karyotypes with greater chromosome numbers, suggesting an evolutionary tendency.

5.5.2 Polyploidy

The information presented in Table 5.3 shows that two basic chromosome numbers are commonly found in the genus: $x = 7$ and $x = 9$, with prevalence of the latter. From these data and without considering the number of accessions analyzed in each species, it became clear that polyploidy predominates in the genus *Brachiaria* (mainly tetraploidy). For example, among 22 accessions of *B. brizantha* analyzed by Mendes-Bonato, et al. (2002a), 1 was diploid, 18 were tetraploid, and 3 were hexaploid. Pentaploids were also detected in this collection (Letteriello et al. 1999). The tendency to tetraploidy was also confirmed by Penteado et al. (2000) in the determination of the ploidy level by flow cytometry in the germplasm collection of *Brachiaria* in Brazil. Among 435 accessions of 13 species analyzed, 13% were diploid, 58% tetraploid, 18% pentaploid, 11% hexaploid, and 0.5% heptaploid; that is, 87% of the accessions maintained in this collection were polyploid.

Polyploidy is very common among grasses. According to Stebbins (1956), 70% of grasses are natural polyploids. In order to explain this high frequency in Gramineae, Stebbins (1985) proposed the "secondary contact hypothesis" by which taxa with "patchy" distributions create frequent opportunities for secondary contact and hybridization between differentiated diploid populations. Following hybridization, highly adapted gene combinations might be buffered and maintained largely by the effects of polyploidy favoring tetrasomic inheritance and by preferential pairing of homologous chromosomes as compared to homoeologous chromosomes.

5.5.3 Meiotic Behavior

Meiotic behavior in *Brachiaria* has been reported in a few cases when compared with cytological studies for chromosome number determinations. Valle and Savidan (1996) reported that natural diploids of *B. brizantha, B. decumbens,* and *B. ruziziensis* exhibit regular meiosis, with chromosome pairing as nine bivalents. Their tetraploid counterparts, however, display irregular meiosis, with univalents and quadrivalents being formed. Sotomayor-Ríos et al. (1960) reported irregular meiosis in tetraploid *B. brizantha.* There were laggards in 69% of cells; 16% of microspore tetrads appeared normal and 14% of pollen stained completely. Analyzing 22 accessions of *B. brizantha* of the Brazilian collection, Mendes-Bonato, et al. (2002a) reported a low frequency of irregular chromosome behavior in the first division, despite normal chromosome pairing at diakinesis in the single diploid accession ($2n = 2x = 18$). The second division was perfectly normal. Among 18 tetraploid accessions ($2n = 4x = 36$), irregular chromosome segregation was common but variable, even among accessions. In six of them, fewer than 16% of tetrads presented micronuclei; in the others, the percentage of abnormal tetrads ranged from 26 to 45%. In the hexaploid accessions ($2n = 6x = 54$), the percentage of abnormal tetrads varied from 13 to 31%.

The most common meiotic abnormalities found in polyploid accessions were related to irregular chromosome segregation in both divisions. Precocious chromosome migration at metaphase plates (Figure 5.14a) and laggards in anaphases (Figure 5.14b) led to micronuclei formation at telophases (Figure 5.14c). Similar observations were also reported in polyploid accessions of several *Brachiaria* species (Pritchard 1967; Sotomayor-Ríos, Schank, and Woodbury 1970; Valle 1986; Basappa, Muniyamma, and Chinnappa 1987). In tetraploid accessions of *B. decumbens,* the meiotic behavior was similar to that observed in *B. brizantha* (Pritchard 1967; Mendes-Bonato et al. 2001a; Mendes-Bonato, 2002b; Junqueira-Filho et al. 2003) and *B. nigropedata* (Utsunomiya, Pagliarini, and Valle 2005). In *B. humidicola,* one tetraploid accession exhibited normal meiosis with 18 bivalents at diakinesis and sexual reproduction (Valle and Glienke 1991),

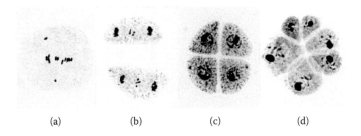

(a)   (b)   (c)   (d)

Figure 5.14 (a) Precocious chromosome migration to the poles in metaphase I; (b) laggards in anaphase II leading to micronuclei formation in (c) tetrad and (d) polyad in *B. brizantha*.

but other tetraploid accessions displayed irregular meiosis (Valle 1986). *Brachiaria mutica* and *B. arrecta* are tetraploid species; the former presented irregular chromosome pairing and apomictic reproduction, and the latter exhibited reasonably regular bivalent pairing and sexual reproduction. A natural hybrid between them was discovered and shown to be completely sterile (Souto 1978).

The most common cytological consequence of irregular chromosome segregation among higher plants is the formation of micronuclei in the tetrads, which remain in the microspore after callose wall dissolution and release. In the polyploid accessions of *B. brizantha* (Mendes-Bonato, et al. 2002b) and in *B. decumbens* (Mendes-Bonato et al. 2001a; Mendes-Bonato, et al. 2002a), the behavior of chromosomes that did not participate in telophase II nuclei was unique. In many cases, they formed micronuclei in one or more microspores of the tetrad; in other cases, these chromosomes were eliminated from the tetrad as microcytes by additional cytokinesis, originating polyads (Figure 5.14d). In a few cases, microspores with micronuclei and microcytes were observed in the same tetrad.

Evidence accumulated in several studies with *Brachiaria* points to the close association of apomixis with polyploidy and irregular meiosis, on the one hand, and sexuality and regular chromosome pairing on the other. This has also been discussed for other genera where polyploidy is frequent.

5.5.4 Unusual Meiotic Behavior

Meiosis is a complex, multistep process of crucial significance in the life cycles of sexually reproducing eukaryotes and marks the transition from sporophyte to gametophyte. It consists of a complex set of processes that involves reduction division (homologous chromosome pairing or synapsis, chiasma formation and crossing over, chromosome segregation, and reductional division) and equational division (mitosis). After two rounds of successive nuclear divisions, the chromosome number is precisely reduced to the haploid level. Successful division requires accurate and synchronized events. First, the DNA is entirely replicated during the S phase of the cell cycle. Next, a bipolar spindle is formed, attached to all chromosomes, and used to separate the chromosomes to opposite poles. Subsequently, the cytoplasm is cleaved between the segregated chromosomes in a plane perpendicular to the spindle. Additionally, the spindle positions itself correctly before cleavage to control the direction of division and to determine whether cytoplasmic components are divided in a symmetric manner. The product of male meiosis is a callose-enclosed tetrad of haploid microspores resulting from the correct partitioning of the cytoplasm.

Analyses on several accessions of different *Brachiaria* species have revealed unusual patterns of microsporogenesis. Studying *B. decumbens* cv. Basilisk, Mendes-Bonato, et al. (2002a) observed that the spindle in metaphase I and anaphase I became heavily stained with propiono-carmine. In telophase I, the interzonal microtubules continued to be intensely stained and, during the phragmoplast formation, the fibers were pushed to the cell wall, persisting until prophase II, even after

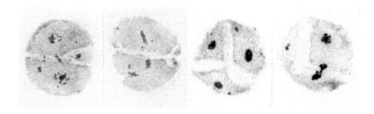

Figure 5.15   Some aspects of irregular cytokinesis in *B. decumbens* leading to cellularization after meiosis I.

cytokinesis. Due to its tetraploid condition, the accession presented many cells with precocious chromosome migration to the poles in metaphase I and laggards in anaphase I that gave rise to micronuclei in telophase I. In other polyploid accessions of *Brachiaria,* micronuclei remained in this condition until the second cytokinesis; however, the micronuclei in this accession organized their own spindle in the second division. In several microsporocytes, the micronuclei with their minispindles were divided further into microcytes by additional cytokinesis. Some curious planes of cytokinesis were found in some cells, with partitioning of cytoplasm into cells of irregular shape. The result consisted of a high frequency of abnormal products of meiosis (Mendes-Bonato 2002a; Figure 5.15).

Another tetraploid ($2n = 4x = 36$) accession of *B. decumbens* analyzed by Risso-Pascotto, Pagliarini, and Valle (2002) revealed an unusual pattern of nucleolar cycle during microsporogenesis. Until late prophase, the nucleolus behavior was normal, when it disappeared at the end of diakinesis. After its dissolution, it was fractioned into multiple micronucleoli of different sizes at metaphase, which persisted during early anaphase. In late anaphase, however, some of the micronuclei entered a process of fusion, thus reducing their number but increasing in size. The fusion process continued throughout the cycle; thus, in late telophase all of them were rejoined into a unique nucleolus in the sister nuclei and persisted in this condition until the end of prophase II, when they disappeared again. This nucleolar behavior was also observed in the second division; at the end of the meiotic process, tetrads presented four microspores with a normal nucleolus (Figure 5.16). Pollen viability was not affected by this anomalous nucleolar behavior.

Cytological characterization of one accession of *B. ruziziensis* ($2n = 2x = 18$) showed a high frequency of abnormal meiotic products, from triads to hexads, and tetrads with micronuclei or microcytes not expected based on its sexual diploid condition (Risso-Pascotto, Pagliarini, and Valle 2003a). Meiosis I showed a low frequency of abnormalities, mainly those related to the chiasma terminalization. In meiosis II, however, the frequency of abnormalities increased exceptionally. Early prophase II was normal, but later, with the nuclear envelope breakdown, the chromosomes were scattered in the cytoplasm. Some chromosomes did not reach the metaphase plate and remained scattered. The behavior of sister cells was diverse. In one cell the chromosomes were totally aligned at the metaphase plate; in the other they were totally scattered, leading to an asynchronous cell division between cells. The cells with scattered chromosomes were unable to progress in meiosis, so anaphase II did not occur and sister chromatids were not released. Cells with nonaligned

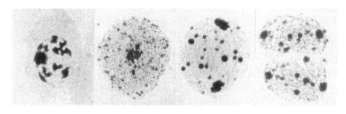

Figure 5.16   Unusual pattern of nucleolar cycle during microsporogenesis in *B. decumbens*.

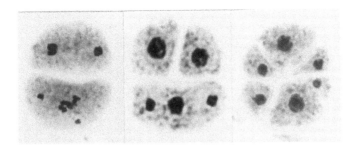

Figure 5.17 Failure in chromosome congression at the metaphase II plate in the diploid accession of *B. ruziziensis,* causing meiosis arrest and leading to polyad formation.

chromosomes at the metaphase plate did not receive a signal to advance to the anaphase, and the scattered chromosomes produced telophase nuclei of different sizes. This asynchronous behavior led to the formation of a wide range of meiotic products (Figure 5.17). The observations suggested that this accession has a mutation affecting the spindle checkpoint arresting the second division.

One accession of *B. brizantha* analyzed by Mendes-Bonato et al. (2001b) presented a severe case of chromosome stickiness during meiosis, impairing chromosome segregation. The accession was tetraploid with chromosomes pairing as bivalents and few quadrivalents at diakinesis. Stages of prophase I were normal and chromosome stickiness became evident from metaphase I on, persisting to microspore stage. Bridges of different thickness were formed in anaphase I and II and the chromosomes did not separate. Some of them persisted until telophase stages. Stickiness was observed also in released microspores after callose dissolution (Figure 5.18).

Cell fusion involving two or several cells has been reported in *B. brizantha* (Mendes-Bonato et al. 2001c; Figure 5.19) and in *B. decumbens* cv. Basilisk (Mendes-Bonato et al. 2001a). In the former, the regular contour of the syncyte suggests that cell fusion took place during premeiotic mitosis by suppression in the cell wall formation, thus being characterized as archesporial syncytes. In *B. decumbens,* the irregular contour indicates fusional syncytes.

An interesting case of cell fusion was found in five tetraploid accessions of *B. jubata* (Mendes-Bonato et al. 2003). Meiosis proceeded normally from prophase I to the end, giving rise to an octad

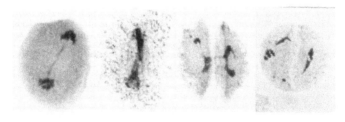

Figure 5.18 Chromosome stickiness in *B. brizantha,* impairing regular chromosome segregation.

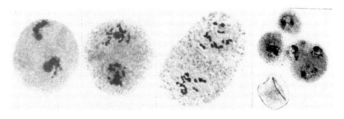

Figure 5.19 Cell fusion in *B. brizantha,* resulting in multinucleated microspores.

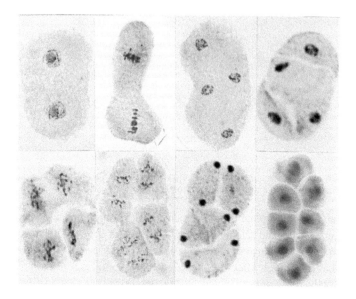

Figure 5.20 Cell fusion in *B. jubata* involving two cells, followed by regular meiosis and normal microspore formation.

with normal microspores that developed into fertile pollen grains (Figure 5.20). Regular octad formation was possible because each chromosome set was maintained in its proper domain, spindles were correctly positioned, and cytokinesis planes were formed in the correct places.

Failure of cytokinesis in the first or second meiotic division leading to dyad and triad formation was also recorded among *B. brizantha* accessions (Figure 5.21; Mendes-Bonato 2000; Risso-Pascotto et al. 2003b). In some cases, restitutional nuclei were observed, producing 2*n* gametes.

Six tetraploid accessions of *B. nigropedata*, among twenty analyzed, presented chromosome transfer among microsporocytes (cytomixis) in early stages of meiosis in frequencies ranging from 0.6 to 9.7% (Utsunomiya, Pagliarini, and Valle 2004). The number of cells involved in the phenomenon varied from two to four, with predominance of the former (Figure 5.22). Transference of a few chromosomes up to the whole genome was found among cells. Prior to cytomixis, chromatin underwent structural alteration similar to chromosome stickiness.

Divergent spindle was another abnormality recorded in some plants of an interspecific hybrid between *B. ruziziensis* and *B. brizantha* (Mendes-Bonato 2006b). In affected plants, prophase I was normal. At metaphase I, however, bivalents were regularly co-oriented at the metaphase plate but distantly positioned and spread over the plate (Figure 5.23a). At anaphase I, chromosomes failed to converge into focused poles due to parallel spindle fibers (Figure 5.23b). As a consequence, an elongated nucleus or several micronuclei were formed in each pole at telophase I (Figure 5.23c). In

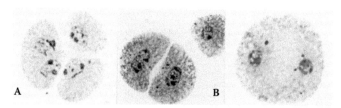

Figure 5.21 Failure of cytokinesis in the first or second meiotic division, leading to (A) a binucleate triad and (B) a dyad with restitutional nuclei in *B. brizantha*.

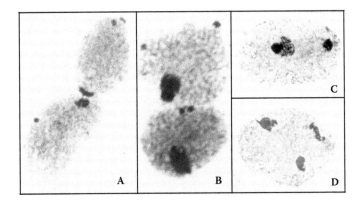

Figure 5.22 Chromosome transfer among microsporocytes in *B. nigropedata*.

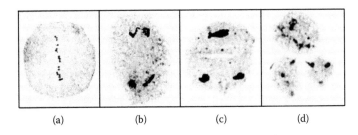

(a)   (b)   (c)   (d)

Figure 5.23 Divergent spindle in a hybrid between *B. ruziziensis* and *B. brizantha*.

the second division, the behavior was the same, leading to the formation of polyads with several micronuclei (Figure 5.23d).

Multiple spindles showing different arrangements inside the meiocyte were recorded in an artificially induced tetraploid accession of *B. ruziziensis* (Risso-Pascotto, Pagliarini, and Valle 2005a). Chromosome pairing at diakinesis ranged from univalents to tetravalents. During first division, multiple spindles separated the genome into several micronuclei (Figure 5.24A). The spindle's position determined the plane of first cytokinesis and the number of chromosomes determined the size of the cell (Figure 5.24B–E). Meiotic products were characterized by polyads with spores of

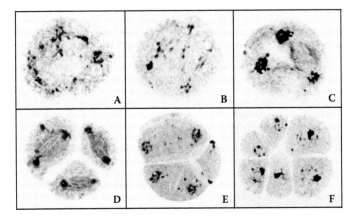

Figure 5.24 A–F. Multiple spindles in an artificially induced tetraploid accession of *B. ruziziensis*, resulting in polyads.

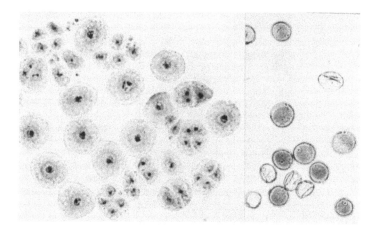

Figure 5.25 Asynchronous meiosis in a hybrid between *B. ruziziensis* and *B. decumbens.*

different sizes and pollen fertility was reduced (Figure 5.24F). These abnormalities have limited the use of this accession as a sexual progenitor in the hybridization program in Brazil.

Asynchronous meiosis was reported in a hybrid between *B. ruziziensis* and *B. decumbens* (Risso-Pascotto et al. 2004a). In young spikelets, 50% of anther meiocytes entered meiosis, exhibiting typical phases of the first and second divisions, while the other 50% showed distinctive features of early prophase. In older spikelets, anthers containing mature pollen grains also displayed meiocytes still undergoing meiosis (Figure 5.25). Pollen fertility was estimated to be 52.8% in dehiscent anthers, and independent genetic control for meiosis synchrony and meiotic stages was suggested.

During cytological analysis in different *Brachiaria* species, some accessions showed some unusual meiotic behavior, and they were analyzed with special attention. Meiotic division and male gametophyte development were analyzed in *B. decumbens* cv. Basilisk ($2n = 4x = 36$), and pollen sterility was detected (Junqueira-Filho et al. 2003). The meiotic process was typical of polyploids with multiple chromosome associations. Precocious chromosome migration to the poles, laggards, and micronucleus formation were abundant in both meiosis I and II, resulting in the tetrads with micronuclei. After callose dissolution, microspores were released into the anther locule and seemed normal. Although each microspore initiated its differentiation by pollen mitosis, in 43% of the microspores, nucleus polarization was not observed, and the typical hemispherical cell plate was not detected. Division was symmetric and microspores lacked differentiation between the vegetative and the generative cell. Both nuclei were of equal size and equal chromatin condensation and had a spherical shape.

After the first pollen mitosis and cytokinesis, each cell underwent a new symmetric mitosis, without nucleus polarization. At the end of the second pollen mitosis, four equal nuclei were observed in each pollen grain. After the second cytokinesis, the cells gave rise to four equal-sized pollen grains with a similar tetrad configuration and remained together. Sterile pollen grains resulted from the abnormal pollen mitosis. Similar abnormal pollen development was recorded in an interspecific hybrid between *B. ruziziensis* and *B. decumbens* (Mendes-Bonato et al. 2004) and in an accession of *B. jubata* (Risso-Pascotto et al. 2005b).

The abnormalities found in the genus *Brachiaria* showed that it is an extraordinary biological material for better understanding the genetic control of meiosis. Some of these also represent important tools for breeding programs such as generation of $2n$ gametes.

5.5.5 Genomic Relationships

Morphological, cytological, and molecular analyses of species contribute to the establishment of the parental relationship. Unfortunately, specific data for genomic relationships in the genus *Brachiaria* are scarce. Many intermediary characteristics were found in *B. brizantha, B. decumbens,* and *B. ruziziensis.* Basappa, Muniyamma, and Chinnappa (1987) stated that the meiotic irregularities present in *B. decumbens* and its strong resemblance to *B. ruziziensis* and *B. brizantha* indicated that this species was a natural hybrid between the latter. Hacker (1988), Valle and Miles (1992), and Lutts, Ndikumana, and Louant (1991, 1994) showed that it is possible to obtain hybrids from crosses between artificially tetraploidized sexual *B. ruziziensis* and the apomictic *B. decumbens* and *B. brizantha.* Later, Tohme et al. (1996) used random amplified polymorphic DNA molecular markers (RAPD) to associate *B. brizantha, B. decumbens,* and *B. ruziziensis* in one group and *B. jubata* and *B. humidicola* in another. An identical grouping based on the most significant morphological characteristics was proposed by Renvoize, Clayton, and Kabuye (1996), who distributed 97 species into nine groups.

A comparison of the morphological and molecular data described in the literature with the karyotype data obtained by Bernini and Marin-Morales (2001) suggests that the chromosome differentiation found in *Brachiaria* accessions was not followed by great morphological changes. The species related by morphological similarities and crossing ability (*B. brizantha, B. decumbens,* and *B. ruziziensis*) had distinct karyotype characteristics. Therefore, the karyotypes described for these species provide additional data in the establishment of relationships among *Brachiaria* species.

5.5.6 Genomic Affinity

Knowledge of genome similarities in different species could provide a means by which evolutionary relationships might be assessed and also represent an important point in alien introgression programs. Gene transfer is facilitated by homologous or homoeologous recombination at meiosis. Chromosome pairing in interspecific hybrids is an excellent method of assessing genomic relationships between species and establishing the phylogeny in the genus. A high pairing affinity of chromosomes presumes that the gene pools of both progenitors are interchangeable.

Interspecific hybridization and meiotic chromosome pairing in the genus *Brachiaria* showed the lack of genome affinity between progenitors of both genomes. Risso-Pascotto et al. (2004b) reported the occurrence of chromosome elimination in an interspecific triploid hybrid between a sexual diploid accession of *B. ruziziensis* ($2n = 2x = 18$) and an apomictic tetraploid accession of *B. brizantha* ($2n = 4x = 36$) by asynchrony during meiosis. The genome of *B. ruziziensis* (R) did not pair with *B. brizantha* (B) and the univalents always remained temporally behind in relation to the *B. brizantha* genome, giving rise to an extra nucleus in each pole (Figure 5.26). Asynchronous meiosis caused the elimination of this very vigorous hybrid from the process of cultivar development because no seed fill was expected and vegetative propagation is unfeasible for pasture establishment in the tropics.

The meiotic behavior in interspecific hybrids is not always identical, even between individuals of the same two species. Studies under way have shown that the meiotic behavior seems to be accession specific. An interspecific tetraploid hybrid between a different R and another accession of B showed that chromosomes associated predominantly as bivalents were equally distributed in two metaphase plates forming two distinct and typical bipolar spindles (Mendes-Bonato, Pagliarini, and

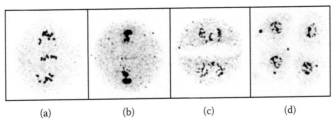

(a)   (b)   (c)   (d)

Figure 5.26 Some aspects of microsporogenesis in a triploid hybrid between *B. ruziziensis* ($2n = 2x = 18$) and *B. brizantha* ($2n = 4x = 36$) showing asynchrony during division.

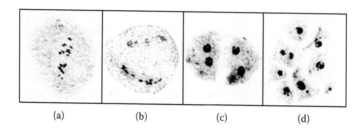

(a)   (b)   (c)   (d)

Figure 5.27 Chromosomes distributed in (a) two metaphase plates and (b, c) forming two distinct and typical bipolar spindles in a hybrid between *B. ruziziensis* and *B. brizantha,* resulting in (d) polyads.

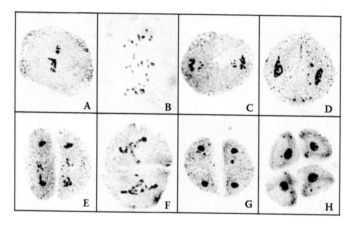

Figure 5.28 A–H. Lack of genome affinity in a pentaploid accession ($2n = 5x = 45$) of *B. brizantha,* suggesting natural hybridization.

Valle 2006a; Figure 5.27). The second division was very irregular, resulting in polyads. This hybrid would also compromise the program and was discarded.

Significant levels of sterility have been found among interspecific hybrids, indicating species incompatibility; however, some interspecific crosses have been successful. Cytological analysis in two other hybrids involving different accessions of *B. ruziziensis* and *B. brizantha* (Risso-Pascotto et al. 2006b) showed irregular meiosis, but abnormalities were related only to its polyploid condition, with both genomes sharing the same metaphase plate and showing the same meiotic rhythm. These abnormalities may not completely compromise fertility and, depending on frequency of occurrence, may allow utilization of the genotype in the breeding program.

Lack of genome affinity was also detected in accessions of *B. brizantha.* Cytological studies showed clear evidence of natural hybridization in two pentaploid accessions ($2n = 5x = 45$) of this species (Mendes et al. 2006). The lack of affinity between parental genomes was clearly evidenced by asynchrony during meiosis. The haploid genome ($n = 9$) showed a unique behavior; it remained as univalent during prophase I (Figure 5.28A), continued separated at the metaphase plate, underwent sister-chromatid segregation, and lagged behind in anaphase I (Figure 5.28B, C). The laggard genome did not always reach the poles on time to be included in the telophase nucleus. However, when the inclusion was effective, this genome was allocated peripherically, changing the otherwise spherical nucleus shape (Figure 5.28D). In the second division, the haploid genome behaved similarly, but because of sister-chromatid segregation during the first division, the chromosomes were slow to reach the poles (Figure 5.28E, F), forming several micronuclei in telophase and tetrad (Figure 5.28G, H). These accessions were characterized as alloautotetraploids.

5.5.7 Chromosome Mapping by Cytogenetics and Molecular Methods

Despite the importance and agronomic interest of the *Brachiaria* species, genetic linkage maps are still not available. The first attempts to construct a molecular map in this genus were made by Pessino et al. (1997, 1998), who tried to locate the apospory gene. Based on the fact that grass genomes display collinearity (i.e., genomic synteny), the authors compared the *Brachiaria* genome with those of maize by restriction fragment length polymorphism (RFLP) and RAPD. The *Brachiaria* genome was systematically scanned using 61 cDNA and genomic maize probes, which detected 65 loci located at 40 cM, on average—one from each other in the maize genome. Two RFLP markers and one RAPD marker co-segregating with apospory were identified. A map of the region was constructed using additional clones that belonged to the same maize linkage group. The results support the hypothesis of a single locus controlling apomixis because that was the only genomic region that presented an apomixis-linked polymorphism.

Also, utilizing synteny among grasses, Arruda et al. (2007) verified the possibility of transferring microsatellites developed for maize, rice, and wheat to species of *Brachiaria*. Their results pointed to the possibility of microsatellite transfer among the several species involved because 13 pairs of primers of maize, 4 of wheat, and 4 of rice amplified putative microsatellite regions for at least one genotype of *Brachiaria*. From the pairs of primers transferred, only one (originally developed for wheat) yielded polymorphic bands among accessions of *B. brizantha* and *B. decumbens*. Therefore, development of specific simple sequence repeat (SSR) markers was undertaken to study diversity in *Brachiaria* germplasm. Genomic libraries were built for five species in the genus (Jungmann et al. 2005) and, recently, SSR markers were developed for *B. brizantha*. Most of the 10 polymorphic loci amplified in *B. ruziziensis* and *B. decumbens*—but not in *B. humidicola* or *B. dictyoneura*— thus supporting the agamic complex structure. Divergence among 23 morphologically contrasting genotypes was estimated, coefficients of similarity calculated, and clustering of genotypes done using the UPGMA method (Jungmann et al. 2007). A well-saturated map for all five important species should aid in the search for markers or quantitative trait loci (QTLs) for traits of agronomic interest, such as insect resistance, tolerance of aluminum, and mode of reproduction that impact the breeding of *Brachiaria* significantly.

5.6 GENE POOLS

At least two gene pools of actual or potential relevance to plant breeding are recognized in *Brachiaria*. Analysis of genetic relationships in the genus, based on morphological characters (Renvoize, Clayton, and Kabuye 1996) or molecular markers (Tohme et al. 1996, Jungmann et al. 2007), confirms the phylogenetic proximity of *B. ruziziensis, B. decumbens,* and *B. brizantha*. Based on hybrid seed set, viability of hybrid seedlings, and chromosome pairing in hybrids, Lutts et al. (1991) found that *B. ruziziensis* is more closely related to *B. decumbens* than to *B. brizantha*. The genetic relationship among *Brachiaria* species was also compared using RAPD primers (Suárez 1994). A total of 58 accessions of six species, including *B. decumbens, B. brizantha, B. ruziziensis, B. jubata, B. humidicola,* and true *B. dictyoneura*, were analyzed. The grouping pattern obtained assigned *B. decumbens, B. brizantha,* and *B. ruziziensis* to one taxonomic group and agamic complex; *B. jubata, B. humidicola,* and true *B. dictyoneura* were assigned to another. Significant levels of sterility are observed in certain interspecific hybrid combinations among *B. ruziziensis, B. decumbens,* and *B. brizantha*, indicating some species incompatibilities (Lutts et al. 1991). However, several interspecific crosses within this agamic complex have been successful. Meiotic chromosome behavior of hybrids suggests that these three species form a more or less coherent gene pool once ploidy barriers are overcome (Lutts et al. 1991).

Brachiaria humidicola is a commercially important species that falls into a distinct taxonomic group from the previously mentioned agamic complex (Renvoize, Clayton, and Kabuye 1996). The two commercial cultivars (Llanero and Tully) are apomictic hexaploid, and hybridization of these lines with existing sexual tetraploid in the *B. ruziziensis* complex has been attempted unsuccessfully (Miles and do Valle 1996). True *B. dictyoneura* is moderately close to *B. humidicola;* hybridization between them may eventually be possible if a closely related sexual hexaploid can be found or synthesized. A natural sexual tetraploid has been identified in *B. humidicola* (Valle and Glienke 1991) and hybridization has finally been obtained (Valle et al. 2008) at the tetraploid level. An approach to introduce sexuality into the hexaploid *B. humidicola* complex includes hybridization of the sexual tetraploid genotype to an apomictic that produces $2n$ gametes identified cytogenetically (Gallo et al. 2007). Hybrids were produced and need to be analyzed for ploidy level and mode of reproduction once they flower.

Another *Brachiaria* species, *B. jubata,* has potentially useful insect resistance, although it is not currently used commercially. Both diploid and tetraploid (sexual and apomictic) genotypes have been identified within the species (Valle 1990). Attempts to hybridize *B. jubata* with sexual material of the *B. ruziziensis* complex have been unsuccessful at the diploid and tetraploid level. Morphological and taxonomic proximity may favor more successful hybridization of *B. jubata* with *B. humidicola* or *B. dictyoneura* than with the *B. brizantha* complex (Renvoize, Clayton, and Kabuye 1996).

5.7 GERMPLASM ENHANCEMENT

5.7.1 Conventional Breeding

Until two decades ago, genetic improvement of *Brachiaria* species depended entirely on selection among naturally existing genotypes, given that genetic recombination was impossible due to apomictic reproduction. Massive importation of seed of the Australian *B. decumbens* cv. Basilisk (a selection from a small collection of natural germplasm) greatly enhanced the productivity of pastures on the vast South American savannas. Although cv. Basilisk is perhaps uniquely adapted and productive on the infertile acid soils typical of tropical America, its deficiencies soon became evident. It is still widely used in the central savannas of Brazil and Latin America; however, in the humid tropics, its susceptibility to spittlebugs, coupled with failure of sward regeneration because of poor seed viability, eliminated this cultivar from tropical forest environments. Programs for producing new *Brachiaria* cultivars were initiated, relying initially on collection and introduction of natural germplasm (Argel and Keller-Grein 1996; Pizarro et al. 1996). Some commercial cultivars—all direct selections from natural germplasm collected in Africa—were listed by Keller-Grein, Maass, and Hanson (1996) and continue to represent the major cultivated pastures throughout this continent. All have limitations (Lapointe and Miles 1992), especially susceptibility to spittlebugs. However, the identification of attributes needed in new cultivars has been relatively straightforward.

In addition to clearly defined objectives, prerequisites for effective plant breeding are (1) an in-depth knowledge of the biology, cytology, and reproductive behavior of the crop; and (2) access to an adequate genetic base. Valle (1990) reported cytology and reproductive behavior in several *Brachiaria* species, providing a solid basis for breeding work when a large new germplasm collection became available in the late 1980s.

At first, genetic improvement based on the combination of attributes from two or more commercial species by conventional hybridization was not directly possible due to apomixis. Despite limitations to the genetic recombination, apomixis has significant potential as a tool in plant breeding, in that even highly heterozygous genotypes breed true through seed. Developing more efficient methods to assess edaphic adaptation, insect resistance, mode of reproduction, and pasture persistence is of particular concern in *Brachiaria* breeding work. Edaphic adaptation is difficult to assess due to the complex set of parameters involved and the time needed for evaluation. Wenzl et al. (2006) developed a greenhouse method to screen for aluminum tolerance and root vigor using rooted vegetative propagules in solution culture. Large numbers of accessions and hybrids were evaluated in each test, and results seemed to correlate well with field data. This approach significantly increased the efficiency in screening genotypes for traits that contribute to edaphic adaptation to infertile and acid soils in the *Brachiaria* breeding program at CIAT.

Recurrent selection was also successfully carried out for six cycles at CIAT to improve resistance to three spittlebug species typical of Colombia (Miles, Cardona, and Sotelo 2006). The population was synthesized by recombining sexual hybrids obtained from crosses between the tetraploid *B. ruziziensis* and nine apomictic genotypes of *B. brizantha* and *B. decumbens*. Selection was based on nymphal survival feeding on rooted tillers in the greenhouse (Cardona, Miles, and Sotelo 1999), and survival dropped from 55.6% at C_2 to 7.0% at C_6. Miles, Cardona, and Sotelo (2006) reported improved resistance in response to intense, recurrent selection on reliable phenotypic data and indicated that resistance to several spittlebug species can be achieved.

Improved methods with the capacity to handle the large volume of segregating genotypes generated in the breeding programs and for the traits of greatest agronomic importance are still in demand. They could further increase efficiency and accuracy in selecting superior genotypes for current breeding schemes.

5.7.2 Exploitation of Apomixis in *Brachiaria* Breeding

Apomixis is generally defined as asexual reproduction through seed. If obligate, apomixis offers a formidable obstacle to conventional genetic recombination. Expression of apomixis is often incomplete, and generally some of the seeds contain embryos resulting from meiosis and fertilization. In most apomictic species, fully sexual genotypes have been found in the same or on closely related species. A frequent pattern is that diploids are obligate sexuals, and different degrees of apomixis are found among related polyploids characterizing facultative apomixis. In species where a mechanism for genetic recombination is available through residual sexuality in facultative apomicts or in cross-compatible sexual species, powerful breeding schemes are potentially available. These would produce novel genetic combinations by conventional hybridization and, furthermore, would fix the resulting apomictic hybrid genotypes, which could be then propagated indefinitely through seed.

To date, only modest plant breeding efforts have been conducted in apomictic species, aside from *Poa,* and primarily in genera of tropical forage grasses in which apomixis occurs naturally. The potential of apomictic reproduction has gained the attention of several research teams and considerable funds have been allocated to determine its genetic control, to isolate the responsible gene(s), and to attempt to transfer apomixis to crops (http://www.apomixis.de/EU-CA.htm, http://www.cimmyt.org/ABC/map/about/Apomixis/Whatisapomixis.htm). Also, cross-compatible sexual genotypes are being sought or produced in apomictic genera/species.

Artificial hybridization in *Brachiaria* has been sought at least since the early 1970s, when Ferguson and Crowder (1974) attempted to produce hybrids by pollinating diploid sexual *B. ruziziensis* with pollen from a tetraploid apomictic *B. decumbens*. This early attempt was unsuccessful, apparently due to the ploidy barrier. The authors suggested artificially doubling the chromosome number of the natural diploid sexual *B. ruziziensis* and attempt crosses with the tetraploid apomicts. Hacker (1988) crossed a diploid sexual *B. decumbens* with a tetraploid apomictic ecotype and obtained a single triploid hybrid that was completely sterile. He suggested (1) colchicine treatment of the sterile triploid hybrid to produce a fertile allohexaploid, and (2) colchicine treatment of the sexual diploid to produce a cross-compatible genotype.

Exploiting apomixis in *Brachiaria* breeding became possible through the work done at the Catholic University of Louvain in Belgium in the early 1980s. An obligate sexual tetraploid was successfully developed by colchicine treatment of sexual diploid *B. ruziziensis* (Gobbe, Swenne, and Louant 1981; Swenne, Louant, and Dujardin 1981). The Belgian material became the basis of the breeding work in *Brachiaria,* although sexual diploid ecotypes of *B. brizantha* (Pinheiro et al. 2000) and *B. decumbens* (Valle et al. 2008) have been successfully tetraploidized to allow for intraspecific crosses.

Ndikumana (1985) reported successful interspecific hybridization in crosses using the tetraploid sexual *B. ruziziensis* and two tetraploid apomictic accessions each of *B. decumbens* or *B. brizantha* as pollinators. Hybrids were confirmed by the presence of parental morphological traits. These results showed close phylogenetic relationships among these three species. Although hybridization resulted in a small population of 35 individuals, analysis of their reproductive mode by embryo-sac morphology suggested simple genetic control of apomixis.

The pioneering Belgian work opened up the possibility of genetic improvement of *Brachiaria.* Applied breeding programs were undertaken after the tetraploid sexual *B. ruziziensis* reached Brazil in 1988 and later at CIAT. At CIAT, the feasibility of large-scale hybridization in the field was demonstrated by Calderón and Agudelo Cortés (1990). They obtained more than 90% hybrid offspring from potted tetraploid sexual plants allowed to open-pollinate in a field of flowering *B. decumbens* or *B. brizantha.* Convincing isozyme evidence proved the hybrid nature of the progenies from this experiment.

The methods currently employed in *Brachiaria* breeding programs are based on general principles of quantitative genetics and on the limited information available on genetic control of the most desirable traits. The objective is to produce and identify apomictic plants with desirable combinations of traits not found in accessions of natural *Brachiaria* germplasm. Due to the narrow genetic base of the tetraploid sexual germplasm, it is reasonable to expect that more than one cycle of hybridization will be needed to achieve the desirable combination of characters. The original Belgian tetraploid sexual not only is extremely susceptible to the spittlebug, but also has poor edaphic adaptation and does not display tolerance to drought or cold. Hence, the purpose is to accumulate favorable alleles over several cycles of recurrent selection and recombination (Miles and Valle 1996; Miles et al. 2004).

Two types of populations are being developed for the breeding of *Brachiaria.* One approach is the improvement of sexual tetraploid populations. A broad-base synthetic in the case of CIAT (Miles et al. 2004) is continuously developed by recurrent selection for agronomic performance, insect resistance, and edaphic adaptation. Progenies are homozygous for sexuality. In Brazil, the strategy involves a crossing block with superior sexual F_1 hybrids in open pollination to produce half-sib families. Inter- and intrafamily selection allows identification of superior sexual progenitors and elite hybrids to participate in the next cycle of crosses. Poor progeny performance indicates progenitors to be discarded. In both programs, the improved populations will subsequently be crossed to elite apomicts to recover apomictic genotypes for development of new cultivars.

A second approach has been to improve both apomicts and sexuals simultaneously in a single population that segregates for reproductive mode. Simple and monogenic control of apomixis makes this scheme feasible, albeit time consuming, because mode of reproduction of many individuals needs to be determined at each cycle. At this point, a molecular marker for apomixis would allow discrimination early, accurately, and quickly. Selected apomictic and sexual plants are intermingled in the field and seeds are harvested only from the sexual plants. The progenies will contain both sexual and apomictic individuals. At flowering, these individuals are sampled for embryo-sac analysis and superior genotypes are selected to enter the next cycle. To complete the cycle, the open-pollinated progeny of these selected sexual and apomictic hybrids is established in the subsequent crossing block. A diagram for the breeding strategy was presented by Miles et al. (2004).

Details of efficient evaluation, selection, and hybridization schemes have yet to be developed. It is necessary to know much more about the genetics of important traits, including a complete genetic model for the control of reproductive mode, before optimal schemes can be designed.

5.7.3 Introgression of Economically Valuable Traits through Wide Crosses

All existing commercial cultivars are either susceptible to biotic stress or have poor edaphic adaptation. Although a desired combination of resistance to these stresses may possibly be identified in naturally occurring ecotypes, it is unlikely that higher forage quality would be included. However, although forage quality is of considerable value in animal production, it does not necessarily improve fitness in nature.

Existing *Brachiaria* breeding programs focus on recognized deficiencies in existing commercial cultivars and tackle the germplasm collection for source of desirable traits. The immediate objective is to develop apomictic cultivars combining the persistence, productivity, and adaptation to infertile acid soils of common *B. decumbens* cv. Basilisk with the durable antibiotic resistance to spittlebugs found in *B. brizantha* cv. Marandu, complemented by the high nutritive value and concentrated flowering of *B. ruziziensis*.

Additional desirable traits in any *Brachiaria* cultivar include resistance to rhizoctonia foliar blight (caused by *Rhizoctonia solani*), rust (caused by *Uromyces setariae-italicae* and/or *Puccinia levis* var. *panici-sanguinalis*), seed-borne diseases such as honey dew (caused by *Claviceps sulcata*), and ergot (caused by *Ustilago operta*). Marked differences in resistance to leaf-cutting ants have been demonstrated among currently available *Brachiaria* cultivars (Valério et al. 1996). Antinutritional quality attributes have been associated with existing *Brachiaria* cultivars (Lascano and Euclides 1996) and should also be selected against.

A wide range in flowering response has been reported in *Brachiaria* species by Hopkinson et al. (1996). It is possible to identify recombinants with any desired photoperiodic response and flowering time in a particular environment. Seed yield and quality (dormancy) and seedling vigor are important for propagation. A nitrification inhibition effect has been identified in *B. humidicola* ecotypes that could greatly benefit tropical systems in avoiding N loss to water tables and to the atmosphere (Subbarao et al. 2007).

5.7.3.1 Sources of Desired Attributes

5.7.3.1.1 The B. ruziziensis/B. decumbens/B. brizantha Complex

A wealth of genetic variation exists within this complex in the germplasm bank. Spittlebug resistance has been identified in several accessions as well as in the commercial *B. brizantha* cv. Marandu and cv. BRS Piatã. Preliminary evidence suggests that resistance is highly heritable. Miles et al. (1995) found significant variation among half-sib families for adult damage scores under natural field infestation ($h^2 = 0.44$) and high correlation of nymphal survival between pollen parent and topcross progeny ($r^2 = 0.897$; $n = 11$). Miles, Cardona, and Sotelo (2006), as mentioned previously, obtained significant progress in insect resistance after six cycles of strong selection against nymphal survival. Common *B. decumbens*, on the other hand, is widely recognized as having broad adaptation to infertile acid soil (Rao, Kerridge, and Macedo 1996).

The genetic control of edaphic adaptation is probably complex and still unknown, but selection as reported previously (Wenzl et al. 2006) has been able to explore the genetic diversity produced in controlled hybridization and select superior hybrids. A wide range of *in vitro* dry matter digestibility has been detected among accessions of natural *Brachiaria* germplasm (Lascano and Euclides 1996). The best genotypes should be used as apomictic progenitors in crosses with elite sexuals. There are interesting features in various accessions that need to be combined to obtain hybrids with superior performance. The ideal cultivar should have the edaphic adaptation found in *B. decumbens*; high nutritive value and rapid pasture establishment found in *B. ruziziensis*; and insect resistance, good dry matter, and seed production common to *B. brizantha*.

5.7.3.1.2 The B. humidicola/B. dictyoneura Complex

There are only two commercial cultivars in the *B. humidicola/B. dictyoneura* group, and they are very vigorous, strongly stoloniferous, and competitive pasture plants. They are both apomictic hexaploid plants. However, nutritional quality, particularly N content, is low. *Brachiaria humidicola* shows poor flowering and seed production at low latitudes. Seed dormancy is very common in these cultivars and may impair good pasture establishment. Cultivars Tully and Llanero are good hosts to spittlebug despite not showing damage under large adult infestation. A reasonable collection of germplasm accessions of *B. humidicola* and *B. dictyoneura* exists, and recently an interesting new variation is being documented by using molecular markers because ecotypes are phenotypically similar (Salgado et al. 2005).

Forage quality is a trait of high priority in the breeding of this complex. Miles and do Valle (1996) pointed out that until a compatible source of sexuality was found for this complex, breeding activities could not be pursued, and this is still true for the hexaploid ecotypes. Breeding at the tetraploid level in *B. humidicola* has just recently produced progenies, and the best 50 hybrids have been planted for agronomic evaluation in Brazil (Valle et al. 2008a). This population will also be used to validate the inheritance of apomixis in *Brachiaria*.

5.7.3.2 Cross-Compatibilities between Species

Interest in wide hybridization in this genus arose because different species have traits of agronomic importance. Once ploidy barriers were overcome, the three most important species (*B. decumbens* [D], *B. brizantha* [B], and *B. ruziziensis* [R]) easily produced hybrids that are mostly or reasonably fertile. These hybrids were generated only through the use of compatible sexual *B. ruziziensis*.

Ndikumana (1985) crossed accessions of these three species and recovered more hybrids in R × D than in R × B crosses. He concluded after examining chromosome pairing that *B. ruziziensis* and *B. decumbens* were closer than *B. ruziziensis* and *B. brizantha*. Ngendahayo (1988), on the other hand, studied self- and cross-compatibility among these species by examining *in vitro* pollen tube growth and fertilization. More than 60% of ovaries contained well-developed pollen tubes in R × D crosses, whereas only 11% of ovaries showed pollen tube growth in R × B crosses. The author concluded that a compatibility barrier did not exist between *B. ruziziensis* and *B. decumbens,* but was strong in interspecific crosses with *B. brizantha*.

Different accessions of the preceding three species have been extensively hybridized in Brazil and in Colombia. Data suggest that compatibility is genotype dependent because many more hybrids have been obtained in first-generation crosses of R × B (cv. Marandu) (227 hybrids) than of R × D (cv. Basilisk) (125 hybrids) in Brazil (Valle and Glienke 1991). Other crosses involving different accessions of B have produced numerous promising sexual and apomictic hybrids, corroborating the genotype-dependent relationship between species in this complex. Crosses at the diploid level between sexual diploids of *B. decumbens* and one accession of *B. brizantha* resulted in three hybrids. All plants were sexual, diploid, vigorous, and stoloniferous.

5.7.4 Potential of Biotechnology in *Brachiaria* Improvement

According to Hayward, Armstead, and Morris (1996), the primary requirements of any breeding program are twofold: a source of variation and the means to manipulate that variation. Considerable variation is still available in the existing germplasm, and much more can be released using sexuality to reach heterozygosity fixed by apomixis. The demand of current plant breeding, however, often requires the incorporation of characteristics that cannot be met by these classical means. Breeders must then explore what new technologies can offer in terms of variation and its effective manipulation.

The use of biotechnology in forage grasses potentially involves a wide range of procedures, such as genome analysis and the construction of genetic maps, somaclonal variation, androgenesis, and transformation. Introducing frase molecular markers may also aid in the identification of hybrids in artificial crosses when emasculation is not feasible. Rapid markers allowed for early discrimination of *B. humidicola* hybrids from self-pollinated plants in Brazil (Valle et al. 2008c). The basic biology of tropical forage grasses is just becoming known and should allow much more effective utilization of biotechnological tools to extend and manipulate available variation. The following topics, discussed by Hayward, Armstead, and Morris (1996), show how these could be or are being employed for *Brachiaria* improvement.

5.7.4.1 Creation of New Variations

Many of the methods developed to create new variations rely heavily on a combination of molecular biology techniques to identify and clone new genes and on effective tissue culture techniques. A wide range of techniques has been developed to allow plant regeneration from diverse tissues, such as embryos, microspores, calli, cell suspensions, and protoplasts. Most of these have been applied to economically important temperate grasses with varying degrees of success. Some of these techniques aim to create or release variation from within the existing genome, such as anther or microspore culture or somaclonal variation linked with *in vitro* selection.

5.7.4.1.1 In Vitro *Selection and Somaclonal Variation*

Tohme et al. (1996) described the first successful protocol for regeneration of genotypes of *B. brizantha, B. dictyoneura* CIAT 6133, *B. decumbens,* and *B. ruziziensis.* The protocol was developed at CIAT. Somaclonal embryogenesis was obtained from mature seeds. Theoretically, all plants regenerated from cells and tissues should be identical to the parental plant, but the plant regeneration process disrupts the genetic stability, creating new genetic variation. The potential for generating and recovering useful somaclonal variation in *Brachiaria* is unknown. No obvious variation was detected in the small populations of regenerated plants in early studies. According to Hayward, Armstead, and Morris (1996), variation will require extremely efficient screening procedures. No obvious variation was also observed in Brazil, when Rodrigues-Otubo, Penteado, and Valle (2000) rescued embryos from interspecific crosses and cultured them *in vitro*, multipropagated, and planted clones in the field to observe somaclonal variation.

5.7.4.1.2 Haploid Production

The artificial production of haploid plants involves growing anthers or free microspores on an artificial medium. The haploid plants require doubling of the chromosome number before they become fertile and can be used in breeding programs. In forage grass breeding, the use of haploids has been somewhat restricted by several factors. Because many forage species are polyploid, reducing the chromosome number does not necessarily lead to the desired homozygosity of the genome. In these species, anther culture may have played a role in revealing variation that would otherwise remain concealed. According to Hayward, Armstead, and Morris (1996), this could be particularly useful in apomictic species like most *Brachiaria,* where a reduced chromosome number complement may lead to the production of sexual as opposed to apomictic forms.

Sexual diploids derived from anther culture would greatly enhance the genetic diversity available at the diploid level in *Brachiaria.* Such material would be extremely useful in genetic mapping. Furthermore, examination of the reproductive mode of an array of random diploid genotypes derived by anther culture from apomictic tetraploid *Brachiaria* could shed light on the relationship between expression of apomixis and ploidy level. The major problem to obtain haploids in the genus *Brachiaria* is that many species are recalcitrant and have not responded to culture.

A project is under way at the Embrapa Genetic Resources and Biotechnology Center to develop an anther culture protocol and obtain haploid and double-haploid plants of diploid *B. brizantha* and

B. decumbens. The major objective is to study the expression of genes related to apomixis. So far, embryogenic calli have been obtained and regeneration protocols are being sought. This type of material could also be potentially used to produce desirable genetic translocations and addition or substitution lines. The production of double haploids is a valuable tool for genetic analysis and for the mapping, isolation, and cloning of specific genes using molecular markers.

5.7.4.1.3 Gene Identification and Insertional Mutagenesis

Molecular approaches based on construction and screening of genomic or cDNA libraries can also be used to clone specific genes. A strategy with great potential for identifying and cloning agronomically important genes is insertional mutagenesis.

Insertion of a foreign gene into the genome of a plant is usually random; hence, such insertions can be at the sites of functional genes and cause disruption. The most widely recognized phenomena of this type are brought about by the germinal or somatic excision of transposons and their reintegration into the genome, often into other genes. A similar system of gene-tagging mutants has also been developed by using insertional mutagenesis, where plasmid or T-DNA is inserted into the genome by transformation, and regenerated plants are screened for mutants. This system is less effective, but probably easier to achieve, than transposon mutagenesis because the inserted plasmid or T-DNA, unlike transposons, cannot be induced to excise and reintegrate germinally. Thus, each transformation event is unique, and a highly efficient system is required to generate the large number of plants needed for identifying mutants. This technique is best applied to species that have a small diploid genome. For this reason, it is not likely to have immediate applicability to *Brachiaria*. However, once isolated from the "model" species, new molecular biology techniques can be used to clone homologous genes from the species of interest.

5.7.4.2 Genetic Transformation by Nonsexual Methods

Genetic transformation and somatic hybridization provide two methods of nonsexual gene transfer. The major difference between conventional plant breeding and genetic transformation is that conventional breeding is concerned with the transfer of large numbers of genes with characteristics only known through phenotypic measurements. In contrast, genetic transformation manipulates small numbers of genes with high precision. Somatic hybridization also involves transfer of whole or partial genomes, but advances in molecular biology now provide more efficient means of handling large numbers of genes.

In plants, successful transfer of foreign genes into a new genetic background requires the production of normal and fertile plants that express the newly inserted genes. Genetic transformation can be done by direct gene transfer to protoplasts or by direct gene transfer to cells by microprojectile bombardment. The protocol for *Brachiaria* transformation using bombardment has been established (Lentini et al. 1999) and this technique is being perfected at Embrapa Genetic Resources and Biotechnology (Carneiro, personal communication 2007). A new approach to transformation is also under way that attempts to use *Agrobacterium* spp. (Cabral et al. 2005). Even though no galls or tumors were formed in *Brachiaria* explants, the possibility of transference of genes from the bacteria is excluded. Tests with a specific vector that has monocot promoters will be used in the next attempts.

The potential of genetic transformation depends fundamentally on the existence of a very limited number of genes (preferably one) having a significant effect on a trait of economic interest. According to Hayward, Armstead, and Morris (1996), some forage-quality, disease-resistant, and developmental attributes in *Brachiaria* may be amenable to genetic improvement through transformation sometime in the future if these genes can be identified.

Somatic hybridization provides a method for nonsexual gene transfer bypassing sexual barriers to combine entire genomes. Species that would normally not be capable of cross-hybridization can form heterokaryons at the protoplast level because there are no initial barriers for cell fusion. Following fusion, protoplasts containing nuclei of the species in common cytoplasm can be induced to regenerate a new cell wall and enter division, where nuclear fusion takes place to form a somatic hybrid cell. Selection and plant regeneration from these somatic hybrids may then be possible. The major difficulty of this method is the genome instability that results in massive or partial chromosome elimination in symmetric somatic hybrids, leading to the development of asymmetric hybridization. Protoplast fusion has conceptual potential in *Brachiaria,* where distantly related species such as *B. humidicola* and *B. decumbens* have complementary desirable attributes that cannot be combined by conventional hybridization.

5.8 PERSPECTIVES AND FINAL COMMENTS

Since their introduction in the mid-1950s, *Brachiaria* grasses have drastically modified the landscape in central Brazil and in most tropical Latin American countries. There is strong evidence that new cultivars will continue to impact production systems and function as agents of technological change in this vast region. These new cultivars (hybrid or not) need to combine adequate forage production with adaptation to soil and climate constraints so as to maintain long-term sustainability of animal and pasture production systems. Thus, varieties should display good nutritive value and resistance to biotic and abiotic stresses, combined with good forage and seed production, in order to ensure the transformation of pasture into valuable animal protein with added economic value.

Research with *Brachiaria* has already yielded important advances, and recently released cultivars are expected to fulfill expectations of increased productivity with sustainable production. However, a wealth of questions must still be answered in the development of improved technologies and methods to work with apomictic tropical forages. Teamwork and multidisciplinary projects are essential to tackle this challenge of working with nondomesticated crops, and each advance is a major contribution to the scientific knowledge of this limited community.

A breeding program can and should take advantage of available technologies, such as molecular, cytogenetics, and mass screening tools, to select elite genotypes more efficiently or to produce new variations amenable to selection. The ideal situation as discussed by Ramalho (2004) is to unite breeders, cytogeneticists, and biotechnologists in a solid program and distribute funds into a coordinated mutual effort. Ramalho expressed his concern over the lack of interest in classic breeding in favor of greater interest in biotechnology or genetic engineering. These new techniques have not yet resulted in significant impact in food production. Therefore, it is premature to overlook traditional methods in breeding and cytology in explaining phenomena observed in progenies.

Brachiaria research is restricted to Brazil and CIAT, despite the enormous economic value as a forage grass for pastures worldwide. This review has attempted to bring together the latest information on the breeding of this genus with the hope of stimulating novel activities and expertise to further promote much needed progress with apomictic grasses.

REFERENCES

Argel, P. J., and G. Keller-Grein. 1996. Regional experience with *Brachiaria*: Tropical America—Humid lowlands. In *Brachiaria: Biology, agronomy, and improvement,* ed. J. W. Miles, B. L. Maass, and C. B. do Valle, 205–224. Cali, Colombia: CIAT/Brasília:EMBRAPA-CNPGC.

Arruda, A., L. Jungmann Cançado, L. Chiari, C. B. do Valle, and L. Jank. 2007. Avaliação da transferência de microssatélites de milho, trigo e arroz para quatro espécies de *Brachiaria* e para *Panicum maximum. Boletim de Pesquisa,* 21, Campo Grande: Embrapa Gado de Corte. 20 pp.

Assis, G. M. L. de, R. F. Euclydes, C. D. Cruz, and C. B. do Valle. 2002. Genetic divergence in *Brachiaria* species. *Crop Breeding Applied Biotechnology,* 2 (3): 327–334.

———. 2003. Discriminação de espécies de *Brachiaria* baseada em diferentes grupos de caracteres morfológicos. *Revista Brasileira de Zootecnia* 32 (3): 576–584.

Barbosa, J. D., C. M. C. Oliveira, C. H. Tokarnia, and P. V. Peixoto. 2006. Hepatogenous photosensitization in horses caused by *Brachiaria humidicola* (Gramineae) in the state of Pará. *Pesquisa Veterinária Brasileira* 26 (3): 147–153.

Basappa, G. P., and M. Muniyamma. 1981. In IOPB chromosome number reports LXXII. *Taxon* 30:694–708.

———. 1983. New taxa in *Brachiaria* Griseb. (Poaceae). *Proceedings of Indian National Science Academy (Part B)* 49:377–384.

Basappa, G. P., M. Muniyamma, and C. C. Chinnappa. 1987. An investigation of chromosome numbers in the genus *Brachiaria* (Poaceae: Paniceae) in relation to morphology and taxonomy. *Canadian Journal of Botany* 65:2297–2309.

Bernini, C., 1997. Análise citogenética e diferenciação cromossômica em espécies do gênero *Brachiaria* Grisebach. MSc thesis, State University of Londrina, Londrina, Brazil, 97 pp.

Bernini, C., and M. A. Marin-Morales. 2001. Karyotype analysis in *Brachiaria* (Poaceae) species. *Cytobios* 104:157–171.

Bir, S. S., and H. S. Chauhan. 1990. SOCGI plant chromosome reports, IX. *Journal of Cytology and Genetics* 25:147–148.

Bir, S. S., and M. Sahni. 1984. SOCGI plant chromosome reports, II. *Journal of Cytology and Genetics* 19:112–113.

Bir, S. S., M. Sahni, S. Kantas, and B. S. Gill. 1987. Cytological investigations on some grasses. *Journal of Cytology and Genetics* 22:23–47.

Brown, W. V. 1951. Chromosome number of some Texas grasses. *Bulletin of Torrey Botany Club* 78:292–299.

Cabral, G. B., V. T. C. Carneiro, C. G. Santana, J. A. Leite, M. V. V. Pires, D. M. A. Dusi. 2005. Prospecção de cepas de *Agrobacterium* spp. para a transformação genética de braquiária. Embrapa Recursos Genéticos e Biotecnologia, *Boletim de Pesquisa,* 115, Brasilia: Embrapa Recursos Genéticos e Biotecnologia, 16p.

Calderón, M. de los A., and J. Agudelo Cortés. 1990. Hibridaciones interespecíficas en el género *Brachiaria*. BS thesis, Universidad Nacional de Colombia, Palmira, Colombia.

Cardona, C., J. W. Miles, and G. Sotelo. 1999. An improved methodology for massive screening of *Brachiaria* spp. genotypes for resistance to *Aeneolamia varia* (Homoptera: Cercopidae). *Journal of Economic Entomology* 92:490–496.

Carnahan, H. L., and H. D. Hill. 1961. Cytology and genetics of forage grasses. *Botanical Review* 27:1–162.

Chatterjee, A. K. 1975. Chromosome studies in some panicoid grasses. *Proceedings of the Indian Science Congress Association* 62:116–119.

Chen, C., and C. Hsu. 1961. Cytological studies in Taiwan grasses. I, tribe Paniceae. *Botany Bulletin Academy Sin* 2:101–110.

———. 1964. The Paniceae (Gramineae) of Formosa. *Taiwania* 9:33–97.

Christopher, J., and A. Abraham. 1976. Studies on the cytology and phylogeny of south Indian grasses III. Subfamily VI: Panicoideae, tribe (i) the Paniceae. *Cytologia* 41:621–637.

CIAT. 1986. Biotechnology. In *Annual Report 1985, Tropical Pastures Program*. Working document no. 17, Cali, Colombia, 1986, 41–57.

Clayton, W. D., and S. A. Renvoize. 1986. Genera Graminium. London: Her Majesty's Stationery Office. 389 pp.

Darlington, C. D., and A. P. Wylie. 1955. *Chromosome atlas of flowering plants*. London: Allen and Unwin. 519 pp.

Davidse, G., and R. W. Pohl. 1972. Chromosome number, meiotic behavior, and notes on some grasses from Central America and the West Indies. *Canadian Journal of Botany* 50:1441–1452.

———. 1974. Chromosome number, meiotic behavior, and notes on tropical America grasses. *Canadian Journal of Botany* 52:317–328.

———. 1978. Chromosome number in tropical American grasses. *Annals of Mississippi Botanical Garden* 65:637–649.

Delay, C. 1951. Nombres chromosomiques chez les Pahnérogames. *Revue Cytologic et Biologique Végétale* 12:161–368.

De Wet, J. M. J. 1958. Additional chromosome numbers in Transvaal grasses. *Cytologia* 23:113–118.

———. 1960. Chromosome numbers and some morphological attributes of various South African grasses. *American Journal of Botany* 47:44–49.

De Wet, J. M. J., and L. Anderson. 1956. Chromosome numbers in Transvaal grasses. *Cytologia* 21:1–10.

Dias Filho, M. B. 2002. Tolerance to flooding in five *Brachiaria brizantha* accessions. *Pesquisa Veterinária Brasileira* 37:439–447.

Dobereiner, J., C. H. Torkania, M. C. C. Monteiro, L. C. H. Cruz, E. G. Carvalho, and A. T. Primo. 1976. Intoxicação de bovinos e ovinos em pastos de *Brachiaria decumbens* contaminados por *Pithomyces chartarum. Pesquisa Veterinária Brasileira* 11:87–94.

Driemeier D., J. Döbereiner, P. V. Peixoto, and M. F. Brito. 1999. Relationship between foamy macrophages in the liver of cattle and the ingestion of *Brachiaria* spp. in Brazil. *Pesquisa Veterinária Brasileira* 19 (2): 79–83.

Dujardin, M. 1978. Étude caryosystématique de quelques espèces de Rottboellinae Africaines (Andropogonae, Gramineae) et réhabilitation du genre Robynsiochloa Jacques. *Bulletin du Jardin Botanique National de Belgique* 48:373–381.

———. 1979. Additional chromosome numbers and meiotic behavior in tropical African grasses from western Zaire. *Canadian Journal of Botany* 57:864–876.

Du Plessis, H., and J. J. Spies. 1988. Chromosome studies in Africa plants, *Bothalia* 18:119–122.

EMBRAPA GADO DE CORTE. 2007. BRS Piatã. Embrapa: Campo Grande, Folder.

Euclides, V. P. B., M. C. M. Macedo, C. B. do Valle, R. Flores, and M. P. Oliveira. 2005. Animal performance and productivity of new ecotypes of *Brachiaria brizantha* in Brazil. In *Proceedings of the International Grassland Congress* 20: 106. Wageningen, the Netherlands: Wageningen Academic Publishers.

Euclides, V. P. B., C. B. do Valle, M. C. M. Macedo, and M. P. Oliveira. 2001. Evaluation of *Brachiaria brizantha* ecotypes under grazing in small plots. In *Proceedings of the International Grassland Congress* 19: 13. Brazil: FEALQ.

Ferguson, J. E. 1974. Method of reproduction in *Brachiaria ruziziensis* Germain at Evrard. PhD thesis, Cornell University, Ithaca, NY. 149 pp.

Ferguson, J. E., and L. V. Crowder. 1974. Cytology and breeding behavior of *Brachiaria ruziziensis. Crop Science* 14:893–895.

Fernandes, C. D., A. T. F. Fernandes, and J. L. Bezerra. 1995. "Mela": Uma nova doença em sementes de *Brachiaria* spp. no Brasil. *Fitopatologia Brasileira* 20:501–503.

Fernandes, C. D., V. F. de Jerba, and J. R. Verzignassi. 2004. Doenças das plantas forrageiras tropicais. In *Proceedings of the VIII Simpósio Brasileiro de Patologia de Sementes*, 51–54, Brazil.

Fischer, M. J., and P. C. Kerridge. 1996. The agronomy and physiology of *Brachiaria* species. In *Brachiaria: Biology, agronomy, and improvement,* ed. J. W. Miles, B. L. Maass, and C. B. do Valle, 43–52. Cali, Colombia: CIAT/Brasília:EMBRAPA-CNPGC.

Gallo, P. H., P. L. Micheletti, K. R. Boldrini, C. Risso-Pascotto, M. S. Pagliarini, and C. B. do Valle. 2007. 2n Gamete formation in the genus *Brachiaria* (Poaceae: Paniceae). *Euphytica* 154:255–260.

Gobbe, J., A. Swenne, and B-P. Louant. 1981. Diploides naturals et autotétraploides induits chez *Brachiaria ruziziensis* Germain et Evrard: Critères d'identification. *L'Agronomie Tropicale* 36:339–346.

Goldblatt, P., ed. 1981. Index to plant chromosome numbers 1975–1978. *Monographs in Systematic Botany*, Missouri Botanical Garden, St. Louis.

Gould, F. W. 1958. Chromosome numbers in south-western grasses. *American Journal of Botany* 45:757–768.

Gould, F. W., and T. R. Soderstrom. 1974. Chromosome numbers of some Ceylon grasses. *American Journal of Botany* 52:1075–1090.

Grof, B., R. P. de Andrade, M. S. França-Dantas, and M. A. de Souza. 1989. Selection of *Brachiaria* spp. for the plateau region of Brazil. In *Proceedings of the International Grassland Congress* 16:267–268. France: Association Française pour la Production Fourragère.

Gupta, B. K. 1971. Cytological investigation in some north Indian grasses. *Genetica Iberica* 23:183–198.

Hacker, J. B. 1988. Sexuality and hybridization in signal grass, *Brachiaria decumbens. Tropical Grasslands* 22:139–144.

Hayward, M. D., I. P. Armstead, and P. Morris. 1996. Theoretical potential of biotechniques in crop improvement. In *Brachiaria: Biology, agronomy, and improvement,* ed. J. W. Miles, B. L. Maass, and C. B. do Valle, 178–195. Cali, Colombia: CIAT/Brasília: EMBRAPA-CNPGC.

Heering, J. H. 1987. An evaluation of *Brachiaria* species from the CIAT-ILCA East Africa collection at the Zwai seed multiplication site. Diploma thesis. University of Wageningen, Wageningen, the Netherlands.

Honfi, A. I., C. L. Quarin, and J. F. M. Valls. 1990. Estudios cariologicos en gramíneas sudamericanas. *Darwiniana* 30:87–94.

Hopkinson, J. M., F. H. D. Souza, S. Diulgheroff, A. Ortiz, and M. Sánchez.1996. Reprodutive physiology, seed production, and seed quality of *Brachiaria*. In *Brachiaria: Biology, agronomy, and improvement,* ed. J. W. Miles, B. L. Maass, and C. B. do Valle, 124–140. Cali, Colombia: CIAT/Brasília: EMBRAPA-CNPGC.

Hoshino, T., and G. Davidse. 1988. Chromosome numbers in grasses (Poaceae) from Southern Africa, I. *Annals of Mississippi Botanical Garden* 75:866–873.

Hsu, C. 1972. Preliminary chromosome studies on the vascular plants of Taiwan, V. Cytotaxonomy on some monocotyledons. *Taiwania* 17:48–53.

Hunter, W. S. 1934. A karyosystematic investigation in the Gramineae. *Canadian Journal of Research* 11:213–241.

Ibrahim, K. M. 1984. *Forage germplasm.* National Agriculture Research Station, Kitale, Kenya. 365 pp.

Jungmann, L., P. M. Francisco, A. C. B. Sousa, J. Paiva, C. B. do Valle, and A. P. Souza. 2005. Development of microsatellites for *Brachiaria brizantha* and germplasm diversity characterization of this tropical grass. In *Molecular breeding for the genetic improvement of forage crops and turf,* p. 19. Wageningen, the Netherlands: Wageningen Academic Publishers.

Jungmann, L., C. B. do Valle, P. R. Laborda, R. M. S. Resende, L. Jank, and A. P. Souza. 2007. Construction of microsatellite-enriched libraries for tropical forage species and characterization of the repetitive sequences found in *Brachiaria brizantha.* In *Molecular breeding for the genetic improvement of forage crops and turf,* ed. M. O. Humphreys, 128. Wageningen, the Netherlands: Wageningen Academic Publishers.

Junqueira Filho, R. G., A. B. Mendes-Bonato, M. S. Pagliarini, N. C. P. Bione, C. B. do Valle, and M. I. O. Penteado. 2003. Absence of microspore polarity, symmetric divisions and pollen cell fate in *Brachiaria decumbens* (Gramineae). *Genome* 46:83–86.

Kammacher, P., G. Anoma, E. Adjanohoune, and A. L. Ake. 1973. Nombres chromosomiques de graminees de Cote-d'Ivoire. *Candollea* 28:191–217.

Keller-Grein, G., B. L. Maass, and J. Hanson. 1996. Natural variation in *Brachiaria* and existing germoplasma collections. In *Brachiaria: Biology, agronomy and improvement,* ed. J. W. Miles, B. L. Maass, and C. B. do Valle, 16–42. Cali, Colombia: CIAT/Brasília: EMBRAPA-CNPGC.

Lapointe, S. L., and J. W. Miles. 1992. Germplasm case study: *Brachiaria* species. In *Pastures for the tropical lowlands: CIAT's contribution,* 43–55. Cali, Colombia: CIAT.

Larsen, K. 1963. Studies in the flora of Thailand, 14. Cytological studies in vascular plants of Thailand. *Dansk Botanical Arkive* 20:211–275.

Lascano, C. E., and V. P. B. Euclides. 1996. Nutritional quality and animal production of *Brachiaria* pastures. In *Brachiaria: Biology, agronomy, and improvement,* ed. J. W. Miles, B. L. Maass, and C. B. do Valle, 106–123. Cali, Colombia: CIAT/Brasília: EMBRAPA-CNPGC.

Lempp, B., C. B. do Valle, M. das G. Morais, R. A. Borges, and E. Detmann. 2005. Physical impediment towards digestive breakdown in leaf blades of *Brachiaria brizantha.* In *Proceedings of the International Grassland Congress* 20:233, Ireland. Wageningen, the Netherlands: Wageningen Academic Publishers.

Lentini, Z., V. T. C. Carneiro, S. J. L. Manzano, L. Galindo. 1999. Processo de Regeneração de Plantas e Transformação Genética de espécies de *Brachiaria*. Brasil: CIAT-Colômbia, Embrapa-Brasil. Patent: PI 9903700-9.

Letteriello, G., C. B. do Valle, D. Christiane, and M. I. O. Penteado. 1999. Citologia e modo de reprodução de acessos pentaplóides de *B. brizantha.* In *Proceedings of the Sociedade Brasileira de Zootecnia*, Brazil, 36, SBZ140:77.

Lutts, S., J. Ndikumana, and B.-P. Louant. 1991. Fertility of *Brachiaria ruziziensis* in interspecific crosses with *Brachiaria decumbens* and *Brachiaria brizantha*: Meiotic behavior, pollen viability and seed set. *Euphytica* 57:267–274.

———. 1994. Male and female sporogenesis and gametogenesis in apomitic *Brachiaria brizantha, B. decumbens* and F1 hybrids with sexual colchicines induced tetraploid *B. ruziziensis. Euphytica* 78:19–25.

Maass, B. L. 1996. Identifying and naming *Brachiaria* species. In *Brachiaria: Biology, agronomy, and improvement,* ed. J. W. Miles, B. L. Maass, and C. B. do Valle, x–xii. Cali, Colombia: CIAT/Brasília: EMBRAPA-CNPGC.

Macedo, M. C. M., J. L. Machado, and C. B. do Valle. 2004. Resposta de cultivares e acessos promissores de *Brachiaria brizantha* ao fósforo em dois níveis de saturação por bases. In *Proceedings of the Sociedade Brasileira de Zootecnia*, 41:339, Brazil, SBZ-Unipress Disc Records. FORR 339.

Mehra, P. N. 1982. *Cytology of East Indian grasses.* Chandigarh, India: P. N. Mehra. 115 pp.

Mehra, P. N., and J. Chowdhury. 1976. In IOPB chromosome number reports LIV. *Taxon* 25:631–649.

Mehra, P. N., and M. L. Sharma. 1975. Cytological studies in some Central and Eastern Himalayan grasses. II, Paniceae. *Cytologia* 40:75–89.

Mehra, P. N., and O. P. Sood. 1974. In IOPB chromosome number reports XLVI. *Taxon* 23:801–812.

Mendes, D. V., K. R. Boldrini, A. B. Mendes-Bonato, M. S. Pagliarini, and C. B. do Valle. 2006. Cytological evidence of natural hybridization in *Brachiaria brizantha* Stapf (Gramineae). *Botanical Journal of the Linnean Society* 150:441–446.

Mendes-Bonato, A. B. 2000. Caracterização citogenética de acessos de *Brachiaria brizantha* (Gramineae). MSc thesis, State University of Maringá, Maringá, Brazil. 59 pp.

Mendes-Bonato, A. B., R. G. Junqueira Filho, M. S. Pagliarini, C. B. do Valle, and M. I. O. Penteado. 2002a. Unusual cytological patterns of microsporogenesis in *Brachiaria decumbens*: Abnormalities in spindle and defective cytokinesis causing precious cellularization. *Cell Biology International* 26:641–646.

Mendes-Bonato, A. B., M. S. Pagliarini, F. Forli, C. B. do Valle, and M. I. O. Penteado. 2002. Chromosome numbers and microsporogenesis in *Brachiaria brizantha* (Gramineae). *Euphytica* 125:419–424.

Mendes-Bonato, A. B., M. S. Pagliarini, N. Silva, and C. B. do Valle. 2001a. Meiotic instability in invader plants of signal grass *Brachiaria decumbens* Stapf (Gramineae). *Acta Scientifica* 23:619–625.

Mendes-Bonato, A. B., M. S. Pagliarini, and C. B. do Valle. 2006a. Abnormal spindle orientation during microsporogenesis in an interspecific hybrid of *Brachiaria* (Gramineae). *Genetics and Molecular Biology* 29 (1): 122–125.

Mendes-Bonato, A. B., M. S. Pagliarini, C. B. do Valle, and L. Jank. 2004. Abnormal pollen mitoses (PM I and PM II) in an interspecific hybrid of *Brachiaria ruziziensis* and *Brachiaria decumbens* (Gramineae). *Journal of Genetics* 83:279–283.

Mendes-Bonato, A. B., M. S. Pagliarini, C. B. do Valle, and M. I. O. Penteado. 2001b. A severe case of chromosome stickiness in pollen mother cells of *Brachiaria brizantha* (Hochst) Stapf (Gramineae). *Cytologia* 66:287–291.

———. 2001c. Archesporial syncytes restricted to male flowers in a hexaploid accession of *Brachiaria brizantha* (Hochst) Stapf (Gramineae). *Nucleus* 44:137–140.

Mendes-Bonato, A. B., M. S. Pagliarini, C. B. do Valle, and C. Risso-Pascotto. 2003. Normal microspore production after cell fusion in *Brachiaria jubata* (Gramineae). *Genetics and Molecular Biology* 26:517–520.

Mendes-Bonato, A. B., C. Risso-Pascotto, M. S. Pagliarini, and C. B. do Valle. 2006b. Cytogenetic evidence for genome elimination during microsporogenesis in interspecific hybrid between *Brachiaria ruziziensis* and *B. brizantha* (Poaceae). *Genetics and Molecular Biology* 29:711–714.

Miles, J. W., C. Cardona, and G. Sotelo. 2006. Recurrent selection in a synthetic Brachiariagrass population improves resistance to three spittlebug species. *Crop Science* 46:1088–1093.

Miles, J. W., and S. L. Lapointe. 1992. Regional germplasm evaluation: A portfolio of germplasm options for the major ecosystems of tropical America. In *Pastures for the tropical lowlands: CIAT's Contribution,* 9–28. Cali, Colombia: CIAT.

Miles, J. W., S. L. Lapointe, M. L. Escandón, and G. Sotelo. 1995. Inheritance of spittlebug resistance in interspecific *Brachiaria* spp. hybrids: Parent–progeny correlation and heritability of field reaction. *Journal of Econic Entomology* 88 (5): 1477–1481.

Miles, J. W., and C. B. do Valle. 1996. Manipulation of apomixis in *Brachiaria* breeding. In *Brachiaria: Biology, agronomy, and improvement,* ed. J. W. Miles, B. L. Maass, and C. B. do Valle, 164–177. Cali, Colombia: CIAT/Brasília: EMBRAPA-CNPGC.

Miles, J. W., C. B. do Valle, I. M. Rao, and V. P. B. Euclides. 2004. Brachiariagrasses. In *Warm-season (C4) grasses,* ed. L. E. Sollenberger, L. Moser, and B. Burson B., 745–783. Agronomy Monograph 45. Madison, WI: ASA-CSSA-SSSA.

Miles, J. W., B. L. Maass, and C. B. do Valle, eds. 1996. *Brachiaria: Biology, agronomy, and improvement.* Cali, Colombia: CIAT/Embrapa, CIAT Publication No. 259, 288 pp.

Moffett, H., and R. Hurcombe. 1949. Chromosome numbers of South African grasses. *Heredity* 3:369–373.

Morrone, O., and F. O. Zoluaga. 1992. Revisíon de las especies sudamericanas nativas e introducidas de los géneros *Brachiaria* e *Urochloa* (Poaceae: Panicoideae: Paniceae). *Darwiniana* 1–4:43–109.

Mulay, B. N., and P. J. Leelamma. 1956. Chromosome number of some desert grasses. *Proceedings of Rajasthan Academy of Science* 6:65–69.

Muniyamma, M. 1976. In IOPB chromosome number reports LIV. *Taxon* 25:155–164.

Muniyamma, M., and K. N. Narayan. 1975. In IOPB chromosome number reports XLVIII. *Taxon* 24:367–372.

Nagabushana, R. S. 1980. In IOPB chromosome number reports LXVIII, *Taxon* 29:545–546.

Narayan, K. N., and M. Muniyamma. 1972. In IOPB chromosome number reports XXXVIII. *Taxon* 12:619–624.

Nassar, N. M. A. 1977. A cytogenetical study on some grasses cultivated in central Brazil. *Ciência e Cultura* 29:9–14.

Nath, J., and M. S. Swaminathan. 1957. Chromosome numbers of some grasses. *Indian Journal of Genetics and Plant Breeding* 17:102.

Nath, J., M. S. Swaminathan, and K. Mehra. 1970. Cytological studies in the tribe Paniceae, Gramineae. *Cytologia* 35:111–131.

Ndikumana, J. 1985. Etude de l'hybridation entre espèces apomictiques et sexuées dans le genre *Brachiaria*. PhD dissertation, Université Catholique de Louvain, Belgium. 210 pp.

Ndikumana, J., and P. N. Leeuw. 1996. Regional experience with *Brachiaria*: Sub-Saharan Africa. In *Brachiaria: Biology, agronomy, and improvement,* ed. J. W. Miles, B. L. Maass, and C. B. do Valle, 247–257. Cali, Colombia: CIAT/Brasília: EMBRAPA-CNPGC.

Ngendahayo, M. 1988. Mechanisms de la réproduction dans le genre *Brachiaria*. PhD dissertation, Université Catholique de Louvain, Belgium. 165 pp.

Nuñez, O. 1952. Investigación carisistematica en las gramíneas argentinas de la tribus Paniceae. *Revista de la Facultad de Agronomia de La Plata* 28:229–255.

Olorode, O. 1975. Additional chromosome count in Nigerian grasses. *Brittonia* 27:63–68.

Oram, R. N. 1990. *Brachiaria*. In *Register of Australian herbage plant cultivars*, 3rd ed., 89–92. Australian Herbage Plant Registration Authority. Div. Plant Industry. CSIRO, East Melbourne. VIC, Australia.

Parodi, L. R. 1946. Gramineas Bonariensis. *Chave para la determinación de los géneros y enumeración de las especies*. Buenos Aires: Ame Agency, 1–112.

Penteado, M. I. O., A. C. M. Santos, I. F. Rodrigues, C. B. do Valle, M. A. C. Seixas, and A. Esteves. 2000. Determinação de ploidia e quantidade de DNA em diferentes espécies do gênero *Brachiaria*. *Boletim de Pesquisa*, 11, Campo Grande: Embrapa Gado de Corte, 32 pp.

Pereira, J. M., C. de P. Rezende, V. P. B. Euclides, C. B. do Valle, and A. M. F. Borges. 2004. Avaliação de novos acessos de *Brachiaria brizantha* no sul da Bahia. 1. Produção animal. In: REUNIÃO ANUAL DA SOCIEDADE BRASILEIRA DE ZOOTECNIA, 41. FORR 270, Campo Grande. Anais. CD-ROM. SBZ-Unipress Disc Records-V2-comunicação.

Pessino, S. C., C. Evans, J. P. A. Ortiz, I. Armstead, C. B. do Valle, and M. D. Hayward. 1998. A genetic map of the apospory region in *Brachiaria* hybrids: Identification of two markers closely associated with the trait. *Hereditas* 128:153–158.

Pessino, S. C., J. P. A. Ortiz, O. Leblanc, C. B. do Valle, C. Evans, and M. D. Hayward. 1997. Identification of maize linkage group related to apomixis in *Brachiaria*. *Theoretical and Applied Genetics* 94:439–444.

Pinheiro, A. A., M. T. Pozzobon, C. B. do Valle, M. I. O. Penteado, and V. T. C. Carneiro. 2000. Duplication of the chromosome number of diploid *Brachiaria brizantha* plants using colchicine. *Plant Cell Reports* 19:274–278.

Pizarro, A. E., C. B. do Valle, G. Keller-Grein, R. Schultze-Kraft, and A. H. Zimmer. 1996. Regional experience with *Brachiaria*: Tropical America—Savannas. In *Brachiaria: Biology, agronomy, and improvement,* ed. J. W. Miles, B. L. Maass, and C. B. do Valle, 225–246. Cali, Colombia: CIAT/Brasília: EMBRAPA-CNPGC.

Pritchard, A. J. 1967. Apomixis in *Brachiaria decumbens* Stapf. *Journal of Australian Agricultural Science* 33:264–265.

Ramalho, M. A. P. 2004. Genetic improvement and agribusiness in Brazil. *Crop Breeding and Applied Biotechnology* 4:127–134.

Rao, I. M., P. C. Kerridge, and M. C. M. Macedo.1996. Nutritional requirements of *Brachiaria* and adaptation to acid soils. In *Brachiaria: Biology, agronomy, and improvement,* ed. J. W. Miles, B. L. Maass and C. B. do Valle, 53–71. Cali, Colombia: CIAT/Brasília: EMBRAPA-CNPGC.

Reeder, J. R. 1967. Notes on Mexican grasses. VI. Miscellaneous chromosome numbers. *Bulletin of Torrey Botany Club* 94:1–17.

———. 1977. Chromosome numbers in western grasses. *American Journal of Botany* 64:102–110.

———. 1984. Chromosome number report LXXXII. *Taxon* 33:26–134.

Renvoize, S. A., W. D. Clayton, and C. H. S. Kabuye. 1996. Morphology, taxonomy, and natural distribution of *Brachiaria* (Trin.) Griseb. In *Brachiaria: Biology, agronomy and improvement,* ed. J. W. Miles, B. L. Maass, and C. B. do Valle, 1–15. Cali: CIAT/Brasília:EMBRAPA-CNPGC.

Risso-Pascotto, C., M. S. Pagliarini, and C. B. do Valle. 2002. Abnormal nucleolar cycle in microsporogenesis of *Brachiaria decumbens* (Gramineae). *Cytologia* 67:355–360.

——. 2003a. A mutation in the spindle checkpoint arresting meiosis II in *Brachiaria ruziziensis*. *Genome* 46:724–728.

——. 2005a. Multiple spindles and cellularization during microsporogenesis in an artificially induced tetraploid accessions of *Brachiaria ruziziensis* (Gramineae). *Plant Cell Reports* 23:522–527.

——. 2006a. Microsporogenesis in *Brachiaria dictyoneura* (Fig. & De Not.) Stapf (Poaceae: Paniceae). *Genetics and Molecular Research* 5:837–845.

——. 2006b. Meiotic behavior in interspecific hybrid between *Brachiaria brizantha* and *Brachiaria ruziziensis* (Gramineae). *Euphytica* 145:155–159.

——. 2006c. A new basic chromosome number for the genus *Brachiaria* (Poaceae: Panicoideae: Paniceae). *Genetic Resources and Crop Evolution* 53:7–10.

Risso-Pascotto, C., M. S. Pagliarini, C. B. do Valle, and L. Jank. 2004a. Asynchronous meiosis in an interspecific hybrid of *Brachiaria ruziziensis* and *B. brizantha*. *Plant Cell Reports* 23:304–310.

——. 2004b. Asynchronous meiotic rhythm as the cause of selective chromosome elimination in an interspecific *Brachiaria* hybrid. *Plant Cell Reports* 22:945–950.

——. 2005b. Symmetric pollen mitosis I (PM I) and suppression of pollen mitosis II (PM II) prevent pollen development in *Brachiaria jubata* (Gramineae). *Brazilian Journal of Medical and Biological Research* 38:1603–1608.

Risso-Pascotto, C., M. S. Pagliarini, C. B. do Valle, and A. B. Mendes-Bonato. 2003b. Chromosome number and microsporogenesis in a pentaploid accession of *Brachiaria brizantha* (Gramineae). *Plant Breeding* 122:136–140.

Rodrigues-Otubo, B., M. I. de O. Penteado, and C. B. do Valle. 2000. Embryo rescue of interspecific hybrids of *Brachiaria* spp. *Plant Cell, Tissue and Organ Culture* 61 (3): 175–182.

Salgado, L. R., L. Chiari, C. B. do Valle, G. O. C. Leguizamon, and J. V. R. Valle. 2005. Diversidade genética de acessos de *Brachiaria humidicola* utilizando a técnica de RAPD. In *Proceedings of Congress Bras,* 664. *Genetics* Brazil, Águas de Lindóia, SBG, Ribeirão Preto.

Sarkar, A. K., N. Datta, V. Chatterjee, and R. Datta. 1973. In IOPB chromosome number report XLII. *Taxon* 22:647–654.

Savidan, Y. H. 1983. Genetics and utilization of apomixis for the improvement of guineagrass (*Panicum maximum* Jacq.). In *Proceedings of the XIV International Grassland Congress,* ed. J. A. Smith and V. W. Hays, 182–184. Lexington, KY.

Schank, S. C., and A. Sotomayor-Ríos. 1968. Cytological studies on *Brachiaria* species. *Soil Crop Science Society of Florida Proceedings* 28:156–162.

Schenk, M. A. M., S. G. Nunes, and J. M. Silva. 1991. Ocorrência de fotossensibilização em eqüinos mantidos em pastagem de *Brachiaria humidicola*. Comunicado Técnico, Embrapa-CNPGC, Campo Grande, MS. 3 pp.

Sharma, A. K., and L. Jhuri.1959. Chromosome analysis of grasses I. *Genetica Iberica* 11:145–173.

Sharma, A. K., and S. Kour. 1980. In chromosome number reports LXIX. *Taxon* 29: 06.

Sharma, M. L., and A. K. Sharma. 1979. Cytological studies in the North Indian grasses. *Cytologia* 44:861–872.

Singh, D. N. 1965. Supernumerary chromosomes in grasses. *Caryologia* 18:547–553.

Singh, D. N., and M. B. E. Godward. 1960. Cytological studies in Gramineae, II. *Heredity* 18:538–540.

Sinha, R. R. P., A. K. Bhardwaj, and R. K. Singh. 1990. SOCGI plant chromosome number reports, IX. *Cytology and Genetics* 25:140–143.

Sinha, R. R. P., and R. P. Jha. 1972. Cytological studies in some grasses of Bihar. *Proceedings of the Indian Science Congress Association* 59:352–353.

Sotomayor-Ríos, A., S. C. Schank, and R. Woodbury. 1970. Cytology and taxonomic description of two *Brachiaria* (Congograss and Tannergrass). *Journal of Agriculture of the University of Puerto Rico* 54:390–400.

Sotomayor-Ríos, A., J. Velez-Fortuno, R. Woodbury, K. F. Schertz, and A. Sierra-Bracero. 1960. Description and cytology of a form of Signalgrass (*Brachiaria brizantha* Stapf.) and its agronomic behavior compared to Guineagrass (*Panicum maximum* Jack.). *Journal of Agriculture of the University of Puerto Rico* 44:208–220.

Souto, S. M. 1978. Híbrido de *Brachiaria* ou Tangola, Niterói, Comunicado Técnico No. 6, Empresa de Pesquisa Agropecuária do Estado do Rio de Janeiro (PESAGRO), Niterói, RJ, Brazil. 2 pp.

Spies, J. J., and H. Du Plessis. 1986. Chromosome studies on African plants, I. *Bothalia* 16:87–88.

———. 1987. Chromosome studies on African plants, III. *Bothalia* 17:131–135.

Spies, J. J., E. van der Merwe, H. Du Plessis, and E. J. L. Saayman. 1991. Basic chromosome numbers and polyploid levels in some South African and Australian grasses (Poaceae). *Bothalia* 21:163–170.

Stebbins, G. L. 1956. Cytogenetics and evolution of the grass family. *American Journal of Botany* 43:891–905.

———. 1985. Polyploidy, hybridization, and the invasion of new habitats. *Annals of Mississippi Botanical Garden* 72:824–832.

Stür, W. W., J. M. Hopkinson, and C. P. Chen. 1996. Regional experience with *Brachiaria*: Asia, the South Pacific, and Australia. In *Brachiaria: Biology, agronomy, and improvement,* ed. J. W. Miles, B. L. Maass, and C. B. do Valle, 258–271. Cali, Colombia: CIAT/Brasília: EMBRAPA-CNPGC.

Suárez, M. M. 1994. Relaciones filogenéticas entre seis especies del género *Brachiaria* utilizando marcadores moleculares RAPD. BS thesis, Departamento de Biología, Facultad de Ciencias Básicas, Pontificia Universidad Javeriana, Santafé de Bogotá, DC, Colombia.

Subbarao, G. V., M. Rondon, O. Ito, T. Ishikawa, I. M. Rao, K. Nakahara, C. Lascano, and W. L. Berry. 2007. Biological nitrification inhibition (BNI)—Is it a widespread phenomenon? *Plant Soil* 294 (1/2): 5–18.

Swenne, A., B.-P. Louant, and M. Dujardin. 1981. Induction par la colchicines de formes autotetraploïdes chez *Brachiaria ruziziensis* Germain et Evrard (Gramineé). *L'Agronomie Tropicale* 36:134–141.

Takeota, T. 1965. Chromosome numbers of some East African grasses. *American Journal of Botany* 52:864–869.

Tohme, J., S. Palacios, S. Lenis, and W. Roca. 1996. Applications of biotechnology to *Brachiaria*. In *Brachiaria: Biology, agronomy, and improvement,* ed. J. W. Miles, B. L. Maass, and C. B. do Valle, 196–204. Cali, Colombia: CIAT/Brasília: EMBRAPA-CNPGC.

Torres-Gonzalez, A. M., and C. M. Morton. 2005. Molecular and morphological phylogenetic analysis of *Brachiaria* and *Urochloa* (Poaceae). *Molecular Phylogenetics and Evolution* 37:36–44.

Trouin, M. 1972. Nombres chromosomiques de quelques gramineae du Soudan. *Adonsia* 12:619–624.

Utsunomiya, K. S., M. S. Pagliarini, and C. B. do Valle. 2004. Chromosome transfer among meiocytes in *Brachiaria nigropedata* (Ficalho & Hiern) Stapf (gramineae). *Cytologia* 69:395–398.

———. 2005. Microsporogenesis in tetraploid accessions of *Brachiaria nigropedata* (Ficalho & Hiern) stapf (Gramineae). *Biocell* 29:295–301.

Valério, J. R., S. L. Lapointe, S. Kelemu, C. D. Fernandes, and F. J. Morales. 1996. Pests and diseases of *Brachiaria* species. In *Brachiaria: Biology, agronomy, and improvement,* ed. J. W. Miles, B. L. Maass, and C. B. do Valle, 87–105. Cali, Colombia: CIAT/Brasília:EMBRAPA-CNPGC.

Valle, C. B. do. 1986. Cytology, mode of reproduction, and forage quality of selected species of *Brachiaria* Griseb. PhD dissertation, Illinois University of Illinois, Urbana, 90 pp.

———. 1990. Coleção de germoplasma de espécies de *Brachiaria* no CIAT: Estudos básicos visando ao melhoramento genético. CNPGC documentos, no. 46, CNPGC/Embrapa, Campo Grande. 33 pp.

Valle, C. B. do. 2001. Genetic resources for tropical areas: achievements and perspectives. In *Proceedings of the International Grassland Congress,* 19, Brazil: FEALQ.v.1, p. 477–482.

Valle, C. B. do, C. Simioni, L. Jank, R. M. S. Resende. 2008a. Polyploidization of sexual diploid *B. decumbens* for intraspecific hybridization. In *Proceedings of the International Grassland Congress,* 21, China: Guangdong People's Publishing House, vol. 2, p. 395.

Valle, C. B. do, L. Chiari, G. A. bitencourt, L. R. Salgado, G. O. C. Leguizamon. 2008c. Use of RAPD markers to identify intraspecific hybrids of *Brachiaria humidicola*. In *Proceedings of the International Grassland Congress,* 21, China: Guangdong People's Publishing House, vol. 2, p. 417.

Valle, C. B. do, G. A. Bitencourt, A. Q. Arce, L. Chiari, L. Jank, R. M. S. Resende, and S. Calixto. 2008b. Preliminary evaluation of *Brachiaria humidicola* hybrids. In *Proceedings of the International Grassland Congress,* 21, China: Guangdong People's Publishing House, vol. 2, p. 418.

Valle, C. B. do, S. Calixto, and M. C. Amézquita. 1993. Agronomic evaluation of *Brachiaria* germplasm in Brazil. In *Proceedings of the International Grassland Congress,* 17: 511–512. Palmerston North: New Zealand: NZGA, TGSA, NZSAP, ASAP-Qld, and NZIAS.

Valle, C. B. do, V. P. B. Euclides, M. C. M. Macedo, J. R. Valério, and S. Calixto. 2001. Selecting new *Brachiaria* for Brazilian pastures. *Proceedings of the International Grassland Congress*, 19, Brazil: FEALQ. ID#13–14.

Valle, C. B. do, V. P. B. Euclides, J. M. Pereira, J. R. Valério, M. S. Pagliarini, M. C. M. Macedo, G. G. Leite, et al. 2004. O capim-xaraés (*Brachiaria brizantha* cv. Xaraés) na diversificação das pastagens de braquiária. *Documentos,* 149: Campo Grande:Embrapa Gado de Corte, 36 p.

Valle, C. B. do, and C. Glienke. 1991. New sexual accessions in *Brachiaria*. *Apomixis Newsletter* 3:11–13.

Valle, C. B. do, C. Glienke, and G. O. C. Leguizamon. 1993. Breeding of apomictic *Brachiaria* through interspecific hybridization. In *Proceedings of the International Grassland Congress*, 17: 427–428. Palmerston North, New Zealand: NZGA, TGSA, NZSAP, ASAP-Qld., and NZIAS, vol. 1.

———. 1994. Inheritance of apomixis *Brachiaria*, a tropical forage grass. *Apomixis Newsletter* 7:42–43.

Valle, C. B. do, G. O. C. Leguizamon, and N. R. Guedes. 1991. Interspecific hybrids of *Brachiaria* (Gramineae). *Apomixis Newsletter* 3:10–11.

Valle, C. B. do, B. L. Maass, C. B. Almeida, and J. C. G. Costa. 1993. Morphological characterization of *Brachiaria* germplasm. In *Proceedings of the International Grassland Congress*, 17: 208–209. Palmerston North, New Zealand: NZGA, TGSA, NZSAP, ASAP-Qld., and NZIAS, vol. 1.

Valle, C. B. do, and J. W. Miles. 1992. Breeding of apomictic species. *Apomixis Newsletter* 5:37–47.

———. 1994. Melhoramento de gramíneas do gênero *Brachiaria*. In Proceedings of the XI Simpósio sobre Manejo da Pastagem, ed. A. M. Peixoto, J. C. de Moura, and V. P. de Faria, 1–23. Piracicaba, FEALQ.

Valle, C. B. do, and Y. H. Savidan. 1996. Genetics, cytogenetics, and reproductive biology of *Brachiaria*. In *Brachiaria: Biology, agronomy, and improvement*, ed. J. W. Miles, B. L. Maass, and C. B. do Valle, 147–163. Cali: CIAT/Brasília:EMBRAPA-CNPGC.

Valle, C. B. do, Y. H. Savidan, and L. Jank. 1989. Apomixis and sexuality in *Brachiaria decumbens* Stapf. In *Proceedings of the International Grassland Congress* 19: 407–408. France: INRA.

Valle, C. B. do, R. J. Singh, and D. A. Miller. 1987. Pachytene chromosomes of *Brachiaria ruziziensis* Germain et Evrard. *Plant Breeding* 98:75–78.

Veldkamp. J. F. 1996. Proposal to conserve the name *Brachiaria* (Trin.) Griseb. (Gramineae) with a conserved type. *Taxon* 45:319–320.

Verzignassi, J. R., A. F. Furben, C. D. Fernandes, and C. B. do Valle. 2001. Ocorrência de *Ustilago operta* em sementes de *Brachiaria brizantha* no Brasil. In *Proc. Cong. Bras. Fitop., Fitopat. Bras.* (Suplemento), Brasil, Soc. Bras. Fitopat., 26:423.

Webster, R. D. 1987. *The Australian Paniceae (Poaceae)*, 228–255. Berlin and Stuttgart, Germany: J. Cramer.

Wenzl, P., A. Arango, A. L. Chaves, M. E. Buitrago, G. M. Patiño, J. Miles, and I. M. Rao. 2006. A greenhouse method to screen Brachiariagrass genotypes for aluminum resistance and root vigor. *Crop Science* 46:968–973.

Zerpa, D. M. de. 1952. Número cromosómico de *Brachiaria decumbens*. *Agronomía Tropical, Venezuela* 2:315.

CHAPTER 6

Birdsfoot Trefoil (*Lotus corniculatus* L.)

William F. Grant and Minoru Niizeki

CONTENTS

6.1 INTRODUCTION

6.1.1 The Genus

The genus *Lotus* contains a heterogeneous assemblage of annual and perennial species distributed widely throughout the world, with the exception of very cold arctic regions and lowland tropical areas of Southeast Asia and South and Central America. A total of 173 species are recognized and provided in a checklist (Kirkbride 1999). The USDA Germplasm Resources Information Network (GRIN) lists 332 species, which includes synonymy (GRIN 2007). A list of *Lotus* species and synonyms may also be found in the Multilingual Multiscript Plant Name Database (http://www.plantnames.unimelb.edu.au/Sorting/Lotus.html). The Food and Agriculture Organization of the United Nations has a "grasslands index" whereby pants may be searched under common names, genus, and Latin names (http://www.fao.org/ag/AGP/AGPC/doc/Gbase/mainmenu.htm).

Lotus species originate from two principal geographic centers of speciation (Meusel and Jager 1962). The most prolific region is centered on the Mediterranean, where the species range throughout Europe, southward around the Sahara desert, and eastward through temperate areas of Asia. Macaronesia (comprising Cape Verde, Canary Islands, Maderia, and the Azores) possesses 44 species, of which 75% are endemic and represent the largest number of *Lotus* species per unit area of land surface (Kirkbride 1999). In Morocco, there are 33 species. Two endemic species are found in Australia. A secondary center of diversity is located in western North America ranging from Mexico to British Columbia, eastward to Manitoba and Arkansas, and in southeastern United States. The largest number of species for a single country is found in the United States (39 species) followed by Mexico (29 species).

The taxa are extremely diverse in form and are adapted to a wide range of ecological habitats. These vary from salt-tolerant annual species that grow at sea level to ones adapted to xerophytic desert conditions and to others that grow at progressively higher elevations until they reach alpine conditions. For more long-term exploitation, *Lotus* species have potential for their use in horticulture as ornamentals, for honey production, for conservation landscaping, for ground covers to prevent soil erosion, for wildlife habitat, for reclamation, and for use under arid, drought, and saline conditions in certain geographic areas (Hardy BBT 1989; Beuselinck and Steiner 1994; Jewett et al. 1996; Rehm et al. 1998; Steiner 1999). Kirkbride, Gunn, and Dallwitz (2004) provide information on the identification of *Lotus* seed using computer technology.

This chapter extends information from past treatments (MacDonald 1946; Henson and Schoth 1962; Seaney and Henson 1970; Duke 1981; Undersander et al. 1993; Beuselinck and Grant 1995; Frame, Charlton, and Laidlaw 1998; Beuselinck 1999). Starting in 2003, the *Lotus Newsletter,* which serves as an informal means of worldwide communication and exchange of materials and research information between those engaged in research and development of *Lotus* species, has been edited by Mónica Rebuffo, INIA La Estanzuela, Uruguay (http://www.inia.org.uy/sitios/lnl/).

6.1.2 The Economic Species

Lotus corniculatus L. is a Eurasian tetraploid ($2n = 4x = 24$) perennial forage legume found in wild and naturalized populations throughout temperate regions of Europe, Asia Minor, North

Lotus corniculatus

Figure 6.1   (See color insert following page 274.) Stem of birdsfoot trefoil. Flowers possess a typical umbel with four to eight florets attached at the end of a relatively long peduncle. Style is attached to developing pod. One seed pod shows dehiscence. (From http://commons.wikimedia.org/wiki/Special:Newimages.)

Africa, North and South America, Australia, and New Zealand (Figure 6.1). The plant is commonly known as birdsfoot trefoil or bird's-foot trefoil. Names in 14 other languages are given at http://www. plantnames.unimelb.edu.au/Sorting/Lotus.html. Birdsfoot trefoil is widely distributed in Europe (Chrtkova-Zertova 1973) and has become extensively colonized in North America (Zandstra and Grant 1968; Table 6.1). The world distribution and references to the history of the species are given in GRIN (2007). In eastern North America, *L. corniculatus* has emerged from an introduced weed in the 1920s to become a successful secondary forage crop (Beuselinck and Grant 1995). It is widely grown in temperate climates in Europe and North and South America for pasture or for hay and silage production (Table 6.2).

Birdsfoot trefoil has distinct advantages for forage production on low-fertile, shallow, poorly drained, acidic and wet soils, and it is tolerant of drought as well as waterlogged conditions (Blumenthal and McGraw 1999; Ayres et al. 2006). On such sites, it outproduces alfalfa and is longer lived than alternative legume crops such as red and white clover. It is compatible with native and introduced grasses. It has high nutritive value because it adds protein to animal diets; has high palatability; does not cause bloat; has a modest requirement for water, calcium, and phosphorous; and has the ability to survive harsh weather conditions that typify northern regions (Bernes, Waller, and Christensson 2000). The plant supplies nitrogen to plants through symbiotic nitrogen fixation. Its chemical attributes are detailed in Duke (1981).

Birdsfoot trefoil is used extensively for soil improvement and on new highway slopes to prevent roadside erosion because it possesses good drought resistance. Knowledge of the agrobiology of this species as well as that of wild species in the genus has been extensively pursued for the introduction of potential desirable characters of economic importance required for the development of plants as new cultivars (Grant 1995, 1996, 1999; Nualsri, Beuselinck, and Steiner 1998; O'Donoughue and Grant 1988; Garcia de los Santos, Steiner, and Beuselinck 2001). However, *L. corniculatus* has a number of undesirable characteristics in need of improvement: its difficulty in establishment, slow regrowth, lack of seedling vigor, pod shattering (dehiscence) caused by indeterminate flowering, and

Table 6.1   Economic Species of *Lotus*[a]

Species[b]	Chromosome Number[c]	Genome Size/1C[d]
Lotus corniculatus L.	$2n = 2x = 24$	1.0 pg[e] (Bennett and Smith 1976)
Diploid forms (see text)	$2n = 2x = 12$	
L. tenuis Waldst. et Kit.	$2n = 2x = 12$	0.5 pg (Cheng and Grant 1973; Grime and Mowforth 1982)
		0.6 pg (Sz.-Borsos 1973)
L. japonicus (Regel) Larsen (*L. corniculatus* var. *japonicus*)	$2n = 2x = 12$	0.5 pg, cv. Gifu B-129 (Cheng and Grant 1973; Sz.-Borsos 1973; Bennett and Smith 1976)
		0.46 pg, 442.8 Mbp (Ito et al. 2000)
		cv. Miyakojima MG-20: 0.49 pg, 472.1 Mbp (Ito et al. 2000)
L. uliginosus Schkuhr (*L. pedunculatus*)	$2n = 2x = 12$ cv. Maku, $2n = 24$	0.55 pg (Grime and Mowforth 1982); 06 pg (Cheng and Grant, 1973; Sz.-Borsos, 1973)
		Genome size not determined
L. subbiflorus Lag. (*L. suaveolens*, *L. hispidus*)	$2n = 4x = 24$	Genome size not determined
L. tetragonolobus L. (*Tetragonolobus purpureus* Moench)	$2n = 2x = 14$	Genome size not determined

[a] For species distribution, see Table 6.2.
[b] See Kirkbride (1999) for a discussion on species nomenclature.
[c] See Grant (1995) for elaboration on chromosome numbers.
[d] Determined by Feulgen microdensitometry.
[e] 1 pg = 965 million base pairs (Mpb).

the presence of hydrocyanic acid and hydrogen cyanide (HCN) in its leaves and stems (Beuselinck and McGraw 1983; Hur and Nelson 1985; O'Donoughue and Grant 1988; Grant 1996; Gebrehiwot and Beuselinck 2001).

Modification of undesirable traits has been hampered by its breeding behavior because of the fact that it is largely an outcrossing species and because of the genetic nature of the species in which the characters are mainly inherited tetrasomically. Although the majority of researchers recognize *L. corniculatus* as a tetraploid species (Beuselinck and Grant 1995), some taxonomists consider that the species also exists in a diploid form (Grant 1995; Steiner 1999). Diploid rhizomatous plants (Beuselinck 1992, 1994) similar to tetraploid rhizomatous plants have been found in Morocco and may lend some credence to the existence of diploid forms.

Two other species, *L. tenuis* Waldst. et Kit. and *L. uliginosus* Schkuhr (*L. pedunculatus* Cav.), are grown for forage but to a lesser extent than birdsfoot trefoil (Beuselinck and Grant 1995). *Lotus tenuis* has fine stems and narrow leaves—hence, its name "narrowleaf birdsfoot trefoil." It is a diploid species (Table 6.1) grown in New York, California, Oregon, Washington, and Argentina. Barufaldi et al. (2007) refer to an induced tetraploid population as Leonel. Narrowleaf trefoil grows on wet soils where *L. corniculatus* is adapted, but it is not as winter hardy and has less seedling vigor (Beuselinck and McGraw 1983). Until 1987, *L. tenuis* was known as *L. tenuis*, at which time it was realized that the name *L. tenuis* took precedence historically over the name *L. tenuis* (Kirkbride 1999).

L. uliginosus is a nonbloating perennial legume with vigorous rhizomes, fine stems, and relatively large leaves—hence, its name "big trefoil." Plants increase their stand via rhizomes. It is a diploid species, with the exception of the tetraploid cultivar Grasslands Maku grown in Australia and New Zealand (Tables 6.1 and 6.2). Similar to *L. corniculatus*, big trefoil is adapted to poorly drained soils and wet sites, but it is not as winter hardy. The nutrient value of big trefoil compares favorably

Table 6.2   Countries with the Greatest Areas of *Lotus* under Cultivation and Commonly Grown Cultivars

Country	Total Area Sown in Hectares	Cultivars
	Lotus corniculatus	
Uruguay	1,100,000	La Estanzuela, San Gabriel
United States	1,000,000	AU Dewey, Dawn, Empire, Norcen, Viking
Argentina	500,000	El Boyero, San Gabriel
Austria	500,000	Leo, Rocco, Oberhaunstadter
Italy	500,000	Franco, Leo
Canada	200,000	Empire, Leo, Upstart, Bull
Germany	200,000	Cornelia, Hoki, Oberhaunstadter, Odenwäider, Rocco
France	180,000	Bonnie, Delpon, Hoki, Madison, Maitland, Oberhaunstadter, Odenwäider, Rodeo, Upstart
	Lotus tenuis	
United States	160,000	Los Banos, New York, Oregon[a]
Argentina	Not known[b]	Tresur Chajá[c,d], Matrero[d], Larrãnaga[d]
	Lotus uliginosus (*L. pedunculatus*)	
Australia	100,000	Grasslands Maku, Sharnae
New Zealand	60,000	Grasslands Maku
United States	28,000	Beaver, Marshfield
	Lotus subbiflorus	
Uruguay	50,000	El Rincón

Source: Blumenthal, M. J., and R. L. McGraw. 1999. In *Trefoil: The Science and Technology of Lotus*, ed. P. R. Beuselinck, 97–119. Madison, WI: American Society of Agronomy and Crop Science Society of America, CSSA Special Publ. No. 28.

[a] Duke, J. A. 1981. *Handbook of Legumes of World Economic Importance*, 125–131. New York: Plenum Press.
[b] Acreage not known. During 2002 and 2003, 230 tons of *L. tenuis* (90% uncertified seed) were grown. Personal communication from María Dubois (csbc@argenseeds.com.ar).
[c] Colares, M., M. M. Mujica, and C. P. Rumi. 2000. *Lotus Newsletter* 31.
[d] Argentina forage trials: http://www.argenseeds.com.ar/pdfs/resultados_03/lotus.pdf

with that of birdsfoot trefoil, but it has a higher level of tannin, which can affect palatability. Data on *L. uliginosus* as forage; on aluminum, pH, water tolerance, wildlife, conservation, and forestry; and in reclamation may be obtained through various links (http://www.bigtrefoil.com/links.html).

A third species, *L. japonicus,* is the target species for genome research to unveil the characteristics of a legume at the molecular level. Prof. Jens Stougaard and colleagues at Aarhus University, Denmark, developed *L. japonicus* suitable for molecular genetics (Handberg and Stougaard 1992). The germplasm (Gifu B-129) has been registered (Stougaard and Beuselinck 1996) and genetic nomenclature guidelines developed (Stougaard et al. 1999). Stougaard carried out the first gene tagging responsible for nodule organogenesis (Schauser et al. 1999), which is also the first gene tagging in a legume. A new accession—Miyaojima MG-20, an early flowering plant—is suitable for indoor handling (Kawaguchi 2000). Both accessions now contribute to the development of the infrastructure of the genetics and genomics of *L. japonicus* (Pedrosa et al. 2002; Akamine et al. 2003; Udvardi, Parniske, and Stougaard 2005; Sandal et al. 2006).

Márquez et al. (2005) have published a text covering all aspects of *L. japonicus* research. *Lotus japonicus* may be accessed also through the Legume Information System (Gonzales et al. 2005), a virtual interface that allows simplified and intuitive transcript comparisons integrating map, genomic, and transcript data from a number of sources via a single but multifaceted Web. The John Innes Center and Sainsbury Laboratory, England, maintain an *L. japonicus* Web site (http://www.lotusjaponicus.org/). Other *L. japonicus* Web sites may be found under "links" on the *Lotus Newsletter* (http://www.inia.org.uy/sitios/lnl). The most recent publications on *L. japonicus* may be found at the National Center for Biotechnology Information PubMed database.

A fourth species is *L. subbiflorus* Lag. A cultivar, El Rincón, is grown in Uruguay for its good growth on poor soil, persistence, productivity, and forage quality (Asuaga 1994; Tables 6.1 and 6.2).

A fifth species, *Lotus burttii* Borsos, which is of no immediate economic value, has been recently introduced as another model species of *Lotus* for molecular genetic studies (Kawaguchi et al. 2005).

6.2 MORPHOLOGY AND DESCRIPTION OF TETRAPLOID *LOTUS CORNICULATUS*

6.2.1 Plant, Stem, and Leaves

Mature plants of birdsfoot trefoil have many well-branched stems, glabrous or sparsely white-hairy, arising from a single crown. Stems are erect, ascending or decumbent, well branched, slender, and moderately leafy. Under favorable growing conditions, the main stem attains a length of 60–90 cm. Stems generally are smaller in diameter than those of alfalfa. Plants can be propagated by means of stem cuttings because roots develop from callous tissue of stem internodes and both roots and shoots develop from axillary buds of the node. There are five broad leaflets (pentafoliolate), with the central three held conspicuously above the others—hence, the use of the name *trefoil*. The lower two leaflets are stipule-like, obliquely ovate, mostly acuminate, 7–10 mm long, and 6–10 mm broad. The upper three leaflets are obovate, rounded at the apex, 10–45 mm long, and 6–10 mm broad. The leaflets of the upper leaves are lanceolate and acuminate. The leaves are compound and are attached alternately on opposite sides of the stem. During darkness, the leaflets close around the petiole and stem. When cut and dried for hay, the leaves fold around the stem and become incon-spicuous, often giving the impression of excessive leaf loss.

6.2.2 Flowers

The inflorescence is a typical umbel having four to eight florets attached in umbels by short pedi-cles at the end of a relatively long peduncle 5–10 cm in length. The calyx is 5–6 mm long, glabrous or finely hairy, with teeth as long as the tube. The flower color varies from light to dark yellow and may be tinged with orange or red stripes. The standard is 10–15 mm long and its broadly rounded limb abruptly passes into a cuneate claw. The wings are 10 mm long, approximately equaling the keel. The keel is incurved at a right angle and is tripped by downward pressure on its petals, exposing the stamens, style, and stigma. The keel color varies from yellow to dark brown and does not influence pollinator foraging behavior or colonization by flowering insects (Jones et al. 1986). The stigma is of the wet type, and the surface is already wet before anthesis and before tripping occurs.

The ontogeny of the inflorescence and flower has been studied by scanning electron microscopy (Prenner 2003). Pollen self-recognition occurs biochemically as a result of protein differences in the ovary prior to fertilization. This suggests that self-incompatibility, a prefertilization event, may be in part responsible for the reduced seed yield of self-pollinated plants (Dobrofsky and Grant 1980). Insects, usually *Hymenoptera,* are required to fertilize flowers. Seed yields are greater when an abundance of insects is present. Significant genetic variation has been found in populations among flowering time and flowering intensity (Kelman and Ayres 2002). Selection for prolific flowering has been associated with an increase in the number of umbels per stem (Kelman and Ayres 2004).

6.2.3 Seed Pods and Seeds

The flowers develop into small pea-like pods or legumes. Seed pods form at right angles at the end of the peduncle in the shape of a bird's foot—hence, the common name *birdsfoot trefoil* or *bird's foot trefoil*. Seed pods are long, cylindrical, brown to almost black, and about 2.5 cm long and 3 mm wide, with 15–20 seeds; seeds are olive-green, brown to almost black, or mottled with black spots; 1.4 mm long; 1.2 mm wide; oval to spherical; and attached to the ventral suture. At maturity, the pods split along both sutures and twist spirally to release seeds. There are about 713–1,210 seeds per gram (varying from 1.02 to 1.41 g/1,000 seeds; Grant 2002). Bullard and Crawford (1996) found that 100-seed weights of 18 different birdsfoot trefoil accessions varied twofold (between 116.2 mg and 239.7 mg). No consistent relationship has been found between seed size and viability or in speed of germination, but seed size is correlated with seedling length and seedling weight (McKersie and Tomes 1982; MacLean and Novak 1997). Good-quality seed is dark brown and shiny. A high proportion of green or dark green or shrunken seed indicates low germination.

Hard seeds are impermeable to water and will not germinate until later in the summer or until the following spring. Hard seeds normally make up 10–30% of the total seed but can reach 50% or more (Hampton et al. 1987). Li and Hill (1989a) found that a seed dormancy of 40–50% occurred when seeds were germinated immediately after harvest. However, this primary dormancy disappeared after drying the seeds under ambient conditions (fewer than 40 days). Hard seeds occur at 30 days after flowering and reach a peak of 98% at the final harvest 42 days after flowering. Germination results on seeds from different maturity stages suggested that "hardseededness" was closely related to the degree of seed maturity (Li and Hill 1989a). Temperatures below 15°C delay germination, but large seeds will germinate at temperatures exceeding 30°C to a greater extent than small seeds will. The minimum acceptable seed germination for Canada Foundation, No. 1 seed, Registered No. 1 seed 85% (http://www1.agric.gov/ab.ca/$department/deptdocs.nsf/all/agdex131).

Seed yield can exceed 600 kg/ha; however, as a result of seed shattering, average yields are much lower (Winch 1976). Seed losses are minimal when crops are harvested approximately 35 days after maximum inflorescence numbers are obtained, coinciding with a change in pod color from dark brown to light tan (Fairey and Smith 1999). Artola, Carrillo-Castañeda, and Garcia de los Santos (2003) have described a strategy to increase seed vigor through hydropriming, in which seeds are hydrated. They have also developed a "birdsfoot trefoil vigor and conductivity test," which is a germination test that is more sensitive and rapid than the standard germination test and has showed a high correlation with seedling emergence (Artola, Garcia de los Santos, and Carrillo-Castañeda 2003; Artola and Carrillo-Castaneda 2005). Nikolic et al. (2006) have found that cytokinins have a positive effect on *in vitro* seed germination.

Seedling vigor is positively correlated with seed size in *L. corniculatus*. Small seed size is considered responsible for poor seedling establishment (references in Grant 2002). The search for larger seed size is feasible; Bullard and Crawford (1996) found a Norwegian ecotype of birdsfoot trefoil to have heavier seed than any other of 18 accessions. Larger seeds produce greater cotyledon area, and the rate of cotyledon expansion is positively correlated with seedling vigor (Hur and Nelson 1995). Hence, a search for greater seed size may offer potential for improvement of seed size in commercial material by interaccessional crossing. Sources for obtaining seed are given in the appendix at the end of this chapter.

6.2.4 Roots

Birdsfoot trefoil has a well-developed, branching taproot, reaching a depth between 1 and 1.1 m, with numerous lateral secondary roots in the upper 30–60 cm of soil. The taproot does not penetrate as deeply as alfalfa, and the distribution of branch roots in the upper soil is more extensive. Roots have the ability to produce new shoots. Segments taken below the crown will develop shoots, and roots and may aid plant survival. Plants generally live for 2–4 years, making reseeding or production of new shoots necessary for persistence.

A form of birdsfoot trefoil that persists by producing rhizomes was found in Morocco (Beuselinck 1992). The rhizomes have potential to promote survival and stand persistence of plants under agricultural conditions (Beuselinck and Steiner 2003; Beuselinck, Peters, and McGraw 1984). The rhizome consists of nodes and internodes, with buds, scale leaves, and adventitious roots occurring at each node (Li and Beuselinck 1996). Introduction of genes for rhizome production into domesticated germplasm has been an important step toward improving survival and persistence. This trait has been transferred into the cultivars Norcen and AU Dewey (Pedersen, Haaland, and Hoveland 1986; Beuselinck, Li, and Steiner 1996). The first rhizomatous cultivar, Steadfast, was released in 1995 (Beuselinck and Steiner 1996). The rhizome trait was verified in the progeny of crosses by random amplified polymorphic DNA (RAPD) analyses (Nualsri et al. 1998). Kallenbach et al. (1998) have shown that strains of rhizobia isolated from Moroccan accessions can effectively nodulate North American cultivars. Clipping a birdsfoot trefoil population with rhizomes during two seasons did not decrease rhizome production (Kallenbach et al. 2001). Nualsri et al. (1998) successfully transferred rhizomatous genes to a birdsfoot trefoil germplasm that sexually reproduces autogamously; they determined that the presence of rhizomes is controlled by a single dominant gene, for which they proposed the symbol R (Nualsri and Beuselinck 1998).

In a study of genotype–environment interaction, ramets of rhizomatous and nonrhizomatous plants were grown at seven locations in the United States. Rhizomes appeared to be beneficial to plant survival and plant growth but did not ensure performance or survival (Beuselinck et al. 2005). Although there was a significant location effect for rhizome mass, there was no evidence to support latitude sensitivity. Variable mortality disclosed the interaction between genotypes and environment that typically goes unnoticed in rhizomatous and nonrhizomatous populations.

6.2.5 Flavonoids

Flavonoids are a large class of secondary plant metabolites of widespread occurrence in higher plants (Harborne and Baxter, 1999). There is a richness and diversity of flavonoid compounds in birdsfoot trefoil. The most common flavonoids are the aglycones found in leaves and flowers and the glycosides composed of monosides found in seeds, leaves, and flowers, and diosides found in seeds, leaves, and aerial parts. Flavonoids are necessary to induce the bacterial production of Nod factors, which are lipochitooligosaccharides that have the capacity to induce the transcription of genes that control the morphological changes observed in plant roots. Nod factors favor mature nodule development and preinfection thread formation (Geurts and Bisseling 2002). Total flavonoid content has been determined in birdsfoot trefoil to range from 50 to 65 mg/g. In contrast, only 15 mg/g is present in *Trifolium pratense* and 1 mg/g for *T. repens* (de Rijke et al. 2004). Sarelli et al. (2003) noted that birdsfoot trefoil contained insignificant amounts of phytoestrogens in contrast to red clover. Flavonoids have also been used to study speciation in the *L. corniculatus* complex and to distinguish diploid and tetraploid populations at different altitudes in the French Alps (Reynaud and Lussignol 2005).

6.3 DAY LENGTH

Photoperiod plays a major role in natural reseeding of pastures. Day length has long been known to influence flowering. A reduction in flowering occurs under a 16-h photoperiod, resulting in plants having a more prostrate, rosette growth habit. Joffe (1958) observed the appearance of normal flowers in 23 days with a photoperiod of either 16 or 18 h. With a photoperiod of 15 h, flowering did not appear for 40 days and, with a 14-h photoperiod, 51 days. Latitude greatly influences flowering in proportion to length of photoperiod (McKee 1963; Steiner 2002). Highest yields occur in northern states because the long photoperiods exceed the minimum required for profuse growth. In northern New South Wales, Australia, day length (~14 h) provides insufficient photoperiod to stimulate flowering and the level of seed-set needed for effective seedling recruitment. Ayres et al. (2007) selected germplasm that flowered strongly in this region so that birdsfoot trefoil might become a perennial legume option for low-day-length and low-latitude regions.

Flower buds may abort even under optimum photoperiods. Factors causing bud abortion or flower drop include adverse temperature, low nutrient level, insects, and diseases. Flower buds have been prolonged from dropping artificially in hybridization studies by spraying the flowers with a hormone (2,4,5 trichlorophenoxypropionic acid) (Grant, Bullen, and de Nettancourt 1962) or applying 1% naphthalene acetic acid in lanolin to the base of the heads (De Lautour, Jones, and Ross 1978).

A photoperiod-insensitive, rapid-flowering, genetic stock has been developed that is a source of genes capable of modifying photoperiod responses (Steiner and Beuselinck 2001; Steiner 2002). Plants cross readily and bidirectionally with other genotypes. This stock may be used in inheritance studies because of its autogamous nature. A greater number of reproductive cycles is obtained than with typical flowering genotypes. There are no other known birdsfoot trefoil genotypes that flower under a 10-h photoperiod. This early flowering genotype may be the result of a single amino acid substitution that influences photoperiodicity, as has been found in *Arabidopsis* (El-Assal et al. 2001).

6.4 CYTOLOGICAL OBSERVATIONS

Birdsfoot trefoil has 24 somatic chromosomes and 12 bivalents at metaphase I (Figure 6.2A, B). Occasionally, a quadrivalent may be observed (Figure 6.2B). The somatic chromosomes are relatively small, with a total complement length (TCL) of 57.50 μm (Cheng and Grant 1973). The chromosomes range in size from 3.86 μm (13.24% TCL) for the longest chromosome to 1.44 μm (5.42% TCL) for the shortest chromosome. The genome size has been determined as 1.0 pg (Table 6.1). The number and size of the chromosomes have discouraged researchers from initiating trisomic studies in birdsfoot trefoil. Trisomics have been produced in the related species *L. tenuis* and *L. uliginosus* (Chen and Grant 1968; St.-Marseille and Grant 1997).

6.4.1 The Cytogenetics of Tetraploid Birdsfoot Trefoil—Autopolyploid or Allopolyploid?

The origin of *L. corniculatus* has been a quandary for many generations because it has a high degree of polymorphism and a number of diploid taxa considered to be involved in its ancestry.

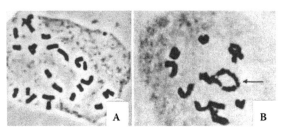

Figure 6.2 Microphotographs of somatic and meiotic chromosomes. A: Somatic chromosome complement of birdsfoot trefoil, 2*n* = 24. (Zandstra, I. I., and W. F. Grant. 1968. *Canadian Journal of Botany* 46:557–583. With permission of the *Canadian Journal of Botany*.) B: Meiotic chromosomes showing 10 bivalents and one quadrivalent indicated by arrow. (Wernsman, E. A., W. F. Keim, and R. L. Davis. 1964. *Crop Science* 4:483–486. With permission of the authors and *Crop Science*.)

Whether tetraploid birdsfoot trefoil is an auto- or allotetraploid is a question that is fundamental to interpreting its ancestry (Grant and Small 1996). Various techniques have been employed to resolve this question, including:

morphology;
cytological analyses;
nuclear DNA comparisons;
interspecific hybridization;
disomic versus tetrasomic inheritance;
comparison of plant secondary products;
seed globulin polypeptides;
chromatography;
isozyme analyses;
random amplified polymorphic DNA markers;
restriction fragment length polymorphic mapping and ribosomal genome DNA sequence data from
    internal transcribed spacer (ITS) sequences; and
phenetic evidence.

Although they are illuminating, the analyses have not led to definitive answers. In most cases, traits used to characterize the species are polymorphic; this has made analyses more difficult and led to conflicting results. Table 6.3 lists more than 50 studies involving a multiplicity of techniques.

The evidence has pointed most strongly to four species as progenitor species: *L. alpinus, L. tenuis, L. japonicus,* and *L. uliginosus.* At present, *L. tenuis* and *L. uliginosus* would appear to be the most likely progenitor species (Ross and Jones 1985; Grant and Small 1996; Steiner and Liston in Steiner 1999). *L. uliginosus* is considered one parent based on ITS analyses (Steiner and Liston in Steiner 1999), its rhizomatous growth habit, and its acyanogenic characteristic, which it shares with birdsfoot trefoil (Grant and Small 1996). From multiple evidence, *L. tenuis* has been suggested as the other parent. Because of the great amount of diversity found among accessions of *L. tenuis,* Steiner (1999) has suggested that *L. tenuis* per se was derived from multiple progenitors. Ribosomal genomic DNA sequence data from ITS sequences showed that the diploid taxa *L. tenuis, L. japonicus,* and *L. uliginosus* have been derived from common ancestors earlier than birdsfoot trefoil (Steiner and Liston in Steiner 1999). This fact has undoubtedly led to previous problems in trying to ascertain the origin of *L. corniculatus.*

6.4.2 Interspecific Hybridization

Some of the goals of interspecific hybridization have been to consider the best manner to transfer genes for superior seedling vigor, for developing a non-pod-shattering cultivar, for developing resistance to root rots, insects, and foliar diseases, development of the rhizomatous character, and male sterility. Grant (1999) has given details on studies utilizing interspecific hybridization within the *L. corniculatus* group. The most successful method for the transfer of genes from diploid species has been to produce interspecific diploid hybrids, double the chromosome number with colchicine, and cross the resulting amphidiploids directly to tetraploid birdsfoot trefoil (Grant 1999). Thus, diploid interspecific hybrids would be useful as a starting point for the improvement of birdsfoot trefoil as a forage species. From a study of 38 accessions identifying plants with low condensed tannins and high *in vitro* digestible dry matter, Kelman (2006b) considered interspecific hybridization a valuable avenue for improvement of agronomic and forage quality characters. A hybrid between *L. uliginosus* and *L. corniculatus* had the desired characters for which Kelman was looking. Beuselinck and McGraw (1983) suggested the feasibility of improving seedling vigor through interspecific crosses using *L. uliginosus, L. tenuis,* and *L. corniculatus.*

Table 6.3   Studies Favoring Auto- or Allotetraploidy

Study	Favoring Autotetraploidy (Au) or Allotetraploidy (Al)	Species Favored	Ref.
L. corniculatus could have arisen from ancestors outside the *L. corniculatus* group	Al		Brand 1898; Pankhurst and Jones 1979
Tetrasomic segregation for cyanide production	Au	*L. tenuis*	Dawson 1941
Induced autotetraploidy of *L. tenuis* and hybridization with *L. corniculatus*	Al		Tome and Johnson 1945
No pod set in crosses between *L. corniculatus* and *L. tenuis* (2n)	Al		Macdonald 1946
Crosses between *L. corniculatus* and *L. tenuis* (2n)	Al		Elliott 1946
Autotetraploid *L. tenuis* did not closely resemble *L. corniculatus*	Al		Guttman 1947
High percentage sterility in crosses between *L. corniculatus* and *L. tenuis* (2n)	Al		McKee 1949
Bivalent pairing and tetrasomic inheritance favoring a segmental allopolyploid	Al		Stebbins 1950
Crosses between *L. corniculatus* and *L. tenuis* (4x)	Al		Keim 1952
Morphological resemblance	Al	*L. alpinus, L. krylovii*	Favarger 1953
Morphological resemblance	Al	*L. alpinus*	Larsen 1954
Crosses between *L. corniculatus* and *L. tenuis* (4x)	Al		Erbe 1955
Crosses between *L. corniculatus* and *L. tenuis* (4x)	Al		Mears 1955
Tetrasomic inheritance for flower color	Au		Bubar 1956
Tetrasomic inheritance for brown keel tip color	Au		Hart and Wilsie 1959
Tetrasomic inheritance for dark green leaf color	Au		Poostchi 1959; Poostchi and MacDonald 1961
Chromosome pairing relationships in the hybrid *L. corniculatus* × *L. uliginosus* (4x)	Al		Gershon 1961a, 1961b
Tetrasomic inheritance for pubescence	Au		Hinckley 1961

Continued

Table 6.3   Studies Favoring Auto- or Allotetraploidy (*Continued*)

Study	Favoring Autotetraploidy (Au) or Allotetraploidy (Al)	Species Favored	Ref.
Induced autotetraploidy of *L. tenuis*, *L. uliginosus*, and *L. japonicus*; hybridization with *L. corniculatus*	Al	*L. japonicus*, *L. uliginosus*	Bent 1962
Interspecific hybrid analyses	Al	*L. alpinus*, *L. filicaulis*, and *L. japonicus*	Grant et al. 1962; Nettancourt and Grant 1963, 1964a, 1964b, 1964c
Crosses between *L. tenuis* (2x) and *L. corniculatus*	Al	*L. tenuis* not a parent	Jaranowski and Wojciechowska 1963
Tetrasomic inheritance for brown keel tip color	Au		Buzzell and Wilsie 1963
Tetrasomic inheritance for a nucleolar body	Au		Wernsman 1963, 1966
Hybrids between *L. corniculatus* and 4x glaber and quadrivalent formation	Au	*L. tenuis*	Wernsman, Keim, and Davis 1964; Wernsman, Davis, and Keim 1965
Tetrasomic inheritance for leaf size	Au		Donovan and McLennan 1964
Phenolic affinities	Al		Harney and Grant 1964a, 1964b, 1965
Tetrasomic inheritance for pubescence, a chlorophyll deficiency, cyanogenesis, lemon yellow flower color, streaks on the corolla, keel tip color, self-incompatibility	Au		Bubar and Miri 1965
Morphology and a chromatographic study of phenolic compounds showed *L. corniculatus* more closely related to *L. tenuis* than *L. uliginosus*	Al		Chrtkova-Zertova 1966; Zandstra and Grant 1968; Grant and Zandstra 1968
Morphology	Al	*L. alpinus, L. krylovii*	Chrtkova-Zertova 1970
Chromosome pairing relationships, crosses between *L. japonicus* (4x), and *L. alpinus* (4x) and *L. corniculatus*	Al	*L. japonicus* and *L. alpinus*	Somaroo and Grant 1971a, 1971b, 1971c, 1972a, 1972b
Tetrasomic inheritance for linamarase production in leaves	Au		Bansal 1971
Karyotype and cytophotometric analyses	Al	*L. alpinus, L. borbasii, L. krylovii*	Cheng and Grant 1973
Cytomorphology	Al	*L. alpinus*	Beuret 1977a, 1977b
Isozyme patterns between *L. uliginosus* (4x) × *L. tenuis* (4x); *L. tenuis* (4x) × *L. corniculatus*	Al		De Lautour et al. 1978
Virescent leaf, unifoliate leaf and chlorotic cotyledon, all segregated tetrasomically	Au		Therrien and Grant 1978

Table 6.3   Studies Favoring Auto- or Allotetraploidy (*Continued*)

Study	Favoring Autotetraploidy (Au) or Allotetraploidy (Al)	Species Favored	Ref.
Tetrasomic inheritance for cyanoglucoside production in leaves	Au		Ramnani 1979
Tetrasomic inheritance for brown keel tip color	Au		Ramnani and Jones 1984a, 1984b
A numerical morphological analysis of characters among *L. alpinus, L. tenuis,* and *L. corniculatus* showed the taxa varied clinally with altitude	Al		Small, Grant, and Crompton 1984
High tannin content is controlled by a single dominant gene with disomic inheritance	Al		Dalrymple, Goplen, and Howarth 1984
Tannin content, phenolic content, cyanide production, morphology, *Rhizobium* specificity, self-incompatibility; tetrasomic inheritance	Al/Au	*L. alpinus* and/or *L. tenuis* (as female parent) with *L. uliginosus* (as male parent)	Ross and Jones 1985
Interspecific hybridization and isoenzymes; duplication of the *Pgi2* loci	Al	*L. tenuis L. alpinus,* or *L. japonicus*	Raelson and Grant 1988; Raelson et al. 1989
Polyploidization process involving 2*n* gametes from *L. tenuis*	Au	*L. tenuis*	Negri and Veronesi 1989; Rim and Beuselinck 1991, 1992; Jay et al. 1991; Reynaud, Jay, and Blaise 1991
Disc electrophoresis of leaf extracts	Al		Grant and Altosaar 1994
Random amplified polymorphic DNA	Al	*L. uliginosus* more distantly related	Campos, Raelson, and Grant 1994
Phenetic analyses	Al	*L. tenuis, L. uliginosus*	Grant and Small 1996
Cyanogenesis controlled by a dominant gene with tetrasomic inheritance	Au		Nikolaichuk and Yalovs'ka 1996
Allozyme makers show tetrasomic inheritance for two loci	Au	*L. alpinus*	Gauthier, Lumaret, and Bedecarrats 1998a, 1998b
Ribosomal genome DNA sequence data from internal transcribed spacer sequences	Al	*L. uliginosus*	Steiner 1999
RFLP markers support tetrasomic inheritance	Au		Fjellstrom, Beuselinck, and Steiner 2001; Fjellstrom et al. 2003

Unreduced gametes from *L. tenuis* intercrossed with *L. corniculatus* offer a means of obtaining progeny exploiting heterotic advantage greater than from tetraploid *L. tenuis* per se (Beuselinck, Steiner, and Rim 2003).

6.4.3 Seed Pod Shattering

L. corniculatus possesses a pod that dehisces and ejects the seeds. No simply inherited nonshattering mutant of *L. corniculatus* has been discovered. The production of an indehiscent cultivar of birdsfoot trefoil would have a significant and positive impact on ruminant animal agriculture as well as for seed producers. The dehiscent pod problem causes high seed prices that discourage farmers from using this pasture species. Despite a number of decades in which attempts have been made to produce a nondehiscent plant, only very limited success has been experienced (Grant 1996). Birdsfoot trefoil seed yield can exceed 600 kg/ha, but as a result of seed shattering, average yields are much lower. Attempts to control seed shattering have been through management practices, such as regulating soil–water relationships (Garcia-Diaz and Steiner 1999, 2000a, 2000b), selection for reduced shattering, and interspecific hybridization of species both within and outside the *L. corniculatus* group (Grant 1996). Techniques for transferring the indehiscent trait into *L. corniculatus* have been discussed, but the present approaches present technical difficulties (Grant 1996).

The molecular approach for indehiscence in *L. corniculatus* has so far not been attempted, but this avenue should lead to an indehiscent plant because a wide array of molecular and genetic techniques is now available (Morris et al. 1999). Both *Agrobacterium tumefaciens* and *A. rhizogenes* can transform *L. corniculatus* (Morris et al. 1999; Webb, Robbins, and Mizen 1994a). These plant pathogens can be modified into vectors for the introduction of a desired gene sequence into the plant genome. It is most likely that indehiscent species such as *L. conimbricensis* can be used as a source of indehiscence genes, transformed, and the indehiscent character transferred to *L. corniculatus*. However, discoveries of SHATTERPROOF MADS-box genes in *Arabidopsis* that control the dehiscence zone and promote the lignification of adjacent cells (Liljegren et al. 2000) led to a new molecular approach. It has been shown that the FRUITFULL MADS-box gene, which is necessary for fruit valve differentiation, is sufficient to prevent formation of the dehiscence zone (Ferrandiz, Liljegren, and Yanofsky 2000). Thus, the MADS-box gene(s) that can control pod shatter would appear to be a major step in producing an indehiscent plant of *L. corniculatus*.

6.4.4 Genetic Linkage Map of *Lotus corniculatus*

Through the use of restriction fragment length polymorphism (RFLP), random amplication of polymorphic DNA (RAPD), inter-simple sequence repeat markers (ISSR), sequence tagged site (STS), and isozyme markers, Fjellstrom, Steiner, and Beuselinck (2003) have developed the first genetic linkage map for *L. corniculatus* with a genome cover of 572.1 cM for 139 loci. Six composite linkage groups were constructed from the combined data of four homologous linkage groups anchored by shared RFLP loci. This linkage map will help establish a foundation for marker-assisted selection for the improvement of desirable characteristics in birdsfoot trefoil. Maps constructed for alfalfa (*Medicago sativa*, $2n = 32$) range from 452 to 709 cM (Sledge, Ray, and Jiang 2005), which is within the range of that for *L. corniculatus*. A genome size for *L. japonicus* is given as 472.1 Mb (Ohmido et al. 2007); for red clover (*T. pratense*, $2n = 14$), it is 468 Mb (Arumuganathan and Earle 1991).

6.5 MUTAGENIC STUDIES

Physical and chemical mutagens have been applied in a range of dosages to birdsfoot trefoil seed to obtain desirable mutants. The mutagenic effects of x-rays (3–12 rad), ethyl methanesulfonate

(EMS), 8-ethoxycaffeine (EOC), *N*-hydroxyurea (HU), and 2-aminopurine (2AP) (concentrations of 0.05–13.2 m*M*) on treated seed (cv. Mirabel) were assessed over four generations (Therrien and Grant 1982). Segregating qualitative mutants were recovered from both selfed and crossed lines. The average mutation frequency in selfed progeny was 1.5 versus 0.8% in crossed progeny from material treated with x-rays and EMS. Material treated with EC, HU, and 2AP had an average mutation rate of 0.05%, compared with <0.001% in controls (Therrien and Grant 1979, 1983). *Vestigial floret* and *chlorotica* mutants were inherited as tetrasomic recessives. EMS was the most effective mutagen from which 0.4–3.1% mutants were obtained in a tetrasomic background (Therrien and Grant 1982).

A sessile inflorescence phenotype arose in progeny from crosses between different cultivars of birdsfoot trefoil (Cree, Leo, Upstart, Viking) (Grant, McDougall, and Coulman 1992). Seed was collected and plants were raised the following year. No plants exhibited the sessile inflorescence character. It is assumed that the plant arose through a physiological condition that was not inherited (Grant, McDougall, and Coulman 1993).

Beuselinck and McGraw (2000) determined the fertility of *L. corniculatus* genotypes expressing the floral mutant, vestigial corolla (*vc*), character. All *vc* genotypes produced some pollen that appeared normal, although the quantity of such pollen was variable. Pollen germination among *vc* genotypes was 22%, compared with 50% for the control. Only 5 of 16 *vc* genotypes produced pods when crossed to controls, and both pod and seed set were less than in control crosses. Reduced fertility of some *vc* genotypes may have resulted from smaller, incompletely developed ovules. Unidirectional reversion of the *vc* genotypes, from abnormal to normal flower morphology, was common. Using the *vc* mutant to eliminate the need for emasculation in hand-pollination programs depended on its fertility. The combination of phenotypic instability and reduced fertility made the *vc* mutant less desirable for use in breeding programs.

It is well known that plants regenerated from protoplasts may induce chromosomal mutations. Of 71 plants regenerated from protoplasts of birdsfoot trefoil cv. Viking, one was octoploid, one mixoploid, and the remainder tetraploid. Chromosome abnormalities such as univalents, lagging chromosomes, fragments, and bridges were seen in PMCs of regenerants (Niizeki 1993a, 1993b).

6.6 ESTABLISHMENT, CULTIVATION, AND UTILIZATION

Birdsfoot trefoil can survive fairly close mowing, grazing, and trampling. However, inoculation is necessary for successful stands in areas where trefoils have not been grown before. Bacteria used to inoculate alfalfa, red clover, and other common legumes are not effective on trefoils. Strains of *Rhizobium* and *Bradyrhizobium* bacteria specific for birdsfoot trefoil are required for effective nodulation. Unlike alfalfa nodules, those on birdsfoot trefoil senesce after each forage harvest, and a new population of nodules must be formed as plants regrow (Vance et al. 1982). A mutational and structural analysis of *R. loti* nodulation showed that, unlike the case with other *Rhizobium* strains, *nodB* is located on an operon separate from *nodACIJ* (Scott et al. 1996). Sequence analysis of the *nodACIJ* and *nodB* operon regions confirmed that *R. loti* common nod genes have a gene organization different from that of other *Rhizobium* spp. A new host-specific nod gene, *nolL*, was identified adjacent to *nodD3*. Mutational analysis showed that *nodD3, nodI, nodJ,* and *nolL* were all essential for *R. loti* strains to nodulate *L. uliginosus* effectively but were not necessary for effective nodulation of the less restrictive host, *L. corniculatus*.

Birdsfoot trefoil is grown commercially in more than 20 countries. The countries with the greatest areas under cultivation for *L. corniculatus, L. tenuis, L. uliginosus,* and *L. subbiflorus* are given in Table 6.2. Birdsfoot trefoil grows naturally in pastures and is used as feed for horses and cattle. Taylor (1984) has given detailed procedures on establishment and management of birdsfoot trefoil and compares cv. Fergus with other grasses. Recommended seeding rates for birdsfoot trefoil range from 5 to 12 kg/ha. In new plantings, seedbeds should be smooth and firm and seed not deeply buried because the seed is small and seedlings grow slowly. Maximum desired planting depth is usually less than 1.3 cm. Rehm et al. (1998) showed that several methods can be used to establish birdsfoot trefoil successfully on sandy soils from the use of a clean, prepared seedbed to no-till seeding. Cattle have the potential to act as a low-cost alternative for seed dissemination, but Doucette, Wittenberg, and McCaughey (2001) consider this a less efficient method for large-scale delivery of viable seed.

Birdsfoot trefoil is adapted to permanent, low-maintenance pastures and has a life span of 5 or 6 years in natural meadows and up to 8 or 9 years in sown pastures (Strelkov 1980). Brummer and Moore (2000) found that kura clover and white clover persisted better in the upper midwestern United States than alfalfa, birdsfoot trefoil, and red clover under continuous grazing by beef cattle. Declining plant density in pure stands can be attributed to competition, high temperature, grass competition, soil acidity, winter killing, and disease (Beuselinck et al. 1984; Hoveland 1994). Broadcast, band, and no-tillage seedlings are used successfully to establish birdsfoot trefoil in new plantings or to interseed into established grass pasture.

Cuomo et al. (2000) and Cuomo, Johnson, and Head (2001) evaluated the effect of sod suppression, interseeding, and planting method on establishment. Methods included glyphosate or no glyphosate treatment, no-till drilling, broadcasting seed, broadcasting seed followed by harrowing, and broadcasting seed followed by a light disking. No differences were detected for planting method. The overriding factor was the suppression of existing vegetation during establishment. Cuomo et al. (2001) found from interseeding kura clover and birdsfoot trefoil into existing cool-season grass pastures that initial emergence and growth were more important than winter survival in establishing legumes in grass sods. Zemenchik, Albrecht, and Shaver (2002) reported improved dry matter production of cool-season grass monocultures by the addition of birdsfoot trefoil. This had the potential to improve milk production of dairy cattle. Under cool summer growing conditions, McKenzie, Papadopoulos, and McRae (2004) have examined the effect of birdsfoot trefoil management on persistence, productivity, and species composition. They found harvesting at 10% bloom to be the best system based on dry matter yield, trefoil content, and stand density.

Robinson et al. (2007) assessed *L. corniculatus* and some native Australian herbaceous legumes for dry matter (DM) digestibility, crude protein (CP), and soluble condensed tannin (CT). Total shoot analyses for *L. corniculatus* for DM was 78%, crude protein 27%, and soluble condensed tannin 2.9 g/kg DM. The nutritive value (DM and CP) was similar to that measured previously (Cassida et al. 2000). The concentration of soluble CT (2.9 g/kg DM) was below that which would cause reduction in forage digestibility and palatability (>55 g CT/kg DM) (Min et al. 2003). The concentration of soluble CT in the shoot (3 g/kg DM) was below the value reported by Min et al. (1999) (17 g/kg DM) and by Kelman and Tanner (1990) (20.9 g/kg DM average). In populations of birdsfoot trefoil in Serbia, Bosnia, and Herzegovina, Vuckovic et al. (2007) determined crude protein content varied between 145.3 and 180.2 g/kg DM and crude fat content between 32.4 and 43.8 g/kg DM. High K concentrations were found that varied between 18.83 and 22.05 g/kg DM, which is above the NRC standard of 10.0 g K/kg DM. Values for P, Ca, Mg, Zn, Cu, Mn, Fe, and Se are given. In comparison with populations from other parts of the world, the authors considered Serbian populations to be superior in chemical properties and nutritional qualities.

In studying the emergence and survival of monoculture and mixtures of birdsfoot trefoil (12%), orchardgrass (18%), white clover (26%), and perennial ryegrass (43%), Skinner (2005) reported that though mixture complexity had significant effects on seedling emergence and mortality, dry matter production in the binary and complex mixtures could be predicted based on emergence, survival, and growth in monocultures. The initial advantage of perennial ryegrass was due to its high emergence rate compared with other species in the mixture. Thus, seedling emergence information gleaned from monocultures is considered a useful tool for predicting initial species composition of more complex mixtures.

Although *L. tenuis* spreads naturally in the grassland of the Flooding Pampa in Argentina, *L. corniculatus* has not spread. It is considered that competitive interaction with other grasses, diseases, and low flood tolerance may be factors for its failure to spread (Vignolo, Fernandez, and Maceira 2002). *L. tenuis* copes better with flooding stress than *L. corniculatus,* even in the presence of natural competitors (Striker et al. 2005). Teakle, Real, and Colmer (2006) reported significant variation for tolerance of combined salinity and water logging in both *L. tenuis* and birdsfoot trefoil. *L. tenuis* cultivars were more tolerant of both treatments than birdsfoot cultivars based on dry matter production relative to aerated treatment. Under aerated NaCl treatment (0–400 mM), *L. tenuis* accumulated half as much Cl⁻ (while Na⁺ was the same) as birdsfoot trefoil, indicating better Cl⁻ "exclusion" as an important trait for salt tolerance in *Lotus* species.

Perennial ryegrass (*Lolium perenne*) in the United Kingdom is generally too competitive and densely tillered to form stable associations with birdsfoot trefoil. When *Agrostis capillaries, Festuca pratensis, Phleum pretense,* and *Poa pratensis* were sown with birdsfoot trefoil, it was found that *P. pratense* depressed *Lotus* dry matter yield and *P. pratensis* gave the lowest overall forage yield (Sheldrick and Martyn 1992). Annual ryegrass (*Lolium* spp.) and festulolium (*Festulolium braunii*) have been found superior as companion crops over oat (*Avena sativa*) and oat + field pea (*Pisum* spp.; Wiersma, Hoffman, and Mlynarek 1999). A reduced seeding rate of the companion crop and early removal reduce competition for light and increase the probability of successful birdsfoot trefoil establishment. Spring-seeded winter grains as a companion crop sometimes do not allow successful establishment of birdsfoot trefoil. Herbicides for weed control are an alternative to a companion crop (Table 6.4). Good stands have also been obtained on high-pH soils by broadcasting a mixture of seed and fertilizer on pastures that cannot be tilled (Winch et al. 1969). Pasture yields have been increased four- to fivefold, and the stocking season has been extended 2–5 months.

Alfalfa, red clover, and birdsfoot trefoil frequently are established in the north central United States with a companion crop to increase establishment and forage yield, reduce erosion, and suppress weeds. In the irrigated steppe of the southern Rocky Mountains, birdsfoot trefoil-tall fescue and kura clover-tall fescue provided more uniform dry matter yield distribution across years than binary mixtures of tall fescue with alfalfa (Lauriault et al. 2006). Legume dry matter yields of birdsfoot trefoil-tall fescue were unaffected by grazing frequency. Woodward et al. (2006) have shown the potential benefit of silage supplementation, particularly with lotus silage, for increased milk solids yield in summer when low pasture growth rates and quality may otherwise limit production. Cows were given lotus silage; the high milk yield resulted from a combination of the higher nutritive value of the silage and possibly the protein-sparing effects of the lotus condensed tannins. Of 13 varieties of *L. corniculatus,* variety Oberhaunstaedter, followed by the variety Lotar, was the most suitable for silage in that both had high plant persistency and higher dry matter yield (Marley, Fychan, and Jones, 2006a). The authors considered that further studies are required on condensed tannins to determine how they affect the ensiling process and, ultimately, protein digestion in the ruminant.

In a study to identify legume species that could regenerate naturally from seed produced the previous year or from residual hardened seed that failed to germinate when first sown, Carr, Poland, and Tisor (2005) found that birdsfoot trefoil has potential to be a self-seeding pasture species in the Great Plains. They found forage dry matter yield ranged from 2 to 5 mg/ha for birdsfoot trefoil and red clover, similar to the yield of alfalfa that persisted in the second year following establishment. Crude protein, acid detergent, and neutral detergent fiber concentrations suggested that forage quality for birdsfoot trefoil was equal or superior to that of alfalfa and red clover.

Living mulches are cover crops planted either before or with a crop and maintained as a living ground cover throughout the growing season (Hartwig and Ammon 2002). Hall, Hartwig, and Hoffman (1984) reported significant soil loss from conventional tilled corn planted into birdsfoot trefoil. However, when corn was planted into a birdsfoot trefoil living mulch on a 14% slope, water runoff, soil loss, and pesticide loss were reduced from 95 to >99%.

Table 6.4　Herbicides That Are or Have Been Used in Birdsfoot Trefoil Stands

Allidochlor	α-chloro-*N*,*N*-diallylacetamide
Aziprotryne	4-azido-*N*-(1-methylethyl)-6-(methylthio)-1,3,5-triazin-2-amine
Benefin	*N*-butyl-*N*-ethyl-2,6-dinitro-4-(trifluoromethyl)benzenamine
Benfluralin	*N*-butyl-N-ethyl-2,6-dinitro-4-(trifluoromethyl)benzenamine
Bentazone	3-(1-methylethyl)-1H-2,1,3-benzothiadiazin-4(3H)-one 2,2-dioxide
Butam	*N*-benzyl-*N*-isopropylpivalamide
Butralin	4-(1,1-dimethylethyl)-*N*-(1-methylpropyl)-2,6-dinitrobenzenamine
Carbofuran	2,3-dihydro-2,2-dimethyl-7-benzofuranyl methylcarbamate
Chloropon	2,2,3-trichloropropionic acid
Chlorsulfuron	2-chloro-*N*-[[(4-methoxy-6-methyl-1,3,5-triazin-2-yl)amino]carbonyl]-benzenesulfonamide
Clopyralid	3,6-dichloro-2-pyridinecarboxylic acid
2,4-D	2,4-dichlorophenoxyacetic acid
2,4-DB	4-(2,4-dichloropheoxy)butyric acid
Dacthal	Chlorthal dimethyl
Dalapon	2,2-dichloropropanoic acid
Dicamba	3,6-dichloro-2-methoxybenzoic acid
Dinoseb	2-(1-methyl-*N*-propyl)-4,6-dinitrophenol
Diphenamid	*N*,*N*-Dimethyldiphenylacetamide
Diquat	1,1′-ethylene-2,2′-bipyridylium ion
Diuron	3-(3,4-dichlorophenyl)-1,1-dimethylurea
Endothal	7-oxabicyclo[2.2.1]heptane-2,3-dicarboxylic acid
EPTC	S-ethyl dipropylthiocarbamate
Ethalfluralin	*N*-ethyl-*N*-(2-methylally1)-2,6-dinitro-4-trifluoromethylaniline
Ethofumesate	2-ethoxy-2,3-dihydro-3,3-dimethyl-5-benzofuranyl methanesulfonate
Fluazifop	(±)-2-[4-[[5-(trifluoromethyl)-2-pridinyl] oxy]phenoxy]propanoic acid
Fluchloralin	*N*-(2-chloroethyl)-2,6-dinitro-*N*-propyl-4-(trifluoromethyl)benzenamine
Glyphosate	*N*-(phosphonomethyl)glycine
Haloxyfop	2-[4-[[3-chloro-5-(trifluoromethyl)-2-pridinyl]oxy]phenoxy]propanoic acid
Harmony	3-[[[(4-methoxy-6methyl-1,3,5, triazine-2-yl)amino]carbonyl]amino]sulfonyl-2-thiophenecarboxylate
Hexazinone	3-cyclohexyl-6-(dimethylamino)-1-methyl-1,3,5-triazine-2,4-(1H,3H)-dione
Imazaquin	2-[4,5-dihydro-4-methyl-4-(1-methylethyl)-5-oxo-1H-imidazol-2-y1]-3-quinolinecarboxylic
Imazetapir	acid (RS)–5-etil-2-(4-isopropil-4-metil-5-oxo-2-imidazolin-2-il)nicotinic)
Isopropalin	4-(1-methylethyl)-2,6-dinitro-*N*,*N*-dipropylbenzenamine
MCPA	(4-chloro-2-methylphenoxy)acetic acid
MCPB	4-(4-chloro-2-methylphenoxy)butyric acid
Methazole	2-(3,4-dichlorophenyl)-4-methyl-1,2,4-oxadiazolidine-3,5-dione
Paraquat	1,1′-dimethyl-4,4′-bipyridinium ion
Picloram	4-amino-3,5,6-trichloro-2-pyridinecarboxylic acid
Profluralin	*N*-(cyclopropylmethyl)-2,6-dinitro-*N*-propyl-4-(trifluoromethyl)benzenamine
Prometryne	2,4-bis(isopropylamino)-6-methylthio-1,3,5-triazine
Pronamide (Propyzamide)	*N*-(1,1-dimethylpropynyl)-3,5-dichlorobenzamide
Sethoxydim	2-[1-(ethoxyimino) butyl]-5-[2-(ethylthio)propyl]-3-hydroxy-2-cyclohexen-1-one
Simazine	2-chloro-4,6-bis(ethylamino)-S-triazine
TCA	trichloroacetic acid
Terbuthylazine	6-chloro-*N*-(1,1-dimethylethyl)-*N*4-ethyl-1,3,5-triazine-2,4-diamine
Terbutryne	(2-tert-butylamino)-4-(ethylamino)-6-(methylthio)-s-triazine
Trifluralin	2,6-dinitro-*N*,*N*-dipropyl-4-(trifluoromethyl)benzenamine

Reproductive performance and wool growth of ewes grazing on birdsfoot trefoil, compared with those of ewes grazing on a perennial ryegrass–white clover pasture, produced 11% more wool and had an ovulation rate up to 14% higher than ewes grazing pasture (Luque et al. 2000). In an experiment of selective grazing preferences by yearling heifers, Poli et al. (2006a) alternated 2.4-m wide strips of a mixture of birdsfoot trefoil with white clover (BW) and red clover (RC). Biting rates were consistently higher on BW than RC, but that changed with time close to 0.50 BW:0.50 RC, reflecting the deliberate exercise of choice by grazing cattle rather than being a chance effect. The distribution of grazing time between sward type and maturity combinations was influenced progressively by herbage mass and height contrasts as strips were grazed down (Poli et al. 2006b).

In a comparison of cattle grazing on rhizomatous versus nonrhizomatous birdsfoot trefoil, Wen et al. (2002) observed that the ability of rhizomatous birdsfoot trefoil to spread by rhizomes did not lead to greater total forage production or livestock performance in the short term. Interseeding birdsfoot trefoil into tall fescue pastures increased total weight gain per hectare of growing ruminants compared to grazing monocultures of tall fescue (Wen et al. 2002). This improvement in gain is usually attributed to the greater crude protein and lower detergent fiber in the mixtures of tall fescue and birdsfoot trefoil compared to a tall fescue monoculture. Although the percentage of birdsfoot trefoil in mixed pastures often declines over time, its value may be underestimated because animals selectively graze birdsfoot trefoil when its proportion in pastures is low (Wen et al. 2004).

Of several forages (lucerne, red and white clover, sulla), Fulkerson et al. (2007) found birdsfoot trefoil to have the highest effective rumen degradable protein (ERDP) value—that is, the portion of rumen degradable protein captured and utilized by rumen microbes. This resulted in a high ERDP to fermentable, metabolizable energy ratio of 29—well above the ratio of 11 required for optimal microbial protein synthesis in rumen of dairy cows.

In a study conducted by White and Scott (1991), birdsfoot trefoil is described as an effective weed suppressor in winter wheat. In control plots with a naturally establishing weed community of white clover (*Trifolium repens*), subclover (*T. subterraneum*), and birdsfoot trefoil, Hiltbrunner et al. (2007) found reduced density of monocotyledonous, dicotyledonous, spring-germinating, and annual weeds by the time of wheat anthesis. Although the grain yield was reduced by 60% or more for all legumes, when compared to control plots kept weed free, a significant negative correlation between the dry matter of the cover crop and the weeds and winter wheat was observed by the time of wheat anthesis. The authors stress that before living legume cover crops can be considered a viable alternative for integrated weed management under organic farming conditions, management strategies need to be identified that maximize the positive effect in terms of weed control at the same time that they minimize the negative impact on growth and yield.

6.7 SOIL AND SOIL FERTILITY

Birdsfoot trefoil is suitable to many types of soil, varying from clays to sandy loams, including saline conditions. It has greater tolerance than alfalfa of acid, infertile, or poorly drained soils. Even though it will grow and persist on such soils, forage and seed yields and winter hardiness are significantly increased by proper applications of lime and fertilizer and are most productive on fertile, moderately well-drained soils (Russelle, McGraw, and Leep 1991). It gives high yields under drained conditions, with water table depths ranging from 60 to 100 cm (Strelkov 1980). It tolerates water logging when actively growing and has moderate tolerance of pH 4.5 (Schachtman and Kelman 1991). At a pH lower than 6.2–6.5, seedling growth and establishment may be slow, and nodulation and N fixation may be retarded or completely inhibited. The cultivar AU Dewey is more tolerant of soil acidity than other cultivars, with good root and forage growth to a pH 4.8 (Hoveland et al. 1990; Voigt and Mosjidis 2002). Nodulation is best with a pH from 6.0 to 6.5.

Birdsfoot trefoil is moderately salt tolerant, showing definite yield reductions only at 6–12 millimhos/cm. Adequate amounts of phosphorus promote early, vigorous seedling growth, thereby increasing the probability of successful establishment (Russelle et al. 1991). It has been shown that birdsfoot trefoil can reduce boron (B) and selenium (Se) concentrations in B- and Se-laden soils. Vegetation management should be considered as a bioremediation tool for removing B and Se from contaminated soils to safe levels (Banuelos et al. 1996). Based on the aluminum activity (μM Al^{3+}) required to reduce yields by 50% (Al_{Rv50}), Wheeler et al. (1992) categorized *L. corniculatus* as a sensitive plant.

Low soil fertility has been associated with high concentrations of condensed tannins, but the results are affected by climatic factors such as moisture stress, temperature, and carbon dioxide concentration (Carter, Theodorou, and Morris 1999). It has been shown that *Lotus* cultivars are responsive to phosphorus and sulfur during early growth and should benefit from superphosphate applications at establishment if good weed control is maintained (Kelman 2006a).

6.8 HYDROGEN CYANIDE CONTENT

Cyanogenic glucosides, generally considered antinutritional factors, are important defense molecules against predators and, in some cases, diseases (Jones 1998; Compton and Jones 1985). In Australia, HCN in *L. cruentus* has been reported to cause poisoning in livestock during drought (Cameron and Prakash 1992). Two independent genes control cyanogenesis: *Ac/ac* for the production of cyanoglycosides and *Li/li* for the production of linamarase. Lotus plants expressing the HCN phenotype are *AcLi* (Bazin, Blaise, and Cartier 1994). Leaves and flowers produce up to five times the concentration of HCN as stems and ripe seed pods. Young leaves contain more cyanide than old leaves. This response has been shown to depend on plant and herbivore response to rising carbon dioxide levels (Bazin et al. 2002). Seeds are acyanogenic, but as seeds germinate and seedlings form cotyledons, HCN is exhibited. Roots are acyanogenic.

HCN has been noted to vary widely in birdsfoot trefoil collections (Grant and Sidhu 1967). Gebrehiwot and Beuselinck (2001) and Gebrehiwot, Beuselinck, and Roberts (2002) observed that HCN concentrations were 50% greater in spring and summer than in fall and winter; winter had the lowest concentration, averaging only 260 µg/g dry matter. Smith et al. (2005) found that survival, plant productivity, and seed predation were not related to cyanogenesis and suggested that phenotypic expression of this gene in a population is not related to short-term adult survival or establishment.

Lotus tenuis has the greatest concentration of HCN, averaging 900 µg/g dry matter. Plants that accumulate >600 µg HCN/g dry matter pose potential damage to livestock (Haskins, Gorz, and Johnson 1987). *L. uliginosus* is acyanogenic. It is possible to select for acyanogenic plants and for cyanogenic content in roots to deal with persistence and disease susceptibility.

6.9 PROANTHOCYANIDINS

Proanthocyanidins (PAs) are a class of polymeric, polyphenolic, plant secondary metabolites that, in high concentrations (>6% of dry matter [DM]), reduce voluntary feed intake, digestibility, and animal performance of ruminants consistent with a defensive role. At lower concentrations, the protein-binding effects protect dietary protein against excessive degradation in the rumen and increase protein utilization, contributing to increases in lactation, wool growth, and live weight gain (Min et al. 2003). Sivakumaran et al. (2006) have examined the PA concentration for several *Lotus* species, including *L. corniculatus* (3.0% DM total PA), *L. uliginosus* (7.2%), and *L. tenuis* (0.8%). PA from *L. corniculatus* is associated with better animal performance than that from *L. uliginosus* (Waghorn, Reed, and Ndlovu 1999), although the PA content of *L. corniculatus* is about half that of

L. uliginosus. PA oligomer and polymer fractions from *L. corniculatus* consisted predominantly of procyanidin units; those of *L. uliginosus* consisted predominantly of prodelphinidin units.

Paolocci et al. (2007) reported the cloning and expression analysis of anthocyanidin reductase (ANR) and leucoanthocyanidin 4-reductase (LAR), two genes encoding enzymes committed to epicatechin and catechin biosynthesis, respectively, in *L. corniculatus.* There are two LAR gene families (LAR1 and LAR2). The steady-state levels of *ANR* and *LAR1* genes correlate with the levels of PA in leaves of wild-type and transgenic plants. *ANR* and *LAR1*, but not *LAR2,* genes produce active proteins following heterologous expression in *Escherichia coli* and are affected by the same basic helix-loop-helix transcription factor that promotes PA accumulation in cells of palisade and spongy mesophyll. This study provided direct evidence that the same subclass of transcription factors can mediate the expression of the structural genes of both branches of PA biosynthesis. Because PAs have the ability to bind reversibly with plant proteins to improve digestion and reduce bloat, engineering this pathway in leaves is a major goal for forage breeders.

6.10 TANNINS

Tannins are phenolic plant secondary metabolites and are involved in plant–pathogen and plant–herbivore interactions (Min et al. 2003). Condensed tannins (CTs) are considered to be the effective antibloat agents in *L. corniculatus* (Jones and Lyttleton 1971). CTs are polymers of flavan-3,4-diols joined by four to eight interflavan bonds with a flavan-3-ol at the 4′ terminal end. CTs of birdsfoot trefoil comprise the flavan-3,4-diols leucocyanidin and leucodelphinidin, which release cyanidin and delphinidin upon acid hydrolysis (Carron et al. 1992; Marles, Ray, and Gruber 2003). At high levels, tannins in forage and fodder are deleterious for use with ruminants and such levels reduce both palatability and nutritive value (Robbins et al. 1998). By definition, CTs bind to protein and are regarded as antinutritional compounds that reduce protein digestibility. Cassida et al. (2000) found trefoil tannin concentrations were positively correlated with undegradable intake protein concentration in cattle diets.

Miller and Ehlke (1997) determined the inheritance of CT concentration in tannin-positive birdsfoot trefoil with high-, medium-, and low-tannin parents. They found primarily additive genetic effects controlled CT concentration, indicating that this character could be used in breeding programs for improved herbage quality. The concentration of CT in the diet and the chemical structure, which affects the activity of the CT, need to be considered when assessing the effects of CT on protein metabolism in ruminants (Aerts et al. 1999). Variations in CT chemistry alter protein-binding capacities. CTs interact with proteins in feed, saliva, and microbial cells; with microbial exoenzymes; and with endogenous proteins or other feed components, which alters digestive processes as compared with diets free of CT. Tannin levels exceeding 40–50 g/kg DM in forages may reduce protein and DM digestibility of the forages by ruminants (McMahon et al. 2000). A species-specific effect of CT occurs on bacteria in the rumen (Min et al. 2002). At low to moderate levels, condensed tannins increase the quantity of dietary protein—especially essential amino acids—flowing to the small intestine. Hedqvist et al. (2000) have determined that the degradability of soluble proteins in birdsfoot trefoil was negatively correlated to tannin concentrations. Unlike alfalfa, legumes that contain CT do not cause bloat. It is well established that tannins that complex with dietary proteins can reduce nitrogen supply to the animal, but the ability of gastrointestinal microorganisms to metabolize these compounds and their effects on microbial populations have received little attention (McSweeney et al. 2001).

In birdsfoot trefoil, CTs are accumulated in large amounts up to 10% dry weight in stems, leaves, roots, and petals (Carron et al. 1992). Levels of CT vary among these tissues and among different genotypes. High tannin content has been associated with prostate growth habit (Kelman, Blumenthal, and Harris 1997). CT concentration is higher in birdsfoot trefoil plants that are rhizomatous (Wen et al. 2003). Häring et al. (2007) investigated the CT concentrations in leaves, stems, and roots and the biomass proportions among these organs. Leaf mass fractions of CT decreased, whereas proportions in stems and roots increased. As a consequence of the unequal distribution of tannins in different plant parts and due to the changing biomass proportions between them, the authors considered that herbivores not only could have different concentrations of CT in their diets but also could be affected by different CT dynamics during the season. Thus, they consider biomass allocation and accumulation of non-CT plant material as important predictors of seasonal variations of CT concentrations. Therefore, knowledge of CT synthesis and its regulation is a necessity.

In a comparison of birdsfoot trefoil and lucerne (*Medicago sativa*) grazed by ewes and lambs, Douglas et al. (1995) found ewe wool production and lamb carcass weight were higher on birdsfoot trefoil. Total CT content was 32–57 g/kg DM for birdsfoot trefoil but negligible for lucerne (less than 2 g/kg DM). In a comparison between birdsfoot trefoil and sulla (*Hedysarum coronarium*), Douglas et al. (1999) reported both species had similar protein concentrations in DM, but sulla had a higher concentration of total CT (88 cf. 50 g CT/kg DM). The CT in sulla reduced carcass weight from 21.2 kg (with PEG drenching) to 18.8 kg. In contrast, CT in birdsfoot trefoil did not affect wool growth, live weight gain, or carcass characteristics. Montossi et al. (2001) reported that CT had no significant effects on diet selection between weaned lambs rotationally grazing swards of annual ryegrass (*Lolium multiflorum*)/white clover (*Trifolium repens*) and Yorkshire fog (*Holcus lanatus*)/*T. repens* with or without birdsfoot trefoil.

In a germplasm collection of 38 accessions of birdsfoot trefoil from nine geographic regions, Kelman (2006b) found CT to be negatively correlated with *in vitro* digestible dry matter and nitrogen. This result helped to identify potentially valuable accessions with low CT (<4% of dry weight) and high *in vitro* digestible dry matter (>70%). Min et al. (2001, 2003) found that birdsfoot trefoil CT increased wool production in sheep through increasing reproductive efficiency and reducing embryonic loss. Condensed tannins of *L. corniculatus* were found to reduce the bacterial proteolysis of the large and small subunit of ribulose-1,5-bisphosphate carboxylate (Rubisco) protein and the growth of proteolytic rumen bacteria (Min et al. 2005).

In Sweden, birdsfoot trefoil was evaluated to assess the possible parasitological benefits when fed as a component in a mixed pasture sward to young sheep. No differential benefit was observed over *Trifolium repens* on the different stages of development for a range of nematode parasite species recovered from lambs (Bernes et al. 2000). It was considered that the low level of CT in the birdsfoot trefoil variety tested and/or the proportion of this plant in the diet did not provide benefits for controlling sheep parasites. Heckendorn et al. (2007) inoculated lambs with larvae of *Haemonchus contortus* 27 days prior to feeding birdsfoot trefoil. When compared to the control, there was a reduction of 63% in total daily fecal egg production and a reduced worm burden of 49%. The concentration of CT in trefoil was 15.2 g CTs/kg DM and is considered the reason for the antiparasitic effect.

Marley et al. (2006b) studied the effects of trefoil and chicory on parasitic nematode development, survival, and migration when compared with perennial ryegrass. When sheep feces containing *Cooperia curticei* eggs were added to pots containing birdsfoot trefoil, chicory, or ryegrass, on a dry matter basis, by day 16 there were 31 and 19% fewer larvae on trefoil and chicory than on ryegrass. In field plots there were a minimum of 8 and 63% fewer infective stage parasitic larvae on trefoil and chicory than on ryegrass. These results indicated that the number of infective stage larvae was reduced by the effect of their sward structure (Marley et al. 2006b). Woodward, Laboyrie, and Jansen (2000) determined that milk yields (kilograms per cow per day) were higher on *L. corniculatus* (21.24) than on *L. corniculatus* + polyethylene glycol by blocking the action of CT (18.61). CT had no effect on intake but contributed to increased milk protein concentration, increased efficiency, and decreased milk fat concentration.

Under dryland farming conditions in New Zealand, Ramirez-Restrepo et al. (2004, 2005, 2006) have shown that the growth of weaned lambs can increase when grazing birdsfoot trefoil (cv. Grasslands Goldie), while reducing the reliance on anthelmintic drenches to control parasites and reducing the amount of insecticide needed to control flystrike. They consider that these effects are due to increased metabolizable protein supply from the protein-binding action of CT, enabling lambs to grow more quickly when carrying a parasite burden, and to birdsfoot trefoil better maintaining its high metabolizable energy value under drought conditions compared to perennial ryegrass (*Lolium perenne*)/white clover (*Trifolium repens*) pasture.

The barnyard or grassy pastoral flavor in pasture-grazed lamb is correlated with indole and ska-tole formed in the rumen from microbial deamination and decarboxylation of tryptophan (Priolo, Micol, and Agabriel 2001). Condensed tannins slow protein degradation in the rumen and will allow for more efficient use of amino acids by rumen microbes and reduce availability of tryptophan for indole and skatole formation. Schreurs et al. (2007) have noted greater indole and skatole concentra-tions when feeding white clover versus birdsfoot trefoil, which they attributed to high solubility and rapid degradation of the forage protein. *L. corniculatus* has a similar nutrient composition to white clover, but the condensed tannins in *L. corniculatus* slow protein degradation and reduce indole and skatole formation. Schreurs et al. (2007) found indole and skatole concentrations peaked in the plasma 1–2 h after the end of feeding, indicating that skatole and indole are rapidly absorbed from the rumen into the blood. They suggest that the white clover component of traditional New Zealand pastures may be the primary cause of undesirable pastoral flavors that result from the presence of indoles in meat and that, to ameliorate undesirable flavors, producers will need to consider using alternative forages such as *L. corniculatus* to reduce protein solubility and degradation rate.

Agrobacterium rhizogenes-transformed birdsfoot trefoil has been used as a model system for the study and genetic manipulation of condensed tannin biosynthesis (Carron, Robbins, and Morris 1994; Morris et al. 1999). A number of CT genotypes of *Lotus* have been selected and clonally micropropagated for transformation.

6.11 NODULATION AND N$_2$ FIXATION

Under nitrogen-limiting conditions, bacteria from the family Rhizobiaceae establish a symbio-sis with leguminous plants to form nitrogen-fixing root nodules. These organs require a coordinated control of the spatiotemporal expression of plant and bacterial genes during morphogenesis. Both plant and bacterial signals are involved in this regulation in the plant host. Plant genes induced during nodule development are the so-called nodulin genes. Nodule morphology of birdsfoot tre-foil is of the desmodioid type—determinate in growth habit and oblate in shape in comparison to *Trifolium*, which is of the crotalarioid type—indeterminate in growth habit, and cylindrical when young, becoming branched and flat (Vance et al. 1982). Unlike alfalfa nodules, those on birdsfoot trefoil senesce after each forage harvest, and a new population of nodules must be formed as plants regrow. Birdsfoot trefoil has a greater nodule mass than does alfalfa but a lower rate of N fixation and lower total seasonal N fixation (Vance et al. 1982). The level of N fixed by pure stands varies with cultural conditions but averages 90 kg/ha annually (Heichel et al. 1985). Strains of bacteria that inoculate birdsfoot and narrowleaf trefoil are generally not effective on big trefoil, which requires a specific inoculum for N fixation. The presence of *Rhizobium loti* in the soil and their densities vary from year to year and within each soil type, supporting the importance of selecting the most efficient and resistant strains to be included in the inoculants (Baraibar et al. 1999).

The expression of plant genes specifically induced during rhizobial infection and the early stages of nodule ontogeny (early nodulin genes) and those induced in the mature, nitrogen-fixing nodule (late nodulin genes) is differentially regulated and tissue or cell specific. De Bruijin et al. (1994) have studied the signal transduction pathway responsible for symbiotic, temporal, and spatial control of expression of an early (*Enod2*) and a late (Leghemoglobin; *lb*) nodulin gene from the stem-nodulated legume *Sesbania rostrata* and in identifying the *cis*-acting elements and *trans*-acting factors involved. By introducing chimeric *S. rostrata lb* promoter-*gus* reporter gene fusions into transgenic *L. corniculatus* plants, they have been able to show that the *lb* promoter directs an infected-cell-specific expression pattern in *Lotus* nodules and have delimited the *cis*-acting element responsible for nodule-infected-cell expression to a 78-pb region of the *lb* promoter.

Arbuscular mycorrhizal fungi (AMF) colonize between 58 and 100% root nodules of birdsfoot trefoil (Scheublin and van der Heijden 2006). However, AMF-colonized nodules never fix N, which shows that these nodules are not functional. The authors suggest that AMF colonize old senescent nodules after nitrogen fixation has stopped, although they could not rule out that AMF colonization inhibited nitrogen fixation. Scheublin, van Logtestijn, and van der Heijden (2007) have shown that AMF influence the competitive relationships between plant species. In competition pairs tested, *L. corniculatus,* the most AMF-dependent competitor, increased relative yield in the presence of AMF, while the relative yield of the less AMF-dependent competitor (*Festuca ovina* or *Plantago lanceolata*) decreased or remained equal. The impact of AMF on plant competition also depended on the identity of the competing plant species. AMF had only a small influence on the competitive relationship between *P. lanceolata* and *L. corniculatus;* in the competition between *F. ovina* and *L. corniculatus,* the shifts were much larger.

Glutamine synthetase is the main enzyme responsible for the assimilation of ammonia in higher plants. The promoter of a nodule-enhanced glutamine synthetase gene from *Phaseolus vulgaris* has been analyzed in transgenic birdsfoot trefoil. *Cis*-acting elements important for nodule-specific expression have been found, and two AT-rich sequences have been implicated in binding nuclear factors (Morris et al. 1999). The protein leghemoglobin is believed to facilitate O_2 diffusion to respiring bacteroids without irreversibly inactivating the O_2-sensitive nitrogenase enzyme complex. A 2-kb leghemoglobin coding sequence forms the nodulin promoter. An organ-specific element is responsible for nodule-specific expression containing two sequences AAAGAT and CTCTT as well as promoters of other nodulin genes. Site-specific mutations in this region have revealed sequences of importance for high-level nodule-specific expression (Morris et al. 1999). The quantification of symbiotic N_2 fixation by legumes is essential to determine their impact on N budgets. Dinitrogen fixation by 2- and 3-year-old stands of birdsfoot trefoil was determined using the ^{15}N isotope dilution method. The percent nitrogen derived from the atmosphere varied during the season, with a peak at the end of the season. Fixed N_2 averaged 145 kg N/ha/year. About 62% of its herbage N was obtained from fixation (Séguin et al. 2000).

Nodulation and nitrogen fixation have been studied under flooded conditions in *L. uliginosus.* More than 99.7% of the nodule biomass and 92% of the nodule number are produced on adventitious, aerial roots in the surface litter and vegetation. This strategy avoided direct contact of most nodulated roots with the anoxic, acidic soil environment and assisted aeration of the root system (Allan et al. 2000). Nitrogenase activity in *L. faba, L. uliginosus,* and *L. subbiflorus* was found to have sixfold the activity of *L. corniculatus* (Gonnet and Diaz 2000).

Lotus japonicus has emerged as a legume in which to study nodulation and nitrogen fixation genes because of the extensive knowledge concerning the genome of this species (Perry et al. 2003).

6.12 DISEASES

All vegetative and reproductive parts of *Lotus* plants are susceptible to species of some 19 fungal genera (listed in Table 7-2 in English 1999). Viruses and bacteria causing disease in *Lotus* under field conditions are rare. Alfalfa mosaic virus, curly top virus, tobacco mosaic virus, and the bacterium *Pseudomonas viridiflava* have been reported (Table 7-3 in English 1999).

Numerous species of nematodes affect root growth and dry matter accumulation. Most reports have focused on two species, *Pratylenchus penetrans* and *Meloidigyne hapla,* the root lesion nematode and the northern root knot nematode, respectively (English 1999). There is wide genetic variability for invasion by root-lesion nematodes (Kimpinski et al. 1999), and symptoms can change with geographic environment (Ostazeski 1967). Plants severely infected with root rot have extensive decay in the central portion of the upper taproot and crown and often fail to regrow after harvest, with stand losses of 68–88%. Severe loss is usually associated with warm weather and high humidity; thus, diseases are of greater importance in southern than in northern areas of the United States. Birdsfoot trefoil is susceptible to parasitism by root-knot (*Meloidogyne* species) and root-lesion (*Pratylenchus penetrans*) nematodes. Synergism between *Fusarium* spp. and *Rhizoctonia solani* with root-lesion nematodes lowers birdsfoot trefoil productivity and increases plant mortality (Thompson and Willis 1975). Fusarium wilt caused by *F. oxysporum* was identified as the primary species contributing to poor stand persistence (Gotlieb and Doriski 1983) and is considered to be the principal cause of the demise of the certified birdsfoot trefoil seed industry in New York state (Bergstrom and Kalb 1995). A greenhouse evaluation method has been developed to screen and characterize birdsfoot trefoil germplasm for reaction to *F. oxysporum* root rot (Altier, Ehlke, and Rebuffo 2000).

Under warm, humid conditions, *Sclerotium rolfsii* and *R. solani* attack the crown and lower foliage, causing leaf blight, which may lead to death of affected plants. The most widespread foliar disease is *Stemphylium loti*. This organism causes reddish brown stem and leaf lesions and results in premature leaf drop or death of stems. English (1992) reported that 90% or more of the leaves of birdsfoot trefoil produced per shoot each year could be destroyed by foliar blight and leaf spot. Immature seed pods may also be attacked, resulting in shriveling and discoloration of seed (Graham 1953). Native fluorescent *Pseudomonas* has been used for controlling seedling diseases (Bagnasco et al. 1998; Pérez et al. 2001). Foliar and stem blights caused by *Phomopsis loti* and *Diaporthe phaseolorurn* are less common but can be severe in some locations. Occasional epidemics of the rust *Uromyces striatus* var. *loti* can cause considerable damage to leaves and stems, but plant death is uncommon (Zeiders 1985).

Hill and Zeiders (1987) and Zeiders and Hill (1988) showed that phenotypic recurrent selection improved the level of disease resistance. This indicates that breeding for resistance to *Fusarium* root rot (*F. oxysporum*) has the potential to increase the persistence of birdsfoot trefoil in the field. However, because this organism is only one of the many species involved in causing root and crown rot, a multiple pest breeding effort is needed to overcome the persistence problem. Papadopoulos et al. (1995) demonstrated that current cultivars contain a wide range of variability for resistance to nematode root-lesion invasion but it is necessary to determine the association between level of root invasion and resistance. Resistance is believed to affect the long-term persistence of birdsfoot trefoil in temperate regions of North America (Willis 1981).

A greenhouse evaluation method has been developed to screen and characterize birdsfoot trefoil germplasm for reaction to *Fusarium* root rot (Altier et al. 2000). A stage-based, matrix population model was developed in relation to clipped and unclipped stands. Plant growth stages represented in the model were seeds, seedlings, and mature vegetative and reproductive plants. Establishment-phase populations were characterized by relatively high mortality and low reproduction (Emery, Beuselinck, and English 1999).

Breeding programs are currently developing selections with improved resistance to root-lesion nematodes (*Pratylenchus penetrans*) and *Fusarium* root rot organisms (Papadopopoulos et al. 1995; Viands et al. 1994) and for susceptibility to foliar and shoot blight caused by *Rhizoctonia* spp. (English and Beuselinck 2000). Emery, Beuselinck, and English (2003) have shown that isolates of *R. solani* are more virulent on leaves and shoots of *L. corniculatus* than were binucleate *Rhizoctonia* isolates. Birdsfoot trefoil offers significant opportunities for investigating temporal and spatial dynamics of genetic structure of *Rhizoctonia* populations in perennial plant populations.

Allelopathic toxins have been implicated in reduction of new forage stands. Such toxins may explain some of the difficulty in establishing birdsfoot trefoil in tall fescue sod (Luu, Matches, and Peters 1982). This suggests that it may be desirable to grow one or more crops between a sod crop and no-till legumes (Vough, Decker, and Taylor 1995). Luu (1980) found that toxins associated with tall fescue swards were largely eliminated by burning the sod prior to seeding birdsfoot trefoil.

6.13 PESTS

A number of insects cause losses of forage and seed (Wipfli et al. 1989; Mackun and Baker 1990). Jones and Turkington (1986) have listed species of phytophagous insects that feed on seeds, buds, leaves, flowers, roots, pods, and shoots. The meadow spittlebug, *Philaenus spumarius,* feeds by sucking plant sap and causes plant stunting and abortion of flower buds. This insect produces a characteristic white foamy mass on stems and leaves. Three plant bugs, *Adelphocoris lineolaris, Lygus lineolaris,* and *Plagiognathus chrysanthemi,* and the potato leafhopper, *Empoasca fabae,* also cause injury (Wipfli, Wedberg, and Hogg 1990a). The plant bugs destroy stem terminals and flowers. The potato leafhopper causes a characteristic yellowing and reddening of leaves and general stunting; heavy infestations can reduce forage yield and quality. Several cultural and chemical control methods have been investigated, such as fiberglass screen barriers (Wipfli, Wedberg, and Hogg 1991), the use of insecticides, rotation (Altier 2003), and burning (Wipfli et al. 1990b; Peterson et al. 1992b).

The trefoil seed chalcid, *Bruchophagus platypterus,* is a small, black, wasp-like, host-specific insect that can greatly reduce seed yields by parasitizing seed. Fertilized females are attracted to seed pods by the presence of volatile compounds that elicit landing and egg-laying behavior (Kamm 1989). Eggs are laid in developing seed pods, and the larvae feed inside the maturing ovule, leaving only hollow, unviable seed. The trefoil seed chalcid differs from other seed chalcids in that it will not seek its host unless it sees yellow-colored flowers (Kamm 1992). It is estimated that commercial seed yields may be reduced as much as 40% or more. In southern Ontario, overall levels of infestation of the trefoil seed chalcid averaged 2–11.7% but were at 50% in some samples (Ellis and Nang'ayo 1992). Peterson, Wedberg, and Hogg (1992a) and Peterson et al. (1992b) considered that applications of malathion at the blossom and pod stage did not provide adequate suppression of *B. platypterus* adults. *Aprostocetus bruchophagi,* a parasitoid of the seed chalcid, was sometimes found in higher numbers in treated plots.

Pesticides do not effectively control the seed chalcid, so growers must rely on cultural practices to reduce seed losses. Large amounts of seed left on the ground after harvest may increase infestations the following year. Chalcid populations can be reduced by avoiding delayed late-season seed harvest, by burning or burying combine trash after harvest to destroy infested seed, and by locating new fields away from old ones.

Different species of mollusks are reported to feed on plants of birdsfoot trefoil (Jones and Turkington 1986).

6.14 WEED CONTROL

Problematic weeds in birdsfoot trefoil stands include:

barnyard grass (*Echinochloa crus-galli* (L.) Beauv.);
Canada thistle (*Cirsium arvense* (L.) Scop.);
cinquefoil (*Potentilla norvegica* L.);
chicory (*Cichorium intybus* L.);
dandelion (*Taraxacum officinale* Weber);
fall panicum (*Panicum dichotomioflorum* Michx.);
field pennycress (*Thlaspi arvense* L.);
fleabane (*Erigeron strigosus* Muhl. ex Willd.);
green foxtail (*Setaria viridis* (L.) Beauv.);
groundsel (*Senecio vulgaris* L.);
hairy vetch (*Vicia villosa* Roth.);
lamb's-quarters (*Chenopodium album* L.);
narrow-leaved hawk's-beard (*Crepis tectorum* L.);

pigweed (*Amaranthus retroflexus* L.);
quackgrass (*Agropyron repens* (L.) Beauv.);
white cockle or white campion (*Silene pratensis* (Rafn) Godron & Gren.; Syn. *S. alba*; *Lychnis alba*);
wild oats (*Avena fatua* L.);
witch grass (*Panicum capillare* L.);
yellow foxtail (*Setaria glauca* (L.) Beauv.);
yellow nutsedge (*Cyperus esculentis* L.); and
yellow rocket (*Barbarea vulgaris* R. Br.).

The particular weeds would depend upon the region where birdsfoot trefoil is grown.

Weed control methods are effective but can be costly in terms of time and factors such as damage to nontarget vegetation and increased soil erosion. During establishment, clipping or mowing can control weeds. Under some conditions, mowing broad-leaved weed species allows weed grasses such as foxtail and fall panicum to grow and become more competitive than the original broad-leaved weeds (Kerr and Klingman 1960). With low soil fertility, weeds can sometimes be controlled and eliminated by liming and fertilization.

Chemicals are available for control of both broad-leaved weeds and grasses. Although the type of chemical for weed control continues to evolve, Table 6.4 gives a record of herbicides that are used or have been used to control weeds. Pre-emergence treatments with EPTC were found to result in a significant reduction in broadleaf weeds and increase in yield (Peters 1964). Treatment with EPTC is effective in suppression of yellow nutsedge, germinating grass, and some broadleaf weeds and has not caused a reduction in yield at the highest rate tested (5 L/ha) (Williams and Adair 1982a). When EPTC or benefin is used, companion grasses cannot be seeded with birdsfoot trefoil. Effective control of clover is provided by TCA applied pre-emergence at 6.7 kg/ha for alsike clover or 13.5 kg/ha for red clover (McGraw, Wyse, and Elling 1983). Lazlo (1992) found pre-emergence treatments with imazetapir superior to imazaquin in controlling weeds.

Postemergence sprays of 2,4-DB or dinoseb can control broad-leaved weeds when birdsfoot trefoil has developed from one to four complete leaves. Seedlings of birdsfoot trefoil are more susceptible to injury from dinoseb than are alfalfa seedlings (Linscott and Hagin 1978). Picloram, clopyralid, and glyphosate control Canada thistle better than MCPA and dicamba (Boerboom and Wyse 1988). Bélanger, Winch, and Townshend (1985) found that cabofuran applied immediately after seeding had a positive effect on plant stand and seedling dry matter production. Haloxyfop alone or combined with dalapon was superior to fluazifop or sethoxydim alone or mixed with dalapon for controlling quackgrass and other grass weeds (Linscott and Vaughan 1989). Over a 3-year period, seed yield averaged 310, 160, 150, and 180 kg/ha after treatment with haloxyfop, fluazifop, sethoxydim, and dalapon, respectively, compared to a control yield of 80 kg/ha.

Wyse and McGraw (1987) found the herbicides trifluralin, profluralin, fluchloralin, ethalfluralin, and benefin effective in controlling white campion during establishment. Trifluralin is reported to be more effective in controlling weeds in birdsfoot trefoil than in alfalfa and red clover (Malik and Waddington 1989). Williams and Adair (1982b) tested 18 herbicides (aziprotryne, benfluralin, bentazone, butam, butralin, dacthal, diphenamid, endothal, EPTC, ethofumesate, isopropalin, MCPA, MCPB, methazole, prometryne, terbuthylazine, terbutryne, and trifluralin) for toxicity on germination and dry weight. Malik and Waddington (1989) found that fluazifop and sethoxydim did not have any adverse effects on yield of forage dry matter, and hexazinone reduced weed content significantly. The herbicide imazethapyr induced injury to *L. corniculatus* at the rate of 70–140 g/ha 20 days after treatment (Wilson 1994). However, it was found to be more effective at reducing vigor in its competitors, resulting in much greater biomass in treated versus untreated plots 28 days after treatment.

Chemicals mentioned here are examples of herbicides that can be used for control of weed competition. Local recommendations should be reviewed each year to determine available herbicides and to ensure their proper use.

Best weed control may possibly be achieved through the development of cultivars resistant to herbicides. Following glyphosate treatment, shoot weights of three C_2 populations were 44–85% greater than their C_0 populations, indicating increased glyphosate tolerance (Boerboom et al. 1991). *In vitro* selection for herbicide tolerance appears practical, and progress has been made for resistant plants for the sulfonylurea herbicides Harmony (Pofelis, Le, and Grant 1992) and chlorsulfuron (MacLean and Grant 1987; Vessabutr and Grant 1995). Dry weight of selected Harmony resistant lines grown from tissue culture increased yield more than 100% (Pofelis et al. 1992). From field selection, birdsfoot trefoil germplasm resistant to sulfonylurea was selected and registered for further studies (Grant and McDougall 1995). Cultivars with resistance to herbicides have not, as yet, been developed for commercial use. It will be necessary to develop this technology in concert with chemical companies, who will need to obtain clearances for use of herbicides on birdsfoot trefoil.

Li and Hill (l989b) obtained increased seed yield from use of the growth regulator paclobutrazol (CAS No.: 76738-62-0) to promote reproductive development and shorten the flowering period.

6.15 BREEDING AND DEVELOPMENT OF NEW CULTIVARS

To date, no cultivar has been released through genetic transformation. Cultivars have been largely the result of breeding and selection within the species rather than as the result of the introduction of foreign genes. Breeding for improved cultivars in birdsfoot trefoil has been by phenotypic recurrent selection with general or specific combining ability to identify desired genetic combinations (Papadopoulos and Kelman 1999). Papadopoulos et al. (1994) developed a technique to assess seeding-year seedling vigor and demonstrated that one cycle of phenotypic recurrent selection was effective in improving seeding-year vigor. Phenotypic recurrent selection can be employed to produce cultivars with more intense expression of winter hardiness (Bubar and Lawson 1959). General combing ability and specific combing ability are important for conditioning winter hardiness in *L. corniculatus* (Papadopoulos and Kelman 1999).

A list of birdsfoot trefoil, big trefoil, and narrow-leaf trefoil cultivars has been published in the *Lotus Newsletter* (http://www.inia.org.uy/sitios/lnl/vol34/grant.pdf; 34:12–26, 2004). More than 200 birdsfoot trefoil cultivars have been named. Information on origin and use for some cultivars may be found in Blumenthal and McGraw (1999) and in GRIN (2007).

6.16 NOVEL APPROACHES TO *LOTUS* SPECIES IMPROVEMENT

6.16.1 Production of Haploid Plants

Evidence for the generation of haploid *Lotus* plants by androgenesis has been presented by Séguin-Swartz and Grant (2006). *L. corniculatus* anthers produced several cell lines from which plants were regenerated. One of the regenerants was morphologically different from the mother plant ($2n = 4x = 24$) for several key traits. Examination of root tip cells revealed a mixoploid condition consisting of a small number of haploid ($2n = 2x = 12$) and octoploid ($2n = 8x = 48$) cells interspersed with tetraploid cells. Although the origin of this plant is uncertain, it is possible that it may have originated from microspore-derived callus tissue that underwent subsequent polyploidization. There is also evidence for the generation of a haploid *L. corniculatus* plant ($2n = 2x = 12$) through parthenogenesis after male sterile tetraploid *L. corniculatus* ($2n = 2x = 24$) was pollinated with *L. tenuis* Wald. et Kit. (Negri and Veronesi 1989).

There is no certainty about why it has been so difficult to generate haploids from legume species. One possible explanation is that the species are outbreeding polyploids that suffer inbreeding depression. Any reduction in ploidy level may result in decreased fitness and the possible unmasking of lethal recessive genes. The fact that there is some evidence for haploid production in *M. sativa* and *L. corniculatus* provides hope for further developments in the future. However, the haploids produced in these species were derived from tetraploid parents and therefore have two sets of chromosomes. Whether viable monoploid plants can be produced remains unknown (Croser et al. 2006).

6.16.2 Somaclonal Variation

Use of tissue and cell culture has significant potential for generating new genetic variability (Bajaj 1990). Niizeki and Grant (1971) were the first to establish plant regeneration using anther culture of *L. corniculatus* calli. *L. corniculatus* is easy to regenerate from callus culture through both organogenesis and somatic embryogenesis. The plant also produces calli from protoplasts with a high potential for regeneration through adventitious buds (Niizeki and Saito 1986).

Plants regenerated from tissue or cell culture are considered as the clones of the tissue or cell donor. Generally, in one or more of the traits among the clones, there is variation that was termed *somaclonal variation* by Larkin and Scowcroft (1981). Somaclonal variation has been found in *L. corniculatus* regenerated from cell cultures (Pezzotti et al. 1985; Vessabutr and Grant 1995). This variation may be used in breeding programs because the variants often occur at higher frequencies than from chemically induced mutagenesis (Gavazzi et al. 1987). Regenerated plants from *L. corniculatus* have proved to be suitable for the evaluation of somaclonal variation in morphological and agronomic traits (Table 6.5).

Niizeki and coworkers have carried out extensive studies on cytogenetics, molecular genetics, and morphological variants derived from single protoplasts in regenerated plants of *L. corniculatus* (Niizeki, Ishikawa, and Saito 1990a; Niizeki et al. 1994b; Niizeki 1996; Niizeki and Kodaira 1994). Southern blot analyses of mitochondrial DNA (mtDNA) using mitochondrial genes have shown novel fragments among protoclones. However, these novel fragments are the same as those fragments found in polymorphisms of seed-derived populations. On the other hand, Southern blot analyses of chloroplast DNA (cpDNA) by using chloroplast genomic DNAs showed few variations among protoclones. In regard to nuclear genes, no variations in Southern blot analyses were detected when using a small subunit of ribulose bisphosphate carboxylase (RuBisCO), phenylalanine ammonia lyase, and ribosomal DNA as probes. However, considerable variation in traits such as plant height and stem diameter was detected, which may be regulated by polygenes. Abnormal meiotic configurations such as univalents, lagging chromosomes, fragments, and bridges were frequently observed in pollen mother cells, and these tended to be related to low pollen fertility.

Because very few physiologically and morphologically abnormal plants were found in the protoclonal populations, mutation of major genes is probably very rare and chromosomal aberrations may occur in that part of the heterochromatin that lacks genetic activity. The possibility still exists that plants containing mutations in major genes or with chromosomal aberrations in part of the euchromatin are eliminated during acclimatization. Variation in inherited quantitative characters was observed in the progeny. The elimination of abnormal chromosome aberrations is most likely the result of plants recovered with high pollen fertility (Niizeki 1996). Therefore, the practicality of a breeding program for quantitative characters including seedling vigor and low HCN by using protoclones in *L. corniculatus* is established and recommended.

One obvious strategy for the use of somaclonal variation is to introduce the best available varieties into cell culture and to select among regenerated plants or their progeny for incremental improvements over existing varieties. Hence, the technique could be used to uncover new variants that retain all the favorable qualities of an existing variety, while adding additional traits such as disease resistance or herbicide resistance. For example, Swanson and Tomes (1983) were able to develop resistance in *L. corniculatus* to the herbicide 2,4-D. Although this technique can be used to select and propagate mutant cells and even give rise to tolerant plants, these are often epigenetic and

Table 6.5 Summary of Studies on Somaclonal Variation in *Lotus corniculatus* L.

Cultivar	Explant or Tissue Used	Somaclonal Variation	Ref.
Leo	Internode-derived calli	2,4-D-tolerant callus, suspension culture lines and regenerated plants	Swanson and Tomes 1983
Leo	Internode-derived calli	Chlorophyllous callus	Swanson, Tomes, and Hopkins 1983
Franco	Calli	Agronomical traits such as plant height, dry matter yield	Damiani et al. 1985
Leo	Hypocotyl-derived calli	Tolerant suspension calli and regenerated plants for 2,4-D and chlorosulfuron (2-chloro-*N*-[[(4-methoxy-6-methyl-1,3, 5-triazine-2-yl) amino] carbonyl]-benzenesulfonamide	MacLean and Grant 1987
Leo	Protoplasts of leaves	Morphological characters	Webb, Woodcock, and Chamberlain 1987
Franco	Leaf-derived calli	Morphological and agronomical traits such as leaflet width and seed yield	Damiani, Pezzotti, and Arcioni 1990
Viking	Protoplasts of calli	Chromosome structure, agronomical traits, and HCN content	Niizeki et al. 1990b
Leo	Protoplasts of root hairs	Plant height and stem number	Rasheed et al. 1990
Leo	Hypocotyl-derived calli	Tolerant plants for sulfonylurea herbicide Harmony {DPX-M6316; 3-[[[(4-methoxy-6- methyl-1,3,5,-triazine-2-yl) amino] carbonyl] amino] sulfonyl-2-thiophenecarboxylate}	Pofelis et al. 1992
Viking	Protoplasts of calli	Mitochondrial and chloroplast genome banding pattern of Southern blot analysis	Niizeki 1996

sometimes unstable. However, in species that are normally vegetatively propagated, this strategy has much to be recommended.

6.16.3 Somatic Cell Hybrids

The hybridization of distantly related species by protoplast fusion has been a practical tool for removing barriers of incompatibility in sexual crossing of agriculturally important plant species. To date, a limited number of reports have been issued on successful hybridization between leguminous species (Sano, Suzuki, and Oono 1988; Niizeki and Saito 1989; Kihara et al. 1992; Kaimori et al. 1998). In addition, agriculturally useful hybrid production has been very difficult because, in most cases, an imbalance in the genomes of the parents results in the rearrangement or partial elimination of the chromosomes of one parent, resulting in an incapability to achieve morphogenesis (Chien, Kao, and Wetter 1982; Sala et al. 1985). Some success in asymmetric protoplast fusion based on the complementation of x-ray- or γ-ray-irradiated and iodoacetamide-treated protoplasts has been reported on leguminous species (Sakai et al. 1996; Liu, Liu, and Li 1999).

Wright, Somers, and McGraw (1987) were the first to produce somatic hybrid plants in the genus *Lotus* by transferring the seed pod indehiscence trait of *L. conimbricensis* Willd. into *L. corniculatus,* two sexually incompatible species. *L. corniculatus* hypocotyl protoplasts were inactivated with IOA to inhibit cell division prior to fusion with suspension-cultured protoplasts *of L. conimbricensis.* Protoplasts of *L. conimbricensis* divided to form callus but did not regenerate plants. Thus, plant regeneration from protoplast-derived calli was used to tentatively identify somatic hybrid cell lines. Plants regenerated from three cell lines exhibited combinations of parental isozymes of phosphoglucomutase and *L. conimbricensis*-specific esterases to indicate that they were somatic hybrids. Unfortunately, both male and female hybrids were sterile.

Aziz et al. (1990) attempted to combine *L. corniculatus* and *L. tenuis*. A cell suspension of *L. tenuis* was established as a source of protoplasts from kanamycin-resistant callus derived from roots transformed by *Agrobacterium rhizogenes*. Such protoplasts were treated with a sublethal dose of sodium iodoacetate prior to their electrofusion with green cotyledon protoplasts of *L. corniculatus*. Putative somatic hybrid colonies were selected on medium containing kanamycin sulphate. The hybridity of plants regenerated from these selected colonies was confirmed by their morphology, esterase-banding patterns, and the presence of condensed tannins in leaves and stems and chromosome complements.

Niizeki and coworkers (Niizeki and Saito 1989; Niizeki, Kihara, and Cai 1994a) have also produced three asymmetric somatic hybrid calli and plants of *L. corniculatus* by protoplast fusion with *Oryza sativa, Glycine max,* and *Medicago sativa* as follows. Asymmetric hybrid calli containing only the nuclei of *L. corniculatus* were produced by protoplast fusion between rice and *L. corniculatus* and analyzed for their mtDNA and cpDNA (Nakajo et al. 1994). In the hybrid calli, novel mtDNA fragments were detected in Southern blot analyses. These results showed that some kind of alteration, such as intergenomic and/or intragenomic recombinations of mtDNA, occurred in the hybrid calli. However, the cpDNA fragment patterns of all hybrid calli lines from Southern blot analyses were found to be identical with those of *L. corniculatus*. Some regenerated plants from hybrid calli were tolerant of low temperatures and low sunlight intensity. In order to produce asymmetric hybrids (cybrids) containing a complete *L. corniculatus* nuclear genome and a small part of a soybean nuclear genome containing the complete *L. corniculatus* nuclear genome, IOA-treated protoplasts of *L. corniculatus* were fused with x-ray-irradiated soybean protoplasts (Kihara et al. 1992; Niizeki et al. 1994a). Peroxidase isozyme and karyotypes of calli obtained from the protoplast fusion clarified the hybridity of the calli. Plant regeneration from the asymmetric hybrid calli was also successful, but regenerated plants did not show hybridity from the analysis of the peroxidase isozyme. The morphology of the regenerated plants resembled that of *L. corniculatus*. The regenerated plants were usually morphologically an erect type in contrast to the creeping type of the parent plants.

In another experiment, donor protoplasts of alfalfa were given lethal doses of x-irradiation and recipient protoplasts of *L. corniculatus* were inactivated with IOA. Donor and recipient protoplasts were fused with polyethylene glycol (Kaimori et al. 1998; Niizeki 2001). Fusion products initiated cell division resulting in calli, some of which had a high capacity for plant regeneration. Many hybrid calli cultured for 1 month were found to have isozymes indicating the banding pattern of one parent for some isozymes and that of another parent in others. These facts suggest that chromosomes or chromosome segments of both parents may be eliminated randomly at an early stage of callus culture. However, after 2 months of culture, most of the banding patterns had altered to those of *L. corniculatus,* indicating that calli cells with *L. corniculatus* genomes were rapidly selected as the dominant calli. Shoot regeneration did not occur from the symmetric somatic hybrid calli of *L. corniculatus* and alfalfa (Niizeki et al. 1989). This fact might be attributed to the imbalance of *L. corniculatus* and the alfalfa nuclear genomes, indicating that most of the alfalfa chromosomes irradiated by x-rays degenerated at some stage in the subcultures.

Accordingly, it may be possible to regenerate shoots from calli derived from asymmetric hybrids with x-ray-irradiated donor protoplasts, whereas regeneration of novel shoots is not likely from symmetrical somatic hybrids carrying complete chromosomes of both parents or some chromosomes of one parent. Isozyme and Southern blot analyses indicated that some hybrid calli had nuclei of *L. corniculatus* and chloroplast genomes of alfalfa. This fact proved that the calli obtained were true asymmetric hybrids. The recalcitrance in shoot regeneration by the calli may be caused by the imbalance in morphogenetic potential of the nucleus and the chloroplast genome of the two species concerned.

Sano et al. (1988) induced protoplast fusion hybrids between *L. corniculatus* and soybean and between *L. corniculatus* and alfalfa. These results, which represent one of the few successful cases of wide hybridization between leguminous species, show the possibility of obtaining somatic hybrids in leguminous species in addition to those reported between species in the Solanaceae and Cruciferae. However, even in species in the Cruciferae, male sterility or low fertility often occurs in interspecific somatic hybridization (Kirti et al. 1992; Lelivelt and Krens 1992).

Hausen and Earle (1997) suggested that one reason for the low fertility might be alloplasmic male sterility caused by incompatibility between the nucleus and the cytoplasm genome. On the other hand, Nothnagel et al. (1997) indicated that backcrosses with one parent might become a useful tool in some cases in the Cruciferae in overcoming male sterility or low fertility. Somatic hybrids of *L. corniculatus* and *L. conimbricensis* also showed male sterility in which no seed was obtained. Thus, from the point of view of plant breeding, it may be worthwhile to attempt to back-cross somatic hybrids with one parent. Progeny of a somatic hybrid of birdsfoot trefoil and rice were obtained when the somatic hybrids were backcrossed with birdsfoot trefoil as the pollen parent (Saito et al. 2005). The progeny plants showed a vigorous increase in plant height and shoot number in the early stage of plant growth; at flowering, the mean flower number of the plants was almost twice as great as that of the plants of the parent cultivar.

6.16.4 Genetic Transformation

That *L. corniculatus* readily regenerates plants in culture and is amenable to transformation by *A. tumefaciens* and *A. rhizogenes* makes it a very suitable species for testing genetic manipulation strategies (Armstead and Webb 1987; Akashi et al. 1998b; Aoki, Kamizawa, and Ayabe 2002; Zhang, Chu, and Fu 2002; De Marchis, Bellucci, and Arcioni 2003). Bacterial genes of *A. rhizogenes* transferred into a plant induce the growth of distinctive roots at the infection site (Tepfer 1984). In culture, these roots continue to grow without a supply of exogenous hormone supplements. They are negatively geotropic, produce opines, and are cytologically stable (Aird, Hamill, and Rhodes 1988). Regenerated transformed plants from hairy roots have the associated hairy root phenotype. In *L. corniculatus,* this transformation results in only minor morphological effects— mainly in flower morphology—but it also significantly reduces fertility (Webb et al. 1990). No adverse effects, however, have been found on the plants' nitrogenase levels or on the ability to fix nitrogen. Such regenerated plants have proved especially valuable for genes involved in nodulation. Nikolic and Mitic (2003) considered that transformed plants regenerated from root sections taken from seedlings transformed by *A. tumefaciens* would be useful to obtain desirable traits and for examining function and expression of transgenes.

A new, continuous culture system of super-growing roots (super roots) has been reported in *L. corniculatus* (Akashi, Hoffmann-Tsay, and Hoffmann 1998a). More than 3 years after initiation, the super roots continue to grow at the initial high rate. They are readily cloned from secondary tips and easily regenerate plants upon transfer to light. The complete system, from primary root culture to plant formation, progresses without need for exogenous hormones. Akashi et al. (2000) reported that regenerated plants from super roots appear morphologically normal and undergo nodulation. Transformation with *nod* genes, combined with the possibility of re-establishing super-growing root cultures from transformed tissues and regenerating plants under hormone-free conditions, may provide a perspective for future nodulation research. In a study of nodulation, Nukui, Ezula, and Minamisawa (2004) produced transgenic *L. japonicus* carrying the mutated melon ethylene receptor gene that confers ethylene insensitivity. When inoculated with *Mesorhizobium loti,* transgenic plants showed a markedly higher number of infection threads and nodule primordia on their roots than wild type. This fact indicates clearly that the ethylene combined receptor gene inhibits the establishment of symbiosis between rhizobia and legumes.

Successful genetic transformation of plants involves not only the production of primary trans-formants showing stable expression of inserted genes but also the inheritance and continued expression of these genes in subsequent generations. However, awareness is increasing that newly inserted genes can be silenced (Finnegan and McElroy 1994; Ulian, Magill, and Smith 1994). *A. rhizogenes* was assessed as a vehicle for transformation of *L. corniculatus* (Webb et al. 1994a, 1994b). Plants were cotransformed using *A. rhizogenes* strain LBA9402, which harbored the bacterial plasmid pRi1855 and the binary transformation vector pJit73. The vector pRi1855 transfers both TL and TR sequences, while pJit73 encodes β-glucuronidase (GUS) as well as two selectable marker genes giving resistance to the antibiotics kanamycin and hygromycin.

Two primary transformants (lines 6 and 12) were resistant to hygromycin and showed a significantly lower GUS activity in line 6 than in line 12. Genetic analysis of progeny indicated that line 6 had one dose of the *uid* gene (GUS gene), and line 12 had two or more independently segregating doses of the gene. Both lines 6 and 12 contained multiple copies of TL-DNA, but only line 6 was TR positive. GUS-positive progeny, which were free of both TL and TR sequences, were identified from both lines. Two out of six progenies of line 6 contained the *uid* gene; however, they did not have detectable GUS activity, although in one of them the tissue was resistant to hygromycin. This suggests silencing of GUS activity in some of the progeny of line 6 (Webb et al. 1994a, 1994b).

Gene silencing was found in *L. corniculatus* plants transformed by using *A. rhizogenes* (Bellucci et al. 1999). Plants with a maize cDNA (G1L) encoding a sulfur-rich γ-zein were obtained by using two fusion genes: one with the CaMV 35S promoter and the other with the RuBisCO small subunit (rbcS) promoter. The highest expression of G1L mRNA was found in plants transformed with G1L under the rbcS promoter. The steady level of G1L mRNA in the leaves was directly correlated with the G1L copy number. Kumagai and Kouchi (2003) artificially induced gene silencing in *L. japonicus*. They investigated the efficacy of self-complementary hairpin RNA expression to induce RNA silencing in the roots and nodules, using hairy root transformation mediated by *A. rhizogenes*. Their results indicated that transient RNA silencing by hairy root transformation provides a powerful tool for loss-of-function analyses of genes that function in roots and root nodules.

Transformation has been frequently used for studies on plant physiological pathways. Higher plants assimilate nitrogen in the form of ammonia through the concerted activity of glutamine synthetase (GS) and glutamate synthase (GOGAT). The enzyme is located in the cytoplasm (GS_1) or in the chloroplast (GS_2). To understand how GS activity affects plant performance, *L. japonicus* plants were transformed with an alfalfa GS_1 gene driven by the CaMV 35S promoter. The transformants showed increased GS activity and an increase in GS_1 polypeptide level in all the organs tested (Ortega et al. 2004). Furthermore, the results consistently showed higher protein concentration, higher chlorophyll content, and higher biomass accumulation in the transgenic plants. The total amino acid content in the leaves and stems of the transgenic plants was 22–24% more than in the tissues of the nontransformed plants. The relative abundance of individual amino acids was similar, except for aspartate/asparagine and proline, which were higher in the transformants.

In vascular plants, glutamine-dependent asparagine synthetase (AS) is the primary source of asparagine. In *E. coli*, asparagine is synthesized by the action of two distinct enzymes: AS-A, which utilizes ammonia as a nitrogen donor, and AS-B, which utilizes both glutamine and ammonia as substrates (Bellucci et al. 2004). The possibility to endow plants with ammonia-dependent AS activity was investigated by heterologous expression of the *E. coli asnA* gene with the aim of introducing a new ammonia assimilation pathway. The bacterial gene was placed under the control of light-dependent promoters and introduced by transformation into *L. corniculatus* plants. Analysis of the transgenic plants, however, indicated transgene silencing, which prevented *asnA* expression in several transgenics. The *asnA*-expressed plants were characterized by premature flowering and reduced growth. A significant reduction of the total free amino acid accumulation was observed in transgenic plants. The content of asparagine in wild-type plants was about 2.5-fold higher than that of transgenic plants. Although glutamine levels in transgenic plants were about three- to fourfold higher than those in wild-type plants, aspartate levels were significantly lower.

On the other hand, a Δ6-fatty-acid desaturase gene, *1S-4*, isolated from the fungus *Mortierella alpina* was introduced into *L. japonicus* by *Agrobacterium*-mediated transformation and constitutively expressed (Chen et al. 2005). Two transgenic *L. japonicus* lines accumulated γ-linolenic acid (GLA) not found in the host at levels of up to 11.34% of the total fatty acids in the leaves. These results indicate that legumes can synthesize GLA by expressing the fungus gene. The ability to synthesize GLA was inherited in the leaves and the seeds. These findings reveal that it is possible to modify the fatty-acid biosynthetic pathway by genetic manipulation in order to produce specific polyunsaturated fatty acids in legume crops.

Root cultures of *L. corniculatus* transformed with *A. rhizogenes* grow rapidly in liquid medium when cultured in the dark, and they produce large numbers of shoots when illuminated (Morris and Robbins 1992). The shoots, which could be regenerated to produce some fertile plants, were maintained in liquid medium as shoot-organ cultures. The accumulation and cellular distribution of condensed tannins increased at a ratio equivalent to the control plants. Condensed tannin accumulation was linearly related to root growth and had a similar spatial distribution in tannin cells in roots and leaves compared to controls.

The first example of the genetic modification of tannin biosynthesis in *L. corniculatus* was the introduction of a heterologous dihydroflavonol reductase gene into hairy root cultures. Transgenic *L. corniculatus* plants harboring antisense dihydroflavonol reductase (AS-DFR) sequences have been produced and analyzed (Carron et al. 1994; Robbins et al. 1994, 1998). The effect of introducing three different antisense *Antirrhinum majus* DFR constructs into a single recipient genotype (S50) had no obvious effects on plant biomass, but levels of condensed tannins showed a statistical reduction in leaf, stem, and root tissues in some of the antisense lines. This is the first report of genetic manipulation of condensed tannin biosynthesis in higher plants (Robbins et al. 1998).

Legume species characteristically accumulate phenylpropanoid phytoalexins under conditions of biological stress such as a pathogen attack or wounding. When *A. rhizogenes*-transformed root cultures of *L. corniculatus* were treated with glutathione, isoflavan phytoalexins accumulated in tissue and in the culture medium (Robbins, Hartnoll, and Morris 1991; Robbins, Thomas, and Morris 1995). This accumulation of phytoalexins was preceded by a transient increase in the activity of phenylalanine ammonia lyase (PAL). Elicitation of PAL occurred throughout the growth curve of *Lotus* hairy roots and in different sectors of transformed root material. Some workers have studied the action of elicitors in experiments on whole plant material; however, few experiments using disorganized legume tissue cultures have been conducted, although they are undoubtedly of value.

6.17 GENE FLOW FROM BIRDSFOOT TREFOIL USING TRANSGENES AS TRACER MARKERS

Pollen dispersal is a major issue in the risk assessment of transgenic crop plants. De Marchis et al. (2003) used transformed plants of *L. corniculatus* as the pollen donors and nontransgenic plants as recipients. The asparagine synthetase gene (*asnA*) was used as a tracer marker. In one location, flow of the *asn*A transgene in nontransgenic plants was detected up to 18 m from a 1.8-m^2 donor plot. In a second location, pollen dispersal occurred up to 120 m from a 14-m^2 pollinating plot. Thus, transgene flow can occur in birdsfoot trefoil, indicating that the size of the recipient "sink" and that of the transgene source plot both influence the degree to which gene flow occurs.

6.18 BIRDSFOOT TREFOIL AND GLOBAL WARMING

Scientists at Agriculture and Agri-Food Canada (Bélanger et al. 2002) predict that perennial forage crops will be at a greater risk of winter injury in the future even though the climate will be warmer. Predicted increases of 2–6°C in minimum temperature during winter months will likely affect survival of perennial forage crops. The loss of snow cover due to warmer winter conditions will increase exposure of plants to subfreezing temperatures and killing frosts and cause loss of cold hardiness during warm periods, soil heaving, and ice encasement—all of which will result in frequent losses of forage stands. Forage crops are also likely to enter the winter in a lower state of cold hardening in the fall due to warmer fall temperatures.

The preceding scenario applies to northeastern Canada and adjacent regions in the United States. It is expected that birdsfoot trefoil cultivars will continue to be developed to meet climatic conditions for winter hardiness as they have in the past. Selection of foreign germplasm shows that winter-hardy birdsfoot trefoil plants may still be obtained (McGraw, Beuselinck, and Marten 1989). Traditional breeding, such as selection from synthetic combinations and phenotypic recurrent selection and progeny testing, can be employed to produce birdsfoot trefoil cultivars with more intense expression of winter hardiness (Papadopoulos and Kelman 1999). New technologies based on laboratory freezing tests and the identification of molecular markers may facilitate the future development of winter-hardy cultivars (Bélanger et al. 2006). The use of mutagens, x-rays, and chemicals has still to be explored to develop winter-hardy plants (Therrien and Grant 1983). Thus, it may be expected that birdsfoot trefoil cultivars will be continually developed for the winter-hardiness trait in northern regions.

6.19 FUTURE OUTLOOK

Birdsfoot trefoil is an excellent perennial, nonbloating, self-seeding forage species for cattle and sheep that has the potential to become a major crop in temperate forage-producing regions of the world. The forage has excellent protein quality that is acceptable even after seed has set, with lower cellulose content and more carbohydrate than alfalfa and most clover. Milk from cows fed on trefoil hay contains more vitamins A and E than that from cows given alfalfa hay. The plant is more tolerant of environmental extremes of drought and moisture and is a suitable low-input substitute for white and red clover on marginal, less fertile soils or free-draining calcareous soils. It will grow on more acid soils than alfalfa but at the same time responds to high fertility and fixes nitrogen for associated grasses.

Plants are slow to become established because of slow seedling growth; therefore, weed control during this phase is very important. Crop competition has been observed to reduce yield. Under warm, humid conditions, fungal diseases such as *Sclerotium rolfsii* and *Rhizoctonia solani* attack the crown and lower foliage, causing leaf blight, which may lead to death of affected plants. However, careful management during and after establishment may greatly reduce or eliminate such problems.

A major drawback is pod shattering (dehiscence) for seed producers. The production of an indehiscent cultivar would have a significant and positive impact for seed producers. Because a wide array of molecular and genetic techniques is now available, this problem should be resolved in the near future.

A wide range of cellular and molecular techniques that offer unlimited opportunity for the plant breeder is now available. Somaclonal variation should become widely used for crop improvement. The techniques are simpler than those of recombinant DNA, and they result in a rich source of genetic variability. In addition, somaclonal variation has been free of the regulatory hurdles that plague products of recombinant DNA. Plants can be transferred directly to the field and evaluated as part of an ongoing breeding program. A tight linkage between tissue culturists/geneticists and breeders is imperative to advance this technology.

The practicality of a breeding program for quantitative characters including seedling vigor and low HCN by using protoclones in *L. corniculatus* is feasible. Birdsfoot trefoil readily regenerates plants in culture and is amenable to transformation by *Agrobacterium tumefaciens* and *A. rhizogenes,* which induce crown gall and hairy root, respectively. These qualities make it a particularly suitable species for testing genetic manipulation strategies. Transformation with *nod* genes, combined with the possibility of re-establishing super growing root cultures from transformed tissues and regenerating plants under hormone-free conditions, may provide a perspective for future nodulation research. Genetic modification of tannin biosynthesis has been possible. It is clear that an integrated approach between the plant breeder and the molecular biologist is necessary.

ACKNOWLEDGMENTS

We thank Dr. Ginette Séguin-Swartz, Agriculture and Agri-Food Canada, Saskatoon, Saskatchewan, for her critical appraisal and constructive comments on the manuscript and also those by the forage specialist Dr. Philippe Séguin, Department of Plant Science, Macdonald Campus, McGill University.

REFERENCES

Aerts, R. J. et al. 1999. Condensed tannins from *Lotus corniculatus* and *Lotus pedunculatus* exert different effects on the in vitro rumen degradation of ribulose-1,5-bisphosphate carboxylase/oxygenase (Rubisco) protein. *Journal of the Science of Food Agriculture* 79:79–85.

Aird, E. L. H., J. D. Hamill, and M. J. C. Rhodes. 1988. Cytogenetic analysis of hairy root cultures from a number of plant species transformed by *Agrobacterium rhizogenes*. *Plant Cell, Tissue and Organ Culture* 15:47–57.

Akamine, S. et al. 2003. cDNA cloning, mRNA expression, and mutational analysis of the squalene synthase gene of *Lotus japonicus*. *Biochimica et Biophysica Acta* 1626:97–101.

Akashi, R., S.-S.Hoffmann-Tsay, and F. Hoffmann. 1998a. Selection of a super-growing legume root culture that permits controlled switching between root cloning and direct embryogenesis. *Theoretical and Applied Genetics* 96:758–764.

Akashi, R. et al. 1998b. High-frequency embryogenesis from cotyledons of bird's-foot trefoil (*Lotus corniculatus*) and its effective utilization in *Agrobacterium tumefaciens*-mediated transformation. *Journal of Plant Physiology* 152:84–91.

Akashi, R. et al. 2000. Plant from protoplasts isolated from a long-term root culture (super root) of *Lotus corniculatus*. *Journal of Plant Physiology* 157:215–221.

Allan, C. E. et al. 2000. Adaptations of *Lotus pedunculatus* Cav. for nitrogen fixation in a riverine wetland plant community. *Botany Journal of Scotland* 52:149–158.

Altier, N. 2003. Caracterización de la población de *Fusarium oxysporum* y potencial patogénico del suelo bajo rotaciones agrícola ganaderas. In Rotaciones. Montevideo, *INIA, Serie Técnica* 134:37–44.

Altier, N. A., N. J. Ehlke, and M. Rebuffo. 2000. Divergent selection for resistance to fusarium root rot in birdsfoot trefoil. *Crop Science* 40:670–675.

Aoki, T., A. Kamizawa, and S. Ayabe. 2002. Efficient *Agrobacterium*-mediated transformation of *Lotus japonicus* with reliable antibiotic selection. *Plant Cell Reports* 21:238–243.

Armstead, I. P., and K. J. Webb. 1987. Effect of age and type of tissue on genetic transformation of *Lotus corniculatus* by *Agrobacterium tumefaciens*. *Plant Cell, Tissue and Organ Culture* 9:95–101.

Artola, A., and G. Carrillo-Castañeda. 2005. The bulk conductivity test for birdsfoot trefoil seed. *Seed Science and Technology* 33:231–236.

Artola, A., G. Carrillo-Castañeda, and G. Garcia de los Santos. 2003. Hydropriming: A strategy to increase *Lotus corniculatus* L. seed vigor. *Seed Science and Technology* 31:455–463.

Artola, A., G. Garcia de los Santos, and G. Carrillo-Castañeda. 2003. A seed vigor test for birdsfoot trefoil. *Seed Science and Technology* 31:753–757.

Arumuganathan, K., and E. D. Earle. 1991. Nuclear DNA content of some important plant species. *Plant Molecular Biology Reporter* 9:208–218.

Asuaga, A. 1994. *Lotus subflorus* cv. El Rincon, a new alternative for extensive improvements of natural pastures. In *Proceedings of the First International Lotus Symposium*, Missouri Botanical Gardens, ed. P. R. Beuselinck and C. A. Roberts, 147–149. Columbia, MO: Univ. of Missouri Ext. Publ., Univ. of Missouri.

Ayres, J. F. et al. 2006. Birdsfoot trefoil (*Lotus corniculatus*) and greater Lotus (*Lotus uliginosus*) in perennial pastures in eastern Australia. 2. Adaptation and applications of Lotus-based pasture. *Australian Journal of Experimental Agriculture* 46:521–534.

Ayres, J. F. et al. 2007. Regeneration characteristics of birdsfoot trefoil (*Lotus corniculatus* L.) in low latitude environments in eastern Australia. *Australian Journal of Experimental Agriculture* 47:833–843.

Ayres, J. F. et al. 2008. Developing birdsfoot trefoil (*Lotus corniculatus* L.) varieties for permanent pasture applications in low latitude regions of eastern Australia. *Australian Journal of Experimental Agriculture* 48:488–498.

Aziz, M. A. et al. 1990. Somatic hybrids between the forage legumes *Lotus corniculatus* L. and *L. tenuis* Waldst et Kit. *Journal of Experimental Botany* 41:471–479.

Bagnasco, P. et al. 1998. Fluorescent *Pseudomonas* spp. as biocontrol agents against forage legume root pathogenic fungi. *Soil Biology and Biochemistry* 30:1317–1322.

Bajaj, Y. P. S. 1990. Somaclonal variation in crop improvement I. Biotechnology in agriculture and forestry, vol. 11. Berlin: Springer–Verlag.

Bansal, R. D. 1971. Inheritance of cyanogenetic enzyme in *Lotus corniculatus*. *Indian Journal of Agricultural Science* 41:67–69.

Bañuelos, G. S. et al. 1996. Accumulation of selenium by different plant species grown under increasing sodium and calcium chloride salinity. *Plant and Soil* 183:49–59.

Baraibar, A. et al. 1999. Symbiotic effectiveness and ecological characterization of indigenous *Rhizobium loti* populations in Uruguay. *Pesquisa Agropecuária Brasileira* 34:1011–1017.

Barufaldi, M. et al. 2007. A systems study of *Lotus*'s leaf area. *Kybernetes* 36:225–235.

Bazin, A., S. Blaise, and D. Cartier. 1994. Polymorphism study of two defense mechanisms in French populations of *Lotus corniculatus* L.: Cyanide and condensed tannins. In *Proceedings of First International Lotus Symposium*, Missouri Botanical Garden, St. Louis, MO, ed. P. R. Beuselinck and C. A. Roberts, 193–198.

Bazin, A. et al. 2002. Influence of atmospheric carbon dioxide enrichment on induced response and growth compensation after herbivore damage in *Lotus corniculatus*. *Ecology and Entomology* 27:271–278.

Bélanger, G., J. E. Winch, and J. L. Townshend. 1985. Carbofuran effects on establishment of legumes in relation to plant growth, nitrogen fixation, and soil nematodes. *Canadian Journal of Plant Science* 65:423–433.

Bélanger, G. et al. 2002. Climate change and winter survival of perennial forage crops in eastern Canada. *Agronomy Journal* 94:1120–1130.

Bélanger, G. et al. 2006. Winter damage to perennial forage crops in eastern Canada: Causes, mitigation, and prediction. *Canadian Journal of Plant Science* 86:33–47.

Bellucci, M. et al. 1999. Transcription of a maize cDNA in *Lotus corniculatus* is regulated by T-DNA methylation and transgene copy number. *Theoretical and Applied Genetics* 98:257–264.

———. 2004. Transformation of *Lotus corniculatus* plants with *Escherichia coli* asparagine synthetase A: Effect on nitrogen assimilation and plant development. *Plant Cell, Tissue and Organ Culture* 78:139–150.

Bennett, M. D., and J. B. Smith. 1976. Nuclear DNA amounts in angiosperms. *Philosophical Transactions of the Royal Society of London, Series B* 274:227–274.

Bent, F. C. 1962. Interspecific hybridization in the genus *Lotus*. *Canadian Journal of Genetics and Cytology* 4:151–159.

Bergstrom, G. C., and D. W. Kalb. 1995. *Fusarium oxysporum* f. sp. *loti*: A specific wilt pathogen of birdsfoot trefoil in New York. *Phytopathology* 85:1555.

Bernes, G., P. J. Waller, and D. Christensson. 2000. The effect of birdsfoot trefoil (*Lotus corniculatus*) and white clover (*Trifolium repens*) in mixed pasture swards on incoming and established nematode infections in young lambs. *Acta Veterinaria Scandinavica* 41:351–361.

Beuret, E. 1977a. Sur la présence dans d'Apennia central d'une race hexaploides de *Lotus corniculatus*. *Bulletin de la Société Neuchâteloise des Sciences Naturelles* 100:107–112.

———. 1977b. Contribution à l'étude de la distribution géographique et de la physiologie de taxons affines di- et polyploïdes. *Bibliotheca Botanica* 133:1–80.

Beuselinck, P. R. 1992. Rhizomes in birdsfoot trefoil: A progress report. In 12th Trifolium Conference, Gainesville, FL, 37–38.

———. 1994. The rhizomes of *Lotus corniculatus* L. In *Proceedings of First International* Lotus *Symposium,* ed. P. R. Beuselinck and C. A. Roberts, 215–219, St. Louis, MO. Columbia: Univ. of Missouri Ext. Publ. MX 411.

———, ed. 1999. *Trefoil: The science and technology of* Lotus, 1–266. Madison, WI: American Society of Agronomy and Crop Science Society of America, CSSA Special Publ. No. 28.

Beuselinck, P. R., and W. F. Grant. 1995. Birdsfoot trefoil. In *Forages. Vol. 1: An introduction to grassland agriculture*, 5th ed., ed. R. F. Barnes, D. A. Miller, and C. J. Nelson, 237–248. Ames: Iowa State Univ. Press.

Beuselinck, P. R., B. Li, and J. J. Steiner. 1996. Rhizomatous *Lotus corniculatus* L: 1. Taxonomic and cytological study. *Crop Science* 36:179–185.

Beuselinck, P. R., and R. L. McGraw. 1983. Seedling vigor of three *Lotus* species. *Crop Science* 23:390–391.

———. 2000. Vestigial corolla in flowers of birdsfoot trefoil. *Crop Science* 40:964–967.

Beuselinck, P. R., E. J. Peters, and R. L. McGraw. 1984. Cultivar and management effects on stand persistence of birdsfoot trefoil. *Agronomy Journal* 76:490–492.

Beuselinck, P. R., and J. J. Steiner. 1994. Registration of CAD birdsfoot trefoil germplasm selected for drought resistance. *Crop Science* 34:543.

———. 1996. Registration of ARS-2620 birdsfoot trefoil. *Crop Science* 36:1414.

———. 2003. Registration of ARS-2622 birdsfoot trefoil germplasm. *Crop Science* 43:1886.

Beuselinck, P. R., J. J. Steiner, and Y. W. Rim. 2003. Morphological comparison of progeny derived from 4x–2x and 4x–4x hybridizations of *Lotus glaber* Mill. and *L. corniculatus* L. *Crop Science* 43:1741–1746.

Beuselinck, P. R. et al. 2005. Genotype and environment affect rhizome growth of birdsfoot trefoil. *Crop Science* 45:1736–1740.

Blumenthal, M. J., and R. L. McGraw. 1999. *Lotus* adaptation, use, and management. In *Trefoil: The science and technology of* Lotus, ed. P. R. Beuselinck, 97–119. Madison, WI: American Society of Agronomy and Crop Science Society of America, CSSA Special Publ. No. 28.

Boerboom, C. M., and D. L. Wyse. 1988. Selective application of herbicides for Canada thistle (*Cirsium arvense*) control in birdsfoot trefoil (*Lotus corniculatus*). *Weed Technology* 2:183–186.

Boerboom, C. M. et al. 1991. Recurrent selection for glyphosate tolerance in birdsfoot trefoil. *Crop Science* 31:1124–1129.

Brand, A. 1898. Monographie der Gattung *Lotus*. *Botanische Jahrbücher für Systematik, Pflanzengeschichte und Pflanzengeographie* 25:166–232.

Brummer, E. C., and K. J. Moore. 2000. Persistence of perennial cool-season grass and legume cultivars under continuous grazing by beef cattle. *Agronomy Journal* 92:466–471.

Bubar, J. S. 1956. Flower color inheritance studies in broadleaf birdsfoot trefoil (*Lotus corniculatus* L.). *Agricultural Institute Review* 11 (3): 37.

Bubar, J. S., and N. C. Lawson. 1959. Note on inheritance of ability to survive winterkilling conditions in birdsfoot trefoil. *Canadian Journal of Plant Science* 39:125–126.

Bubar, J. S., and R. K. Miri. 1965. Inheritance of self-incompatibility and brown keel tip in *Lotus corniculatus* L. *Nature* 205:1035–1036.

Bullard, M. J. and T. J. Crawford. 1996. Seed yield, size and indeterminacy in diverse accessions of *Lotus corniculatus* L. grown in the UK. *Plant Varieties and Seeds* 9:21–28.

Buzzell, R. I., and C. P. Wilsie. 1963. Genetic investigations of brown keel tip color in *Lotus corniculatus* L. *Crop Science* 3:128–130.

Cameron, B. G., and N. Prakash. 1992. The floral and seed structure of a poisonous species of *Lotus* from Australia. *Lotus Newsletter* 23:1–5.

Campos, L. P., J. V. Raelson, and W. F. Grant. 1994. Genome relationships among *Lotus* species based on random amplified polymorphic DNA (RAPD). *Theoretical and Applied Genetics* 88:417–422.

Carr, P. M., W. W. Poland, and L. J. Tisor. 2005. Forage legume regeneration from the soil seed bank in western North Dakota. *Agronomy Journal* 97:505–513.

Carron, T. R., M. P. Robbins, and P. Morris. 1994. Genetic modification of condensed tannin biosynthesis in *Lotus corniculatus*. 1. Heterologous antisense dihydroflavonol reductase down-regulates tannin accumulation in "hairy root" culture. *Theoretical and Applied Genetics* 87:1006–1015.

Carron, T. R. et al. 1992. Condensed tannin levels in different tissues and different developmental stages of transformed and nontransformed *Lotus corniculatus*. *Lotus Newsletter* 23:49–52.

Carter, E. B., M. K. Theodorou, and P. Morris. 1999. Responses of *Lotus corniculatus* to environmental change. 2. Effect of elevated CO_2, temperature and drought on tissue digestion in relation to condensed tannin and carbohydrate accumulation. *Journal of the Science of Food Agriculture* 79:1431–1440.

Cassida, K. A. et al. 2000. Protein degradability and forage quality in maturing alfalfa, red clover, and birdsfoot trefoil. *Crop Science* 40:209–215.

Chen, C.-C., and W. F. Grant. 1968. Morphological and cytological identification of the primary trisomics of *Lotus pedunculatus* (Leguminosae). *Canadian Journal of Genetics and Cytology* 10:161–179.

Chen, R. et al. 2005. Production of the γ-linolenic acid in *Lotus japonicus* and *Vigna angularis* by expression of the Δ6-fatty-acid desaturase gene isolated from *Mortierella alpina*. *Plant Science* 169:599–605.

Cheng, R. I.-J., and W. F. Grant. 1973. Species relationships in the *Lotus corniculatus* group as determined by karyotype and cytophotometric analyses. *Canadian Journal of Genetics and Cytology* 15:101–115.

Chien, Y. C., K. N. Kao, and L. R. Wetter. 1982. Chromosomal and isozyme studies of *Nicotiana tabacum-Glycine max* hybrid cell lines. *Theoretical and Applied Genetics* 62:301–304.

Chrtkova-Zertova, A. 1966. Bemerkungen zur Taxonomie von *Lotus uliginosus* Schkuhr und *L. pedunculatus* Cav. *Folia Geobotanica et Phytotaxonomica* 1:78–87.

———. 1970. Bemerkungen zur Taxonomie und Nomenklatur von *Lotus krylovii* Schischk. et Serg. und *L. corniculatus* L. subsp. *frondosus* Freyn. *Folia Geobotanica et Phytotaxonomica* 5:89–97.

———. 1973. A monographic study of *Lotus corniculatus* L. I. Central and northern Europe. *Rozpravy Ceskoslovenské Akademie Ved. Rada Matematickych a Prírodn i ch Ved. Academia, Praha* 83 (4): 1–94.

Colares, M., M. M. Mujica, and C. P. Rumi. 2000. Analysis of the early expression characters for the cultivars differentiation of *Lotus glaber* Mill. (= *Lotus tenuis* (Waldst. et Kit.) *Lotus Newsletter* 31.

Compton, S. G., and D. A. Jones. 1985. An investigation of the responses of herbivores to cyanogenesis in *Lotus corniculatus* L. *Biological Journal of Linnean Society* 26:21–38.

Croser, J. S. et al. 2006. Toward doubled haploid production in the Fabaceae: Progress, constraints, and opportunities. *Critical Reviews in Plant Sciences* 25:139–157.

Cuomo, G. J., D. G. Johnson, and W. A. Head. 2001. Interseeding kura clover and birdsfoot trefoil into existing cool-season grass pastures. *Agronomy Journal* 93:458–462.

Cuomo, G. J. et al. 2000. Pasture renovation and grazing management impacts on cool-season grass pastures. *Journal of Production Agriculture* 12:564–569.

Dalrymple, E. J., B. P. Goplen, and R. E. Howarth. 1984. Inheritance of tannins in birdsfoot trefoil. *Crop Science* 24:921–923.

Damiani, F., M. Pezzotti, and S. Arcioni. 1990. Somaclonal variation in *Lotus corniculatus* L. in relation to plant breeding purposes. *Euphytica* 46:35–41.

Damiani, F. et al. 1985. Variation among plants regenerated from tissue culture of *Lotus corniculatus* L. *Zeitschrift für Pflanzenzüchtung* 94:332–3395.

Dawson, C. D. R. 1941. Tetrasomic inheritance in *Lotus corniculatus* L. *Journal of Genetics* 42:49–72.

De Bruijin, F. J. et al. 1994. Regulation of nodulin gene expression. *Plant and Soil* 161:59–68.

De Lautour, G., W. T. Jones, and M. D. Ross. 1978. Production of interspecific hybrids in *Lotus* aided by endosperm transplants. *New Zealand Journal of Botany* 16:61–68.

De Marchis, F., M. Bellucci, and S. Arcioni. 2003. Measuring gene flow from two birdsfoot trefoil (*Lotus corniculatus*) field trials using transgenes as tracer markers. *Molecular Ecology* 12:1681–1685.

de Rijke, E. et al. 2004. Flavonoids in Leguminosae: Analysis of extracts of *T. pratense* L., *T. dubium* L., *T. repens* L., and *L. corniculatus* L. leaves using liquid chromatography with UV, mass spectrometric and fluorescence detection. *Analytical and Bioanalytical Chemistry* 378:995–1006.

Dobrofsky, S., and W. F. Grant. 1980. Electrophoretic evidence supporting self incompatibility in *Lotus corniculatus*. *Canadian Journal of Botany* 58:712–716.

Donovan, L. S., and H. A. McLennan. 1964. Further studies on the inheritance of leaf size in broadleaf birdsfoot trefoil, *Lotus corniculatus* L. *Canadian Journal of Genetics and Cytology* 6:164–169.

Doucette, K. M., K. M. Wittenberg, and W. P. McCaughey. 2001. Seed recovery and germination of reseeded species fed to cattle. *Journal of Range Management* 54:575–581.

Douglas, G. B. et al. 1995. Liveweight gain and wool production of sheep grazing *Lotus corniculatus* and lucerne (*Medicago sativa*). *New Zealand Journal of Agricultural Research* 38:95–104.

———. 1999. Effect of condensed tannins in birdsfoot trefoil (*Lotus corniculatus*) and sulla (*Hedysarum coronarium*) on body weight, carcass fat depth, and wool growth of lambs in New Zealand. *New Zealand Journal of Agricultural Research* 42:55–64.

Duke, J. A. 1981. *Handbook of legumes of world economic importance,* 125–131. New York: Plenum Press.

El-Assal, S. E. et al. 2001. A QTL for flowering time in *Arabidopsis* reveals a novel allele of *CRY2*. *Nature* 29:435–440.

Elliott, F. C., 1946. The inheritance of self-incompatibility in *Lotus tenuis* Wald. et Kit. MSc thesis, Iowa State Univ., Ames.

Ellis, C. R., and F. L. O. Nang'ayo. 1992. The biology and control of the trefoil seed chalcid, *Bruchophagus platypterus* (Walker) (Hymenoptera: Eurytomidae) in Ontario. *Proceedings of the Entomology Society of Ontario* 123:111–122.

Emery, K. M., P. Beuselinck, and J. T. English. 1999. Evaluation of the population dynamics of the forage legume *Lotus corniculatus* using matrix population models. *New Phytology* 144:549-560.

———. 2003. Genetic diversity and virulence of *Rhizoctonia* species associated with plantings of *Lotus corniculatus*. *Mycological Research* 107:183–189.

English, J. T. 1992. Modular demography of *Lotus corniculatus* infected by *Rhizoctonia*. *Phytopathology* 82:1104.

———. 1999. Diseases of *Lotus*. In *Trefoil: The science and technology of* Lotus, ed. P. R. Beuselinck, 121–131. Madison, WI: American Society of Agronomy and Crop Science Society of America, CSSA Special Publ. No. 28.

English, J. T., and P. R. Beuselinck. 2000. Methods for evaluating birdsfoot trefoil for susceptibility to foliar and shoot blight caused by *Rhizoctonia* spp. *Crop Science* 40:841–843.

Erbe, L. W. 1955. A study of the fertility relationships of *Lotus corniculatus,* tetraploid *L. tenuis* and their F_1 hybrid progeny. MSc thesis, Univ. Vermont, Burlington.

Fairey, D. T., and R. R. Smith. 1999. Seed production in birdsfoot trefoil, *Lotus* species. In *Trefoil: The science and technology of* Lotus, ed. P. R. Beuselinck, 145–166. Madison, WI: American Society of Agronomy and Crop Science Society of America, CSSA Special Publ. No. 28.

Favarger, C. 1953. Notes de caryologie alpine II. *Bulletin de la Société Neuchâteloise des Sciences Naturelles* 76:137–139.

Ferrandiz, C., S. J. Liljegren, and M. F. Yanofsky. 2000. Negative regulation of the SHATTERPROOF genes by FRUITFULL during *Arabidopsis* fruit development. *Science* 289:436–438.

Finnegan, J., and D. McElroy. 1994. Transgene inactivation: Plants fight back. *Bio/Technology* 12:883–888.

Fjellstrom, R. G., P. R. Beuselinck, and J. J. Steiner. 2001. RFLP marker analysis supports tetrasomic inheritance in *Lotus corniculatus* L. *Theoretical and Applied Genetics* 102:718–725.

Fjellstrom, R. G., J. J. Steiner, and P. R. Beuselinck. 2003. Tetrasomic linkage mapping of RFLP, PCR, and isozyme loci in *Lotus corniculatus* L. *Crop Science* 43:1006–1020.

Frame, J., J. F. L. Charlton, and A. S. Laidlaw. 1998. *Temperate forage legumes*. Wallingford, U.K.: CAB International, 327 pp.

Fulkerson, W. J. et al. 2007. Nutritive value of forage species grown in the warm temperate climate of Australia for dairy cows: Grasses and legumes. *Livestock Science* 107:253–264.

Garcia de los Santos, G., J. J. Steiner, and P. R. Beuselinck. 2001. Adaptive ecology of *Lotus corniculatus* L. genotypes: II. Crossing ability. *Crop Science* 41:564–570.

Garcia-Diaz, C. A., and J. J. Steiner. 1999. Birdsfoot trefoil seed production: I. Crop-water requirements and response to irrigation. *Crop Science* 39:775–783.

———. 2000a. Birdsfoot trefoil seed production: II. Plant-water status on reproductive development and seed yield. *Crop Science* 40:449–456.

———. 2000b. Birdsfoot trefoil seed production: III. Seed shatter and optimal harvest time. *Crop Science* 40:457–462.

Gauthier, P., R. Lumaret, and A. Bedecarrats. 1998a. Genetic variation and gene flow in alpine diploid and tetraploid populations of *Lotus* (*L. alpinus* (D.C.) Schleicher *L. corniculatus* L.). II. Insights from morphological and allozyme markers. *Heredity* 80:683–693.

———. 1998b. Genetic variation and gene flow in Alpine diploid and tetraploid populations of *Lotus* (*L. alpinus* (D.C,) Schleicher *L. corniculatus* L.). II. Insights from RFLP of chloroplast DNA. *Heredity* 80:694–701.

Gavazzi, G. et al. 1987. Somaclonal variation versus chemically induced mutagenesis in tomato (*Lycopersicon esculentum* L.). *Theoretical and Applied Genetics* 74:733–738.

Gebrehiwot, L., and P. R. Beuselinck. 2001. Seasonal variations in hydrogen cyanide concentration of three *Lotus* species. *Agronomy Journal* 93:603–608.

Gebrehiwot, L., P. Beuselinck, and C. A. Roberts. 2002. Seasonal variations in condensed tannin concentration of three *Lotus* species. *Agronomy Journal* 94:1059–1065.

Gershon, D. 1961a. Breeding for resistance to pod dehiscence in birdsfoot trefoil, (*Lotus corniculatus* L.), and some studies of the anatomy of pods, cytology and genetics of several *Lotus* species and their interspecific hybrids. PhD thesis, Cornell University, Ithaca, NY.

———. 1961b. Genetic studies of effective nodulation in *Lotus* spp. *Canadian Journal of Microbiology* 7:961–963.

Geurts, R., and T. Bisseling. 2002. *Rhizobium* Nod factor perception and signaling. *Plant Cell (Suppl.)* 14:S239–S249.

Gonnet, S., and P. Diaz. 2000. Glutamine synthetase and glutamate synthase activities in relation to nitrogen fixation in *Lotus* spp. *Revista Brasileira de Fisiologia Vegetal* 12:195–202.

Gonzales, M. D. et al. 2005. The legume information system (LIS): An integrated information resource for comparative legume biology. *Nucleic Acids Research* 33 (database issue): D660–D665.

Gotlieb, A. R., and H. Doriski. 1983. Fusarium wilt of birdsfoot trefoil in Vermont and New York. *Plant Disease Report* 46:655–656.

Graham, J. H. 1953. A disease of birdsfoot trefoil caused by a new species of *Stemphylium*. *Phytopathology* 43:577–579.

Grant, W. F. 1995. A chromosome atlas and interspecific–intergeneric index for *Lotus* and *Tetragonolobus* (Fabaceae). *Canadian Journal of Botany* 73:1787–1809.

———. 1996. Seed pod shattering in the genus *Lotus* (Fabaceae): A synthesis of diverse evidence. *Canadian Journal of Plant Science* 76:447–456.

———. 1999. Interspecific hybridization and amphidiploidy of *Lotus* as it relates to phylogeny and evolution. In *Trefoil: The science and technology of Lotus*, ed. P. R. Beuselinck, 43–60. Madison, WI: American Society of Agronomy and Crop Science Society of America, CSSA Special Publ. No. 28.

———. 2002. Seed size and seed weight in some *Lotus* (Fabaceae) species. *Seed Technology* 24:119–121.

Grant, W. F., and I. Altosaar. 1994. Acrylamide gel electrophoresis in the separation of soluble leaf proteins in *Lotus*. *Lotus Newsletter* 25:47–49.

Grant, W. F., M. R. Bullen, and D. de Nettancourt. 1962. The cytogenetics of *Lotus*. I. Embryo-cultured interspecific diploid hybrids closely related to *L. corniculatus*. *Canadian Journal of Genetics and Cytology* 4:105–128.

Grant, W. F., and R. B. McDougall. 1995. Registration of H401-4-4-2 birdsfoot trefoil germplasm resistant to sulfonylurea. *Crop Science* 35:286–287.

Grant, W. F., R. B. McDougall, and B. Coulman. 1992. Sessile inflorescence—A putative new mutant in birdsfoot trefoil. *Lotus Newsletter* 23:11–13.

———. 1993. Sessile inflorescence—An environmental morphological abnormality. *Lotus Newsletter* 24:43.

Grant, W. F. and B. S. Sidhu. 1967. Basic chromosome number, cyanogenetic glucoside variation, and geographic distribution of *Lotus* species. *Canadian Journal of Botany* 45:639–647.

Grant, W. F., and E. Small. 1996. The origin of the *Lotus corniculatus* (Fabaceae) complex: A synthesis of diverse evidence. *Canadian Journal of Botany* 74:975–989.

Grant, W. F., and I. I. Zandstra. 1968. The biosystematics of the genus *Lotus* (Leguminosae) in Canada. II. Numerical chemotaxonomy. *Canadian Journal of Botany* 46:585–589.

Grime, J. P., and M. A. Mowforth. 1982. Variation in genome size—An ecological interpretation. *Nature* 299:151–153.

GRIN. 2007. USDA Germplasm Resources Information Network (http://www.ars-grin.gov/cgi-bin/npgs/html/tax_search.pl); list of *Lotus* species; http://www.ars-grin.gov/cgi-bin/npgs/html/taxon.pl?300317; world distribution, history).

Gruber, M. et al. 2008. Variation in morphology, plant habit, proanthocyanidins, and flavonoids within a *Lotus* germplasm collection. *Canadian Journal of Plant Science* 88:121–132.

Guttman, R. H. 1947. Relationships between chromosome number and some morphological characteristics in a sample from a native population and artificial polyploids of *Lotus corniculatus* L. MSc thesis, Cornell University, Ithaca, NY.

Hall, J. K., N. L. Hartwig, and L. D. Hoffman. 1984. Cyanazine losses in runoff from no-tillage corn in "living" and dead mulches vs. unmulched, conventional tillage. *Journal of Environmental Quality* 13:105–110.

Hampton, J. G. et al. 1987. Temperature effects on the germination of herbage legumes in New Zealand. *Proceedings of New Zealand Grasslands Association* 48:177–183.

Handberg, K., and J. Stougaard. 1992. *Lotus japonicus*, an autogamous, diploid legume species for classical and molecular genetics. *Plant Journal* 2:487–496.

Harborne, J. B., and H. Baxter. 1999. *The handbook of natural flavonoids.* 2 vols. New York: John Wiley & Sons, 1768 pp.

Hardy BBT Limited. 1989. *Manual of plant species suitability for reclamation in Alberta,* 2nd ed. *Alberta Land Conservation and Reclamation Council Report* No. RRTAC 89–4, 1989.

Häring, D. A. et al. 2007. Biomass allocation is an important determinant of the tannin concentration in growing plants. *Annals of Botany* 99:111–120.

Harney, P. M., and W. F. Grant. 1964a. A chromatographic study of the phenolics of species of *Lotus* closely related to *L. corniculatus* and their taxonomic significance. *American Journal of Botany* 51:621–627.

———. 1964b. The cytogenetics of *Lotus* (Leguminosae). VII. Segregation and recombination of chemical constituents in interspecific hybrids between species closely related to *L. corniculatus* L. *Canadian Journal of Genetics and Cytology* 6:140–146.

———. 1965. A polygonal presentation of chromatographic investigations on the phenolic content of certain species of *Lotus. Canadian Journal of Genetics and Cytology* 7:40–51.

Hart, R. H., and C. P. Wilsie. 1959. Inheritance of a flower character, brown keel tip, in *Lotus corniculatus* L. *Agronomy Journal* 51:379–380.

Hartwig, N. L., and H. U. Ammon. 2002. Cover crops and living mulches. *Weed Science* 50:688–699.

Haskins, F. A., H. J. Gorz, and B. E. Johnson. 1987. Seasonal variation in leaf hydrocyanic acid potential of low- and high-dhurrin sorghums. *Crop Science* 27:903–906.

Hausen, L. N., and E. D. Earle. 1997. Somatic hybrids between *Brassica oleracea* L. and *Sinapis alba* L. with resistance to *Alternaria brassicae* (Berk.) Sacc. *Theoretical and Applied Genetics* 94:1078–1085.

Heckendorn, F. et al. 2007. Individual administration of three tanniferous forage plants to lambs artificially infected with *Haemonchus contortus* and *Cooperia curticei. Veterinary Parasitology* 146:123–134.

Hedqvist, H. et al. 2000. Characterization of tannins and in vitro protein digestibility of several *Lotus corniculatus* varieties. *Animal Feed Science and Technology* 87:41–56.

Heichel, G. H. et al. 1985. Dinitrogen fixation, and N and dry matter distribution during 4 year stands of birdsfoot trefoil and red clover. *Crop Science* 25:101–105.

Henson, P. R., and H. A. Schoth. 1962. *The trefoils—Adaptation and culture.* Agric. Handb. U.S. Dept. Agriculture, agricultural handbook, 223; 1–16.

Hill, R. R., Jr., and K. T. Zeiders. 1987. Among- and within-population variability for forage yield and *Fusarium* resistance in birdsfoot-trefoil. *Genome* 29:761–764.

Hiltbrunner, J. et al. 2007. Legume cover crops as living mulches for winter wheat: Components of biomass and the control of weeds. *European Journal of Agronomy* 26:21–29.

Hinckley, R. A. 1961. The inheritance of pubescence in *Lotus corniculatus* L. MSc thesis, Purdue Univ., W. Lafayette, IN.

Hoveland, C. S. 1994. Birdsfoot trefoil management problems in a stressful environment. In *Proceedings of the First International Lotus Symposium,* St. Louis, Missouri, ed. P. R. Beuselinck and C. A. Roberts, 142–146. Columbia: Univ. of Missouri Ext. Publ.

Hoveland, C. S. et al. 1990. Birdsfoot trefoil research in Georgia. *Georgia Agricultural Experiment Station Research Bulletin* 396:1–22.

Hur, S. N., and C. J. Nelson. 1985. Temperature effects on germination of birdsfoot trefoil and seombadi. *Agronomy Journal* 77:557–560.

———. 1995. Cotyledon and leaf development associated with seedling vigor of six forage legumes. *Journal of Korean Society of Grasslands Science* 15:19–23.

Hwang, S.-F. et al. 2006. Effect of seed treatments and root pathogens on seedling establishment and yield of alfalfa, birdsfoot trefoil and sweetclover. *Plant Pathology* 5:322–328.

Ito, M. et al. 2000. Genome and chromosome dimensions of *Lotus japonicus. Journal of Plant Research* 113:435–442.

Jaranowski, J., and B. Wojciechowska. 1963. Cytological studies in the genus *Lotus.* Part II. Embryology of the interspecific cross *L. corniculatus* L. x *L. tenuifolius* L. *Genetica Polonica* 4:277–292.

Jay, M. et al. 1991. Evolution and differentiation of *Lotus corniculatus/Lotus alpinus* populations from French southwestern Alps. III. Conclusions. *Evolutionary Trends in Plants* 5:157–160.

Jewett, J. G. et al. 1996. A survey of CRP land in Minnesota. 1. Legume and grass persistence, *Journal of Production Agriculture* 9:528–534.

Joffe, A. 1958. The effect of photoperiod and temperature on the growth and flowering of birdsfoot trefoil *Lotus corniculatus* L. *South African Journal of Agricultural Science* 1:435–450.

Jones, D. A. 1998. Why are so many food plants cyanogenic? *Phytochemistry* 7:155–162.

Jones, D. A., and R. Turkington. 1986. *Lotus corniculatus* L. *Journal of Ecology* 74:1185–1212.

Jones, D. A. et al. 1986. Variation of the color of the keel petals in *Lotus corniculatus* L. 3. Pollination, herbivory and seed production. *Heredity* 57:101–112.

Jones, W. T., and J. W. Lyttleton. 1971. Bloat in cattle. XXXIV. A survey of forages that do and do not produce bloat. *New Zealand Journal of Agricultural Research* 13:101–107.

Kaimori, N. et al. 1998. Asymmetric somatic cell hybrids between alfalfa and birdsfoot trefoil. *Breeding Science* 48:29–34.

Kallenbach, R. L. et al. 1998. Rhizomatous birdsfoot trefoil exhibits a unique response to nitrogen-free conditions. *Agronomy Journal* 90:709–713.

———. 2001. Summer and autumn growth of rhizomatous birdsfoot trefoil. *Crop Science* 41:149–156.

Kamm, J. A. 1989. In-flight assessment of host and nonhost odors by alfalfa seed chalcid (Hymenoptera: Eurytomidae). *Environmental Entomology* 18:56–60.

———. 1992. Influence of celestial light on visual and olfactory behavior of seed chalcids (Hymenoptera: Eurytomidae). *Journal of Insect Behavior* 5:273–287.

Kawaguchi, M. 2000. *Lotus japonicus* 'Miyakojima' MG-20: An early-flowering accession suitable for indoor handling. *Journal of Plant Research* 113:507–509.

Kawaguchi, M. et al. 2005. *Lotus burttii*, takes a position of the third corner in the *Lotus* molecular genetics triangle. *DNA Research* 12:69–77.

Keim, W. F. 1952. Interspecific hybridization studies in *Trifolium* and *Lotus* utilizing embryo culture techniques. PhD thesis, Cornell University, Ithaca, NY.

Kelman, W., and J. Ayres. 2002. Genetic analysis of seed components in birdsfoot trefoil (*Lotus corniculatus*). In *Plant Breeding for the 11th Millenium*, ed. J. A. McComb, 504–506. *Proceedings 12th Australasian Plant Breeding Conference*, Perth.

Kelman, W. M., and J. F. Ayres. 2004. Genetic variation for seed yield components in birdsfoot trefoil (*Lotus corniculatus* L.). *Australian Journal of Experimental Agriculture* 44:259–263.

Kelman, W. M. 2006a. The interactive effects of phosphorus, sulfur and cultivar on the early growth and condensed tannin content of greater *Lotus* (*Lotus uliginosus*) and birdsfoot trefoil (*L. corniculatus*). *Australian Journal of Experimental Agriculture* 46:53–58.

———. 2006b. Germplasm sources for improvement of forage quality in *Lotus corniculatus* L. and *L. uliginosus* Schkuhr (Fabaceae). *Genetic Resources and Crop Evolution* 53:1707–1713.

Kelman, W. M., M. J. Blumenthal, and C. A. Harris. 1997. Genetic variation for seasonal herbage yield, growth habit, and condensed tannins in *Lotus pedunculatus* Cav. and *Lotus corniculatus* L. *Australian Journal of Agricultural Research* 48:959–968.

Kelman, W. M., and G. J. Tanner. 1990. Foliar condensed tannin in *Lotus* species growing on limed and unlimed soils in south-eastern Australia. *Proceedings of the New Zealand Grasslands Association* 52:51–54.

Kerr, H. D., and D. L. Klingman. 1960. Weed control in establishing birdsfoot trefoil. *Weeds* 8:157–1670.

Kihara, M. et al. 1992. Asymmetric somatic hybrid calli between leguminous species of *Lotus corniculatus* and *Glycine max* and regenerated plant from the calli. *Japanese Journal of Breeding* 42:55–64.

Kimpinski, J. et al. 1999. Invasion and reproduction of *Pratylenchus penetrans* in birdsfoot trefoil cultivars. *Phytoprotection* 80:179–184.

Kirkbride, J. H., Jr. 1999. Checklist of *Lotus* species. In *Trefoil: The science and technology of* Lotus, ed. P. R. Beuselinck, 229–230. Madison, WI: American Society of Agronomy and Crop Science Society of America, CSSA Special Publ. No. 28.

Kirkbride, J. H., Jr., C. R. Gunn, and M. J. Dallwitz. 2004. Worldwide identification of legume (Fabaceae) seeds using expert computer technology. *Seed Science Technology* 32:53–68.

Kirti, P. B. et al. 1992. Production and characterization of intergeneric somatic hybrids of *Trachystoma ballii* and *Brassica juncea*. *Plant Cell Reports* 11:90–92.

Kumagai, H., and H. Kouchi. 2003. Gene silencing by expression of hairpin RNA in *Lotus japonicus* roots and root nodules. *Molecular Plant–Microbe Interactions* 16:663–668.

Larkin, P. J., and W. R. Scowcroft. 1981. Somaclonal variation—A novel source of variability from cell cultures for plant improvement. *Theoretical and Applied Genetics* 60:197–214.

Larsen, K. 1954. Chromosome numbers of some European flowering plants. *Botanisk Tidsskrift* 50:163–174.

Lauriault, L. M. et al. 2006. Performance of irrigated tall fescue-legume communities under two grazing frequencies in the southern Rocky Mountains, USA. *Crop Science* 46:330–336.

Lazlo, L. N. 1992. The effects of herbicides formulated of imidazolinone on *Lotus corniculatus* L. (c.v. G kes-kenylevelu). *Lotus Newsletter* 23:47–48.

Lelivelt, C. L. C., and F. A. Krens. 1992. Transfer of resistance to the beet cyst nematode (*Heterodera schachtii* Schm.) into the *Brassica napus* L. gene pool through intergeneric somatic hybridization with *Raphanus sativus* L. *Theoretical and Applied Genetics* 83:887–894.

Li, B., and P. R. Beuselinck. 1996. Rhizomatous *Lotus corniculatus* L. II. Morphology and anatomy. *Crop Science* 36:407–411.

Li, Q., and M. J. Hill. 1989a. Seed development and dormancy characteristics in *Lotus corniculatus* L. *New Zealand Journal of Agricultural Research* 32:333–336.

———. 1989b. Effect of the growth regulator PP 333 (Paclobutrazol) on plant growth and seed production of *Lotus corniculatus* L. *New Zealand Journal of Agricultural Research* 32:507–514.

Liljegren, S. J. et al. 2000. SHATTERPROOF MADS-box genes control seed dispersal in *Arabidopsis. Nature* 404:766–770.

Linscott, D. L., and R. D. Hagin. 1978. Weed control during establishment of birdsfoot trefoil (*Lotus corniculatus*) and red clover (*Trifolium pratense*) with EPTC and dinoseb. *Weed Science* 26:497–501.

Linscott, D. L., and R. Vaughan. 1989. Quackgrass (*Agropyron repens*) control in birdsfoot trefoil (*Lotus corniculatus*) seed production. *Weed Technology* 3:102–104.

Liu, B., Z. L. Liu, and X. W. Li. 1999. Production of a highly asymmetric somatic hybrid between rice and *Zizania latifolia* (Griseb): Evidence for intergenomic exchange. *Theoretical and Applied Genetics* 98:1099–1103.

Luque, A. et al. 2000. The effect of grazing *Lotus corniculatus* during late summer–autumn on reproductive efficiency and wool production in ewes. *Australian Journal of Agricultural Research* 51:385–391.

Luu, K. T. 1980. Characterization of allelopathic effects of tall fescue on birdsfoot trefoil. PhD thesis, Univ. of Missouri, Columbia.

Luu, K. T., A. G. Matches, and E. J. Peters. 1982. Allelopathic effects of tall fescue on birdsfoot trefoil as influenced by N fertilization and seasonal changes. *Agronomy Journal* 74:805–808.

MacDonald, H. A. 1946. Birdsfoot trefoil (*Lotus corniculatus* L.)—Its characteristics and potentialities as a forage legume. *Cornell Agricultural Experiment Station Memoir* 261:1–182.

Mackun, I. R., and B. S. Baker. 1990. Insect populations and feeding damage among birdsfoot trefoil–grass mixtures under different cutting schedules. *Journal of Economic Entomology* 83:260–267.

MacLean, N. L., and W. F. Grant. 1987. Evaluation of birdsfoot trefoil (*Lotus corniculatus*) regenerated plants following *in vitro* selection for herbicide tolerance. *Canadian Journal of Botany* 65:1275–1280.

MacLean, N. L., and J. Novak. 1997. In vitro selection for improved seedling vigour in birdsfoot trefoil (*Lotus corniculatus* L.). *Canadian Journal of Plant Science* 77:385–390.

Malik, N., and J. Waddington. 1989. Weed control strategies for forage legumes. *Weed Technology* 3:288–296.

Marles, M. A. S., H. Ray, and M. Y. Gruber. 2003. New perspectives on proanthocyanidin biochemistry and molecular regulation. *Phytochemistry* 64:367–383.

Marley, C. L., R. Fychan, and R. Jones. 2006a. Yield, persistency and chemical composition of *Lotus* species and varieties (birdsfoot trefoil and greater birdsfoot trefoil) when harvested for silage in the UK. *Grass and Forage Science* 61:134–145.

Marley, C. L. et al. 2006b. The effects of birdsfoot trefoil (*Lotus corniculatus*) and chicory (*Cichorium intybus*) when compared with perennial ryegrass (*Lolium perenne*) on ovine gastrointestinal parasite development, survival and migration. *Veterinary Parasitology* 138:280–290.

Márquez, A. J. et al., eds. 2005. Lotus japonicus handbook. New York: Springer, 384 pp.

Marshall, A. et al. 2008. A high-throughput method for the quantification of proanthocyanidins in forage crops and its application in assessing variation in condensed tannin content in breeding programs for *Lotus corniculatus* and *Lotus uliginosus. Journal of Agriculture and Food Chemistry* 56:974–981.

McGraw, R. L., P. R. Beuselinck, and G. C. Marten. 1989. Agronomic and forage quality attributes of diverse entries of birdsfoot trefoil. *Crop Science* 29:1160–1164.

McGraw, R. L., D. L. Wyse, and L. J. Elling. 1983. Tolerance of alsike clover (*Trifolium hybridum*), red clover (*T. pratense*), and birdsfoot trefoil (*Lotus corniculatus*) to TCA and dalapon. *Weed Science* 31:100–102.

McKee, G. W. 1963. Influence of daylength on flowering and plant distribution in birdsfoot trefoil. *Crop Science* 3:205–208.

McKee, R. 1949. Fertilization relationships in the genus *Lotus. Agronomy Journal* 41:313–316.

McKenzie, D. B., Y. A. Papadopoulos, and K. B. McRae. 2004. Harvest management affects yield and persistence of birdsfoot trefoil (*Lotus corniculatus* L.) in cool summer climates. *Canadian Journal of Plant Science* 84:525–528.

McKersie, B. D., and D. T. Tomes. 1982. A comparison of seed quality and seedling vigor in birdsfoot trefoil. *Crop Science* 22:1239–1241.

McMahon, L. R. et al. 2000. A review of the effects of forage condensed tannins on ruminal fermentation and bloat in grazing cattle. *Canadian Journal of Plant Science* 80:469–485.

McSweeney, C. S. et al. 2001. Microbial interactions with tannins: Nutritional consequences for ruminants. *Animal Feed Science and Technology* 91:83–93.

Mears, K. A. 1955. Studies in species hybridization in the genus *Lotus*. MSc thesis, University of Vermont, Burlington.

Meusel, H., and E. Jager. 1962. Über die Verbreitung einiger Papilionaceen-Gattungen. In *Die Kulturpflanze, Part 3*, ed. H. Stubbe, 249–262. Berlin: Rudolf Mansfeld zum Gedachtnis, Akademie-Verlag.

Miller, P. R., and N. J. Ehlke. 1997. Inheritance of condensed tannins in birdsfoot trefoil. *Canadian Journal of Plant Science* 77:587–593.

Min, B. R. et al. 1999. The effect of condensed tannins in *Lotus corniculatus* upon reproductive efficiency and wool production in sheep during late summer and autumn. *Journal of Agricultural Science* 132:323–334.

———. 2001. The effect of condensed tannins in *Lotus corniculatus* upon reproductive efficiency and wool production in ewes during autumn. *Animal Feed Science and Technology* 92:185–202.

———. 2002. *Lotus corniculatus* condensed tannins decrease in vivo populations of proteolytic bacteria and affect nitrogen metabolism in the rumen of sheep. *Canadian Journal of Microbiology* 48:911–921.

———. 2003. The effect of condensed tannins on the nutrition and health of ruminants fed fresh temperate forages: A review. *Animal Feed Science and Technology* 106:3–19.

———. 2005. The effect of condensed tannins from *Lotus corniculatus* on the proteolytic activities and growth of rumen bacteria. *Animal Feed Science and Technology* 121:45–58.

Montossi, F. et al. 2001. A comparative study of herbage intake, ingestive behavior and diet selection, and effects of condensed tannins upon body and wool growth in lambs grazing Yorkshire fog (*Holcus lanatus*) and annual ryegrass (*Lolium multiflorum*) dominant swards. *Journal of Agricultural Science* 136:241–251.

Monza, J. et al. 2008. Characterization of an indigenous population of rhizobia nodulating *Lotus corniculatus*. *Soil Biology and Biochemistry* 24:241–247.

Morris, P., and M. P. Robbins. 1992. Condensed tannin formation by *Agrobacterium rhizogenes* transformed root and shoot organ cultures of *Lotus corniculatus*. *Journal of Experimental Botany* 43:221–231.

Morris, P. et al. 1999. Application of biotechnology to *Lotus* breeding. In *Trefoil: The science and technology of* Lotus, ed. P. R. Beuselinck, 199–228. Madison, WI: American Society of Agronomy and Crop Science Society of America, CSSA Special Publ. No. 28.

Nakajo, S. et al. 1994. Somatic cell hybridization in rice (*Oryza sativa* L.) and birdsfoot trefoil (*Lotus corniculatus* L.). *Breeding Science* 44:79–81.

Negri, V., and F. Veronesi. 1989. Evidence for the existence of 2n gametes in *Lotus tenuis* Wald. et Kit. (2n = 2x = 12): Their relevance in evolution and breeding of *Lotus corniculatus* L. (2n = 4x = 24). *Theoretical and Applied Genetics* 78:400–404.

Nettancourt, D. de, and W. F. Grant. 1963. The cytogenetics of *Lotus* (Leguminosae). II. A diploid interspecific hybrid between *L. tenuis* and *L. filicaulis*. *Canadian Journal of Genetics and Cytology* 5:338–347.

———. 1964a. La cytogénétique de *Lotus* (Leguminosae). III. Un cas de cytomixie dans un hybride interspécifique. *Cytologia* 29:191–195.

———. 1964b. The cytogenetics of *Lotus* (Leguminosae). VI. Additional diploid species crosses. *Canadian Journal of Genetics and Cytology* 6: 29–36.

———. 1964c. Gene inheritance and linkage relationships in interspecific diploid hybrids closely related to *Lotus corniculatus* L. *Canadian Journal of Genetics and Cytology* 6:277–287.

Niizeki, M. 1993a. Chromosomal mutations induced by protoplast culture in *Lotus corniculatus* L. *Lotus Newsletter* 24:44.

———. 1993b. Regeneration of plants from protoplasts of *Lotus* spp. (birdsfoot trefoil). In *Biotechnology in agriculture and forestry*, vol. 22. *Plant protoplasts and genetic engineering III*, ed. Y. P. S. Bajaj, 69–78. Berlin: Springer–Verlag.

———. 1996. Somaclonal variation in *Lotus corniculatus* L. (Birdsfoot trefoil). In *Biotechnology in agriculture and forestry,* vol. 36. *Somaclonal variation in crop improvement II,* ed. Y. P. S. Bajaj, 146–159. Berlin: Springer–Verlag.

———. 2001. Somatic hybridization between *Medicago sativa* L. (alfalfa) and *Lotus corniculatus* L. (birdsfoot trefoil). In *Biotechnology in agriculture and forestry,* vol. 49. *Somatic hybridization in crop improvement II,* ed. T. Nagata and Y. P. S. Bajaj, 341–355. Berlin: Springer–Verlag.

Niizeki, M., and W. F. Grant. 1971. Callus, plantlet formation and polyploidy from cultured anthers of *Lotus* and *Nicotiana. Canadian Journal of Botany* 49:2041–2051.

Niizeki, M., and T. Kodaira. 1994. Stability of somaclonal variation in *Lotus corniculatus* L. *Lotus Newsletter* 25:8–9.

Niizeki, M., and K. Saito. 1986. Plant regeneration from protoplasts of birdsfoot trefoil, *Lotus corniculatus* L. *Japanese Journal of Breeding* 36:177–180.

———. 1989. Callus formation from protoplasts fusion between leguminous species of *Medicago sativa* and *Lotus corniculatus. Japanese Journal of Breeding* 39:373–377.

Niizeki, M., R. Ishikawa, and K. Saito. 1990a. Variation in a single protoplast- and seed-derived population of *Lotus corniculatus* L. *Theoretical and Applied Genetics* 80:732–736.

Niizeki, M., M. Kihara, and K.-N. Cai. 1994a. Somatic hybridization between birdsfoot trefoil (*Lotus corniculatus* L.) and soybean (*Glycine max* L.). In *Biotechnology in agriculture and forestry,* vol. 27. *Somatic hybridization in crop improvement I,* ed. Y. P. S. Bajaj, 132–144. Berlin: Springer–Verlag.

Niizeki, M. et al. 1989. Somatic cell hybridization among gramineous and leguminous species. In *Proceedings of 6th International Congress of Society for the Advancement of Breeding Research in Asia and Oceania,* Tsukuba, Japan, 501–504.

———. 1990b. Somatic cell hybrids between birdsfoot trefoil and soybean. *Lotus Newsletter* 21:14–17.

———. 1994b. Cytogenetical and molecular genetical analysis on somaclonal variation in *Lotus corniculatus* L. In *Proceedings of 1st International Lotus Symposium,* St. Louis, MO, ed. P. R. Beuselinck and C. A. Roberts, 109–111. Columbia, MO: Univ. of Missouri Ext. Publ.

Nikolaichuk, V. I., and H. I. Yalovs'ka. 1996. A character of inheritance of the synthesis of cyanoglucosides in birdsfoot trefoil. *Dopovidi Natsional'Noyi Akademiyi Nauk Ukrayiny* 0 (4): 140–142 [Russian].

Nikolic, R., and N. Mitic. 2003. Morphological changes in atypical bird's foot trefoil plants obtained during genetic transformation by *Agrobacterium. Acta Biologica Jugoslavica Series F Genetics* 35:177–185.

Nikolic, R. et al. 2006. Effects of cytokinins on in vitro seed germination and early seedling morphogenesis in *Lotus corniculatus* L. *Journal of Plant Growth Regulation* 25:187–194.

Nothnagel, T. et al. 1997. Successful backcrosses of somatic hybrids between *Sinapis alba* and *Brassica oleracea* with the *Brassica oleracea* parent. *Plant Breeding* 116:89–97.

Nualsri, C., and P. R. Beuselinck. 1998. Rhizomatous *Lotus corniculatus:* IV. Inheritance of rhizomes. *Crop Science* 38:1175–1179.

Nualsri, C., P. R. Beuselinck, and J. J. Steiner. 1998. Rhizomatous *Lotus corniculatus* L.: III. Introgression of rhizomes into autogamous germplasm. *Crop Science* 38:503–509.

Nukui, N., H. Ezula, and K. Minamisawa. 2004. Transgenic *Lotus japonicus* with an ethylene receptor gene *Cm-ERS1/H70A* enhances formation of infection threads and nodule primordial. *Plant and Cell Physiology* 45:427–435.

O'Donoughue, L. S., and W. F. Grant. 1988. New sources of indehiscence for birdsfoot trefoil (*Lotus corniculatus,* Fabaceae) produced by interspecific hybridization. *Genome* 30:459–468.

Ohmido, N., S. Sato, S. Tabata, and K. Fukui. 2007. Chromosome maps of legumes. *Chromosome Research* 15:97–103.

Ortega, J. L. et al. 2004. Biochemical and molecular characterization of transgenic *Lotus japonicus* plants constitutively over-expressing a cytosolic glutamine synthetase gene. *Planta* 219:807–818.

Ostazeski, S. A. 1967. An undescribed fungus associated with a root and crown rot of birdsfoot trefoil (*Lotus corniculatus*). *Mycologia* 59:970–975.

Pankhurst, C. E., and W. T. Jones. 1979. Effectiveness of *Lotus* root nodules. III. Effect of combined nitrogen on nodule effectiveness and flavolan synthesis in plant roots. *Journal of Experimental Botany* 30:1109–1118.

Paolocci, F. et al. 2007. Ectopic expression of a basic helix-loop-helix gene transactivates parallel pathways of proanthocyanidin biosynthesis. Structure, expression analysis, and genetic control of *leucoanthocyanidin 4-reductase and anthocyanidin* reductase genes in *Lotus corniculatus. Plant Physiology* 143:504–516.

Papadopoulos, Y. A., and W. M. Kelman. 1999. Traditional breeding of *Lotus* species. In *Trefoil: The science and technology of* Lotus, ed. P. R. Beuselinck, 187–198. Madison, WI: American Society of Agronomy and Crop Science Society of America, CSSA Special Publ. No. 28.

Papadopoulos, Y. A. et al. 1994. Assessment of phenotypic recurrent selection techniques for improving vigor in birdsfoot trefoil. In *Proceedings of First International Lotus Symposium,* ed. P. R. Beuselinck and C. A. Roberts, 225–228, St. Louis, MO. Columbia: Univ. of Missouri Ext. Publ.

———. 1995. Susceptibility of birdsfoot trefoil varieties to invasion of root-lesion nematodes. In *Proceedings of 11th Eastern Forage Improvement Conference,* 45–46. Ottawa, ON. Agric. and Agri-Food, Canada, Ottawa, ON.

Pedersen, J. F., R. L. Haaland, and C. S. Hoveland. 1986. Registration of 'Au Dewey' birdsfoot trefoil. *Crop Science* 26:1081.

Pedrosa, A. et al. 2002. Chromosomal map of the model legume *Lotus japonicus. Genetics* 161:1661–1672.

Pérez, C. et al. 2001. Uso de *Pseudomonas fluorescentes* nativas para el control de enfermedades de implantación en *Lotus corniculatus* L. *Agrociencia* 5:41–47.

Perry, J. A. et al. 2003. A TILLING reverse genetics tool and a web-accessible collection of mutants of the legume *Lotus japonicus. Plant Physiology* 131:866–871.

Peters, E. J. 1964. Pre-emergence, preplanting and postemergence herbicides for alfalfa and birdsfoot trefoil. *Agronomy Journal* 56:415–419.

Peterson, S. S., J. J. Wedberg, and D. B. Hogg. 1992a. Sweep counts of the trefoil seed chalcid, *Bruchophagus-platypterus* (Walker) (Hymenoptera: Eurytomidae) and two of its parasitoids in malathion-treated plots of birdsfoot trefoil, and predicted damage based on sweep sampling. *Journal of Agricultural Entomology* 9:25–35.

Peterson, S. S. et al. 1992b. Plant bug (Hemiptera: Miridae) damage to birdsfoot trefoil seed production. *Journal of Economic Entomology* 85:250–255.

Pezzotti, M. et al. 1985. Somaclonal variation and its potential use in plant breeding of *Lotus corniculatus* L. *Italian Society of Agriculture Genetics Meeting 1985,* 465–466.

Pofelis, S., H. Le, and W. F. Grant. 1992. The development of sulfonylurea herbicide-resistant birdsfoot trefoil (*Lotus corniculatus*) plants from in vitro selection. *Theoretical and Applied Genetics* 83:480–488.

Poli, C. H. E. C. et al. 2006a. Selective behavior in cattle grazing pastures of strips of birdsfoot trefoil and red clover. 1. The effects of relative sward area. *Journal of Agricultural Science* 144:165–171.

———. 2006b. Selective behavior in cattle grazing pastures of strips of birdsfoot trefoil and red clover. 2. The effects of sward maturity and structure. *Journal of Agricultural Science* 144:173–181.

Poostchi, I. 1959. Agronomic, genetic and cytological investigation on vigor of establishment, growth form and other morphological characters in birdsfoot trefoil, *Lotus corniculatus* L. PhD thesis, Cornell Univ., Ithaca, NY.

Poostchi, I., and H. A. MacDonald. 1961. Inheritance of leaf color in broadleaf birdsfoot trefoil. *Crop Science* 1:327–328.

Prenner, G. 2003. A developmental analysis of the inflorescence and the flower of *Lotus corniculatus* (Fabaceae-Loteae). *Mitteilungen des Naturwissenschaftlichen Vereines für Steiermark* 133:99–107.

Priolo, A., D. Micol, and J. Agabriel. 2001. Effects of grass feeding systems on ruminant meat color and flavor. A review. *Animal Research* 50:185–200.

Raelson, J. V., and W. F. Grant. 1988. An isoenzyme study in the genus *Lotus* (Fabaceae). Evaluation of hypotheses concerning the origin of *L. corniculatus* using isoenzyme data. *Theoretical and Applied Genetics* 76:267–276.

Raelson, J. V. et al. 1989. An isoenzyme study in the genus *Lotus* (Fabaceae). Segregation of isoenzyme alleles in synthetic allo- and autotetraploids, and in *L. corniculatus. Theoretical and Applied Genetics* 77:360–368.

Ramirez-Restrepo, C. A. et al. 2004. Use of *Lotus corniculatus* containing condensed tannins to increase lamb and wool production under commercial dryland farming conditions without the use of anthelmintics. *Animal Feed Science and Technology* 117:85–105.

———. 2005. Use of *Lotus corniculatus* containing condensed tannins to increase summer lamb growth under commercial dryland farming conditions with minimal anthelmintic drench input. *Animal Feed Science and Technology* 122:197–217.

———. 2006. Production of *Lotus corniculatus* L. under grazing in a dryland farming environment. *New Zealand Journal of Agricultural Research* 49:89–100.

Ramnani, A. D. 1979. Studies on cyanogenesis and the colour of the keel petals of *Lotus corniculatus* L. PhD thesis, Univ. Hull, U.K.

Ramnani, A. D., and D. A. Jones. 1984a. Genetics of cyanogenesis, cyanoglucoside and linamarase production in the leaves of *Lotus corniculatus* L. *Pakistan Journal of Botany* 16:145–154.

———. 1984b. Inheritance of the brown keel tip character in *Lotus corniculatus* L. *Bangladesh Journal of Botany* 13:52–59.

Rasheed, J. H. et al. 1990. Root hair protoplasts of *Lotus corniculatus* (birdsfoot trefoil) express their totipotency. *Plant Cell Reports* 8:565–569.

Real, D. et al. 1992. Waterlogging tolerance and recovery of 10 *Lotus* species. *Australian Journal of Experimental Agriculture* 48:480–487.

Rehm, G. W. et al. 1998. Methods for establishing legumes on sandy soils. *Journal of Production Agriculture* 11:108–112.

Reynaud, J., M. Jay, and S. Blaise. 1991. Evolution and differentiation of populations of *Lotus corniculatus* L. (Fabaceae) from the southern French Alps (Massif du Ventoux and Montagne de Lure). *Canadian Journal of Botany* 69:2286–2290.

Reynaud, J., and M. Lussignol. 2005. The flavonoids of *Lotus corniculatus*. *Lotus Newsletter* 35:75–82.

Rim, Y. W., and P. R. Beuselinck. 1991 2*n* Pollen production in diploid *Lotus* species. *Agronomy Abstracts Annual Meeting of the American Society of Agronomy* 208.

———. 1992. Verification of crossing between *Lotus corniculatus* ($2n = 4x = 24$) and 2*n* pollen producing *L. tenuis* ($2n = 2x = 12$). *Agronomy Abstracts Annual Meeting of the American Society of Agronomy* 112.

Robbins, M. P., J. Hartnoll, and P. Morris. 1991. Phenylpropanoid defense responses in transgenic *Lotus corniculatus*. 1. Glutathion elicitation of isoflavan phytoalexins in transformed root cultures. *Plant Cell Reports* 10:59–62.

Robbins, M. P., B. Thomas, and P. Morris. 1995. Phenylpropanoid defense responses in transgenic *Lotus corniculatus*. II. Modeling plant defense responses in transgenic root cultures using thiol and carbohydrate elicitors. *Journal of Experimental Botany* 46:513–524.

Robbins, M. P. et al. 1994. A study of the genetic manipulation of flavonoids and condensed tannins in the *Lotus corniculatus* using antisense technology. In *Proceedings of 1st Lotus Symposium*, St. Louis, MO, ed. P. R. Beuselinck and C. A. Roberts, 118–122. Columbia: University of Missouri Experimental Publication.

———. 1998. Genetic manipulation of condensed tannins in higher plants. II. Analysis of birdsfoot trefoil plants harboring antisense dihydroflavonol reductase constructs. *Plant Physiology* 116:1133–1144.

Robinson, K. et al. 2007. Perennial legumes native to Australia—A preliminary investigation of nutritive value and response to cutting. *Australian Journal of Experimental Agriculture* 47:170–176.

Ross, M. D., and W. T. Jones. 1985. The origin of *Lotus corniculatus*. *Theoretical and Applied Genetics* 71:284–288.

Russelle, M. P., R. L. McGraw, and R. H. Leep. 1991. Birdsfoot trefoil response to phosphorus and potassium. *Journal of Production Agriculture* 4:114–120.

Saito, M. et al. 2005. Characteristics of progeny plants originating from asymmetric somatic cell hybridization of birdsfoot trefoil (*Lotus corniculatus* L.) and rice (*Oryza sativa* L.). *Breeding Science* 55:379–382.

Sakai, T. et al. 1996. Introduction of a gene from fertility restored radish (*Raphanus sativa*) into *Brassica napus* by fusion of x-irradiated protoplasts from a radish restorer line and iodoacetamide-treated protoplasts from a cytoplasmic male-sterile cybrid of *B. napus. Theoretical and Applied Genetics* 93:373–379.

Sala, C. et al. 1985. Selection and nuclear DNA analysis of cell hybrids between *Daucus carota* and *Oryza sativa. Journal of Plant Physiology* 118:409–419.

Sandal, N. et al. 2006. Genetics of symbiosis in *Lotus japonicus*: Recombinant inbred lines, comparative genetic maps, and map position of 35 symbiotic loci. *Molecular Plant–Microbe Interactions* 19:80–91.

Sano, H., Y. Suzuki, and K. Oono. 1988. Somatic cell hybridization of hyacinth bean and soybean. *Plant Tissue Culture Letters* 5:11–14.

Sarelli, L. et al. 2003. Phytoestrogen content of birdsfoot trefoil and red clover: Effects of growth stage and ensiling method. *Acta Veterinaria Scandinavica Section A, Animal Science* 53:58–63.

Schachtman, D. P., and W. M. Kelman. 1991. Potential of *Lotus* germplasm for the development of salt, aluminium and manganese tolerant pasture plants. *Australian Journal of Agricultural Research* 42:139–149.

Scharenberg, A. et al. 2007. Palatability in sheep and in vitro nutritional value of dried and ensiled sainfoin (*Onobrychis viciifolia*) birdsfoot trefoil (*Lotus corniculatus*), and chicory (*Cichorium intybus*). *Archives of Animal Nutrition* 61:481–496.

Schauser, L. et al. 1999. A plant regulator controlling development of symbiotic root nodules. *Nature* 402:191–195.

Scheublin, T. R., and M. G. A. van der Heijden. 2006. Arbuscular mycorrhizal fungi colonize nonfixing root nodules of several legume species. *New Phytology* 172:732–738.

Scheublin, T. R., R. S. P. van Logtestijn, and M. G. A. van der Heijden. 2007. Presence and identity of arbuscular mycorrhizal fungi influence competitive interactions between plant species. *Journal of Ecology* 95:631–638.

Schreurs, N. M. et al. 2007. Concentration of indoles and other rumen metabolites in sheep after a meal of fresh white clover, perennial ryegrass or *Lotus corniculatus* and the appearance of indoles in the blood. *Journal of the Science of Food and Agriculture* 87:1042–1051.

———. In press. Pastoral flavor in meat products from ruminants fed fresh forages and its amelioration by forage condensed tannins. *Animal Feed Science and Technology.*

Scott, D. B. et al. 1996. Novel and complex chromosomal arrangement of *Rhizobium loti* nodulation genes. *Molecular Plant–Microbe Interactions* 9:187–197.

Seaney, R. R., and P. R. Henson. 1970. Birdsfoot trefoil. *Advances in Agronomy* 22:119–157.

Séguin, P. et al. 2000. Dinitrogen fixation in Kura clover and birdsfoot trefoil. *Agronomy Journal* 92:1216–1220.

Séguin-Swartz, G., and W. F. Grant. 2006. Evidence for androgenesis in the genus *Lotus* (Fabaceae). *Lotus Newsletter* 26:4–8.

Sheldrick, R. D., and T. M. Martyn. 1992. Further developments with *Lotus* screening in the U.K. *Lotus Newsletter* 23:37–40.

Sivakumaran, S. et al. 2006. Variation of proanthocyanidins in *Lotus* species. *Journal of Chemical Ecology* 32:1797–1816.

Skinner, R. H. 2005. Emergence and survival of pasture species sown in monocultures or mixtures. *Agronomy Journal* 97:799–805.

Sledge, M. K., I. M. Ray, and G. Jiang. 2005. An expressed sequence tag SSR map of tetraploid alfalfa (*Medicago sativa* L.). *Theoretical and Applied Genetics* 111:980–992.

Small, E., W. F. Grant, and C. W. Crompton. 1984. A taxonomic study of the *Lotus corniculatus* complex in Turkey. *Canadian Journal of Botany* 62:1044–1053.

Smith, B. M. et al. 2005. The effect of provenance on the establishment and performance of *Lotus corniculatus* L. in a re-creation environment. *Biological Conservation* 125:37–46.

Somaroo, B. H. and W. F. Grant. 1971a. Interspecific hybridization between diploid species of *Lotus* (Leguminosae). *Genetica* 42:353–367.

———. 1971b. Meiotic chromosome behavior in induced autotetraploids and amphidiploids in the *Lotus corniculatus* group. *Canadian Journal of Genetics and Cytology* 13:663–671.

———. 1971c. An interspecific diploid hybrid of *Lotus* (Leguminosae) with a B chromosome. *Canadian Journal of Genetics and Cytology* 13:158–160.

———. 1972a. Chromosome differentiation in diploid species of *Lotus* (Leguminosae). *Theoretical and Applied Genetics* 42:34–40.

———. 1972b. Crossing relationships between synthetic *Lotus* amphidiploids and *L. corniculatus*. *Crop Science* 12:103–105.

Stebbins, G. L., Jr. 1950. *Variation and evolution in plants,* 1–643. New York: Columbia Univ. Press.

Steiner, J. J. 1999. Birdsfoot trefoil origins and germplasm diversity. In *Trefoil: The science and technology of Lotus*, ed. P. R. Beuselinck, 81–96. Madison, WI: American Society of Agronomy and Crop Science Society of America, CSSA Special Publ. No. 28.

———. 2002. Birdsfoot trefoil flowering response to photoperiod length. *Crop Science* 42:1709–1718.

Steiner, J. J., and P. R. Beuselinck. 2001. Registration of RG-BFT photoperiod insensitive and rapid-flowering autogamous birdsfoot trefoil genetic stock. *Crop Science* 41:607–608.

St.-Marseille, P., and W. F. Grant. 1997. Segregation by morphological analyses of trisomy types in *Lotus tenuis* (Fabaceae). *Canadian Journal of Botany* 75:1209–1214.

Stougaard, J., and P. R. Beuselinck. 1996. Registration of GIFU-129-S9 *Lotus japonicus* germplasm. *Crop Science* 36:476.

Stougaard, J. et al. 1999. Genetic nomenclature guidelines for the model legume *Lotus japonicus*. *Trends in Plant Science* 4:300–301.

Strelkov, V. 1980. Biological and ecological properties of *Lotus corniculatus* L. (wild-growing birdsfoot tre-foil). In *Proceedings of XIII International Grasslands Congress, Leipzig, 1977,* ed. E. Wojahn and H. Thons, 280–285.

Striker, G. G. et al. 2005. Physiological and anatomical basis of differential tolerance to soil flooding of *Lotus corniculatus* L. and *Lotus* glaber Mill. *Plant and Soil* 276:301–311.

Swanson, E. B., and D. T. Tomes. 1983. Evaluation of birdsfoot trefoil regenerated plants and their progeny after in vitro selection for 2,4-dichlorophenoxyacetic acid. *Plant Science Letters* 29:19–24.

Swanson, E. B., D. T. Tomes, and W. Hopkins. 1983. Modifications to callus culture characteristics and plastid differentiation by the formation of an albino callus of *Lotus corniculatus. Canadian Journal of Botany* 61:2500–2505.

Sz.-Borsos, O. 1973. Cytophotometric studies on the DNA contents of diploid *Lotus* species. *Acta Botanica Academy of Science Hungary* 18:49–58.

Tanaka, H. et al. 2008. Transgenic superroots of *Lotus corniculatus* can be regenerated from superroot-derived leaves following *Agrobacterium*-mediated transformation. *Journal of Plant Physiology.* 165:1313–1316

Taylor, T. H. 1984. AGR-104 'Fergus' bird's-foot trefoil. Issued 3/84. Revised: www.ca.uky.edu/agc/pubs/agr/agr104/agr104.htm

Teakle, N. L., D. Real, and T. D. Colmer. 2006. Growth and ion relations in response to combined salinity and waterlogging in the perennial forage legumes *Lotus corniculatus* and *Lotus tenuis. Plant and Soil* 289:369–383.

Tepfer, D. 1984. Transformation of several species of higher plants by *Agrobacterium* rhizogenes; sexual trans-mission of the transformed genotype and phenotype. *Cell* 37:959–967.

Therrien, M. C., and W. F. Grant. 1978. EMS-induced mutants in birdsfoot trefoil *Lotus corniculatus. Lotus Newsletter* 9:20.

———. 1979. Physical and chemical mutagenesis in birdsfoot trefoil (*Lotus corniculatus*). *Mutation Breeding Newsletter* 14:3-4.

———. 1982. Induction of two mutants in birdsfoot trefoil (*Lotus corniculatus*) by x-rays and chemical muta-gens. *Canadian Journal of Plant Science* 62:957–963.

———. 1983. Induced quantitative variation for agronomic and related characters in bird's-foot trefoil (*Lotus corniculatus*). *Canadian Journal of Plant Science* 63:649–658.

Thompson, L. S., and C. B. Willis. 1975. Influence of fensulfothion and fenamiphos on root lesion nematode numbers and yield of forage legumes. *Canadian Journal of Plant Science* 55:727–735.

Tome, G. A., and I. J. Johnson. 1945. Self- and cross-fertility relationships in *Lotus corniculatus* L. and *Lotus tenuis* Wald. et Kit. *Journal of the American Society of Agronomy* 37:1011–1023.

Udvardi, M. T., M. Parniske, and J. Stougaard. 2005. *Lotus japonicus*: Legume research in the fast lane. *Trends in Plant Science* 10:222–228.

Ulian, E. C., J. M. Magill, and R. H. Smith. 1994. Expression and inheritance pattern of two foreign genes in petunia. *Theoretical and Applied Genetics* 88:433–440.

Undersander, D. et al. 1993. Birdsfoot trefoil for grazing and harvested forage. Madison, WI: University Wisconsin Extension Service, North Central Region Extension Publication 474, 15 pp.

Vance, C. P. et al. 1982. Birdsfoot trefoil (*Lotus corniculatus*) root nodules: Morphogenesis and the effect of forage harvest on structure and function. *Canadian Journal of Botany* 60:505–518.

Vessabutr, S., and W. F. Grant. 1995. Isolation, culture and regeneration of protoplast from birdsfoot trefoil (*Lotus corniculatus*). *Plant Cell, Tissue and Organ Culture* 49:9–15.

Viands, D. R. et al. 1994. Cooperative project to develop birdsfoot trefoil with multiple disease resistance. *Lotus Newsletter* 25:45–46.

Vignolio, O. R., O. N. Fernandez, and N. O. Maceira. 2002. Biomass allocation to vegetative and reproductive organs in *Lotus glaber* and *L. corniculatus* (Fabaceae). *Australian Journal of Botany* 50:75–82.

Voigt, P. W., and J. A. Mosjidis. 2002. Acid-soil resistance of forage legumes as assessed by a soil-on-agar method. *Crop Science* 42:1631–1639.

Vough, L. R., A. M. Decker, and T. H. Taylor. 1995. Forage establishment and renovation. In *Forages. Vol. 2: The science of grassland agriculture,* 5th ed., ed. R. F. Barnes, A. Miller, and C. J. Nelson, 29–43. Ames: Iowa State Univ. Press.

Vuckovic, S. et al. 2007. Morphological and nutritional properties of birdsfoot trefoil (*Lotus corniculatus* L.) autochthonous populations in Serbia and Bosnia and Herzegovina. *Genetic Resources and Crop Evolution* 54:421–428.

Waghorn, G. C., J. D. Reed, and L. R. Ndlovu. 1999. Condensed tannins and herbivore nutrition. *Proceedings of XVIII International Grasslands Congress* 3:153–166.

Webb, K. J., M. P. Robbins, and S. Mizen. 1994a. Expression of GUS in primary transformants and segregation patterns of GUS, T_L- and T_R-DNA in the T_1 generation of hairy root transformants of *Lotus corniculatus*. *Transgenic Research* 3:232–240.

———. 1994b. Segregation of *Agrobacterium rhizogenes* T-DNA from other inserted genes in the T1 progeny of *Lotus corniculatus*. In *Proceedings of First International Lotus Symposium*, St. Louis, MO, ed. P. R. Beuselinck and C. A. Roberts, 114–117. Columbia, MO: University of Missouri Extension Publication.

Webb, K. J., S. Woodcock, and D. A. Chamberlain. 1987. Plant regeneration from protoplasts of *Trifolium repens* and *Lotus corniculatus*. *Plant Breeding* 98:111–118.

Webb, K. J. et al. 1990. Characterization of transgenic root cultures of *Trifolium repens, T. pretense,* and *Lotus corniculatus* and transgenic plants of *Lotus corniculatus*. *Plant Science* 70:243–254.

Wen, L. et al. 2002. Performance of steers grazing rhizomatous and nonrhizomatous birdsfoot trefoil in pure stands and in tall fescue mixtures. *Journal of Animal Science* 80:1970–1976.

———. 2003. Condensed tannin concentration of rhizomatous and nonrhizomatous birdsfoot trefoil in grazed mixtures and monocultures. *Crop Science* 43:302–306.

———. 2004. Cattle preferentially select birdsfoot trefoil from mixtures of tall fescue and birdsfoot trefoil. *Forage and Grazinglands*, online: doi:10.1094/FG-2004-0924-01-RS.

Wernsman, E. A. 1963. Occurrence of a nucleolar-like body in *Lotus corniculatus* L. *Proceedings of Iowa Academy of Science* 70:135–138.

———. 1966. Behavior and inheritance of a nucleolar-like body in *Lotus*. *Canadian Journal of Genetics and Cytology* 8:737–743.

Wernsman, E. A., R. L. Davis, and W. F. Keim. 1965. Interspecific fertility of two *Lotus* species and their F_1 hybrids. *Crop Science* 5:452–454.

Wernsman, E. A., W. F. Keim, and R. L. Davis. 1964. Meiotic behavior in two *Lotus* species. *Crop Science* 4:483–486.

Wheeler, D. M. et al. 1992. Effect of aluminium on the growth of 34 plant species: A summary of results obtained in low ionic strength solution culture. *Plant and Soil* 146:61–66.

White, J. G., and T. W. Scott. 1991. Effects of perennial forage-legume living mulches on no-till winter wheat and rye. *Field Crop Research* 28:135–148.

Wiersma, D. W., P. C. Hoffman, and M. J. Mlynarek. 1999. Companion crops for legume establishment: Forage yield, quality, and establishment success. *Journal of Production Agriculture* 12:116–122.

Williams, G. H., and F. Adair. 1982a. Effect of herbicides on *Lotus corniculatus*. *Lotus Newsletter* 13:26–27.

———. 1982b. Effect of herbicides on *Lotus corniculatus*. *Annals of Applied Biology*, Suppl. 3, Tests of Agrochemicals and Cultivars 100:82–83.

Willis, C. B. 1981. Reaction of 5 forage legumes to *Meloidogyne hapla*. *Plant Disease* 65:149–150.

Wilson, R. G. 1994. Effect of imazethapyr on legumes and the effect of legumes on weeds. *Weed Technology* 8:536–540.

Winch, J. E. 1976. Varieties, seed and mixtures of bird's-foot trefoil. *Ontario Ministry of Agriculture of Food No. 76-067, AGDEX 122/30*, 1–2.

Winch, J. E. et al. 1969. The use of mixtures of granular dalapon, birdsfoot trefoil seed and fertilizer for rough-land pasture renovation. *Journal of Brazilian Grasslands Society* 24:302–307.

Wipfli, M. S., J. L. Wedberg, and Hogg. 1990a. Damage potentials of three plant bug (Hemiptera: Heteroptera: Miridae) species to birdsfoot trefoil grown for seed in Wisconsin, USA. *Journal of Economic Entomology* 83:580–584.

———. 1990b. Cultural and chemical control strategies for three plant bug *Heteroptera miridae* pests of birds-foot trefoil in northern Wisconsin USA. *Journal of Economic Entomology* 83: 2086–2091.

———. 1991. Screen barriers for reducing interplot movement of three adult plant bug *Hemiptera miridae* species in small plot experiments. *Great Lakes Entomology* 24:169–172.

Wipfli, M. S. et al. 1989. Insect pests associated with birdsfoot trefoil *Lotus corniculatus* in Wisconsin, USA. *Great Lakes Entomology* 22:25–34.

Woodward, S. L., P. J. Laboyrie, and E. B. L. Jansen. 2000. *Lotus corniculatus* and condensed tannins—Effects on milk production by dairy cows. *Asian-Australasian Journal of Animal Science* 13 (Suppl. Vol. A): 521–525.

Woodward, S. L. et al. 2006. Supplementing fresh pasture with maize, lotus, sulla and pasture silages for dairy cows in summer. *Journal of the Science of Food and Agriculture* 86:1263–1270.

Wright, R. L., D. A. Somers, and R. L. McGraw. 1987. Somatic hybridization between birdsfoot trefoil *Lotus corniculatus* L. and L. conimbricensis Willd. *Theoretical and Applied Genetics* 75:151–156.

Wurst, S. et al. 2008. Earthworms counterbalance the negative effect of microorganisms on plant diversity and enhance the tolerance of grasses to nematodes. *Oikos* 117:711–718.

Wyse, D. L., and R. L. McGraw. 1987. Control of white campion (*Silene alba*) in birdsfoot trefoil (*Lotus corniculatus*) with dinitroaniline herbicides. *Weed Technology* 1:34–36.

Zandstra, I. I., and W. F. Grant. 1968. The biosystematics of the genus *Lotus* (Leguminosae) in Canada. I. Cytotaxonomy. *Canadian Journal of Botany* 46:557–583.

Zeiders, K. E. 1985. First report of rust caused by *Uromyces* species on birdsfoot trefoil in the United States. *Plant Disease* 69:727.

Zeiders, K. E., and R. R. Hill Jr. 1988. Measurement of resistance to fusarium wilt/root and crown rot in birdsfoot trefoil populations. *Crop Science* 28:468–474.

Zemenchik, R. A., K. A. Albrecht, and R. D. Shaver. 2002. Improved nutritive value of kura clover- and birdsfoot trefoil-grass mixtures compared with grass monocultures. *Agronomy Journal* 94:1131–1138.

Zhang, Z. X., C. C. Chu, and Y. K. Fu. 2002. Transformation of leguminous forage *Lotus corniculatus* by *Agrobacterium* mediated GA *20-oxidase* gene. *Acta Prataculturae Sinica* 11:97–100.

APPENDIX

Sources of Seed Samples and Information on *Lotus* Species

Seed samples may be obtained from the Margot Forde Forage Germplasm Center, Palmerston North, New Zealand (http://www.agresearch.co.nz/seeds/). By selecting genus and species for *Lotus*, 64 species, three hybrids, and one tetraploid (*L. pedunculatus*) are listed.

National Plant Germplasm System of the USDA, Agricultural Research Service (http://www.ars-grin.gov/cgi-bin/npgs/html/tax_acc.pl): lists 451 accessions of *L. corniculatus* from 48 countries.

Millennium Seed Bank located at Wakehurst Place in West Sussex, Royal Botanic Gardens, Kew (http://data.kew.org/sid/SidServlet?Clade=&Order=&Family=&APG=off&Genus=Lotus&Species=&StorBehav=0&WtFlag=on): lists 60 species with seed weights present for each species.

OH-Seed Identification, Ohio State Seed Identification (http://www.oardc.ohio-state.edu/seedid/): seed of *L. corniculatus* and *L. uliginosus* listed, but number of accessions not given.

USDA ARS National Seed Herbarium, Systematic Botany and Mycology Laboratory, Beltsville, MD (http://www.ars.usda.gov/is/np/systematics/seeds.htm); (http://www.ars-grin.gov/cgi-bin/npgs/html/tax_search.pl); (http://www.ars-grin.gov/npgs/acc/acc_queries.html). The latter URL lists 883 accessions of *Lotus* (restricted to active and available accessions), 500 accessions of *L. corniculatus,* 64 accessions of *L. tenuis,* and 101 accessions of *L. uliginosus.* Complete accession information is provided. Accessions of other species may be obtained by entering the species name.

USDA ARS Systematic Botany and Mycology Laboratory, Beltsville, MD: fruits and seeds (http://nt.ars-grin.gov/SBMLWeb/OnlineResources/Fabaceae/).

World Forage Gene Banks (http://www.forage.prosser.wsu.edu/Links.html): lists sources from Africa, Asia, Australia and the Pacific, Europe and Northern Asia, and the Americas.

USDA Natural Resources Conservation Service (http://plants.usda.gov/java/nameSearch?keywordquery=Lotus&mode=sciname&submit.x=13&submit.y=7): lists 152 taxa of *Lotus,* including synonyms.

Plant Gene Resources of Canada, Agriculture and Agri-Food Canada, Saskatoon Research Center, Saskatchewan, Canada (http://pgrc3.agr.gc.ca/cgi-bin/npgs/html/acc_search.pl?accid=Lotus): lists 171 accessions and their origins, 114 accessions of *L. corniculatus*, 0 accessions of *L. tenuis*, 6 accessions of *L. uliginosus,* and 51 accessions of other species. The species may be determined through the preceding URL.

Note: All URLs cited in this chapter were retrievable as of August 2008.

CHAPTER 7

Clover

J. B. Morris, G. Pederson, K. Quesenberry, and M. L. Wang

CONTENTS

7.1 INTRODUCTION

Clover (*Trifolium* spp.) is an important forage crop worldwide; it is used not only as a forage species in the United States but also as a sustainable crop to manage agricultural pests. Research has been devoted recently to alternative uses, including the preference of white clover (*T. repens*) during the morning hours for Holstein-Friesian dairy cows (Rutter et al. 2004). White clover has also been found to be a low-cost method to remove diesel from contaminated soils in subarctic regions (Palmroth, Puhakka, and Pichtel 2002). An edible vaccine has been fused in transgenic white clover for shipping fever (Lee et al. 2003). Crimson clover in combination with hairy vetch used as a cover crop was successful in a no-tillage system without nitrogen fertilization and dryland conditions for seedless watermelon production (Rangappa, Bhardwaj, and Hamama 2002). Cabbage rotation with red clover (*T. pretense*) was effective in controlling weed (*Ambrosia artemisiifolia*) infestations

(Dillard, Shah, and Bellinder 2004). Red clover was discovered to inhibit cyclooxygenase (COX) activity associated with the reduction in the incidence of a range of cancers (Lam et al. 2004). Red clover hay significantly reduced *Escherichia coli* in cattle (Jacobson et al. 2002).

Clovers (*Trifolium* spp.) are leguminous plants characterized by having a five-cleft persistent calyx with bristleform teeth. The corolla is persistent but withering, and the petal claws (except for the oblong or ovate standard) are more or less united below with the stamen tube consisting of a short and obtuse keel. The tenth stamen is more or less separate, and the fruit is small and membranous, often included in the calyx, one to six seeded, indehiscent, operculate or opening by one of the sutures. The leaves are mostly palmate (sometimes pinnate) and trifololiate (hence the name *Trifolium—Tri* means three, and *folium* means a leaf) with toothed leaflets. Stipules are united with the petiole and flowers in heads or head-like racemes. Flower colors range from white to purple (Fernald 1950).

7.2 GERMPLASM COLLECTIONS

The *Trifolium* collection contains more than 5,000 accessions, representing about 200 species collected or donated from more than 90 countries (NPGS 2003; www.ars-grin.gov/npgs/searchgrin. html). The National Plant Germplasm System (NPGS) of the *Trifolium* collection consists of accessions that meet the objectives of plant introduction, preservation of cultivated germplasm, and crop improvement (Greene 1998). Sixty percent of the collection represents cultivated species and 40% consists of breeding materials, landraces, and wild types. The majority of clover cultivated species are represented by *T. pratense* L. (35% of the accessions), followed by *T. repens* L. (21% of the accessions) and other minor cultivated species (18% of the accessions). Nearly 5% of the collection are related to red and white clover. About 88% of the clover species were acquired and added to the USDA collection during the last two decades (Taylor, Gibson, and Knight 1977). The genus *Trifolium* comprises 225–250 species (Zohary and Heller 1984; Cleveland 1985; NPGS 2003).

7.3 MAINTENANCE

Quality maintenance of clover germplasm is essential for long-term preservation and future utilization. Seed viability through standardized germination testing (Association of Official Seed Certification Agencies 2000) is required. Clover seeds are cleaned and dried at 21°C, 25% relative humidity (RH) for 2–4 weeks prior to storage in sealed bags at –18°C. The annual clover species are preserved at the USDA, ARS, Plant Genetic Resources Conservation Unit (1109 Experiment St., Griffin, GA, 30223; www.ars-grin.gov/ars/SoAtlantic/Griffin/pgrcu/). The perennial clover species are preserved at the USDA, ARS National Temperate Forage Legume Germplasm Resources Center (24106 North Bunn Road, Prosser, WA, 99350; www.forage.prosser.wsu.edu/).

Clover seed regeneration is a component of maintenance and is a very important curation responsibility. The self-pollinated species are regenerated in normal field conditions at a rate of 50 plants per accession and planted in 6-m rows. Cross-pollinated species are planted in two rows consisting of 100 plants within required pollination cages. Honeybee hives are placed in each cage at 50% bloom. All clover species are harvested beginning about 30 days after the first flower; the process continues until 10,000 seeds are harvested.

7.4 EVALUATION

Regenerating clover species in the field or greenhouse requires scientists to observe and record characterization data. Characterization is the observation and documentation of plant morphological traits. These traits are assigned common terms and scored using agreed-upon conventions. These data show which plants are in a collection and provide an array of information for use in plant development and breeding (NPGS 2003). Additionally, clover morphological traits are usually highly heritable and observed in any environment where the accession is grown. Genetic markers, including amplified fragment length polymorphism (AFLP), have been identified in subterranean clover for identifying genetic variation, core collections, and redundancy among accessions. Historically, clovers have been evaluated for soil enrichment and livestock herbage (Taylor et al. 1977), bee plants and ornamentals (Duke 1981), erosion control, cover crops (Hanson 1974), and as gene sources for pest resistance (Cope and Taylor 1985). Clovers have also been used in the past for human food (Glob 1970).

7.5 DISSEMINATION

Another very important component of curation involves the dissemination of seed to researchers throughout the world. Many of the accessions are sent to scientists conducting research on forage, cover crop, and seed production. However, during the last decade, clover accessions have been sent to scientists in many other areas of research, including evolution, saline soil tolerance, fodder, variety testing, archaeology, medicinal plants, phylogenetic relationships, ozone, and demonstration plots. The Plant Genetic Resources Conservation Unit (PGRCU) disseminated 3,608 clover accessions to U.S. public and foreign public researchers between 1995 and 2006.

7.6 TAXONOMY

Although a recent molecular phylogenetic analysis of the genus *Trifolium* has created two subgenera and revised sections of the genus (Ellison et al. 2006), we have followed the treatment of Zohary and Heller (1984) in the following discussions. Ranked in general descending order of agricultural importance, the major taxa used as forage are *T. pratense* and *T. repens* followed by *T. incarnatum*, *T. vesiculosum*, *T. hybridum*, *T. subterraneum*, *T. hirtum*, *T. alexandrinum*, *T. medium*, *T. lappaceum*, *T. nigrescens*, *T. fragiferum*, *T. ambiguum*, *T. resupinatum*, *T. glomeratum*, and *T. dubium* (Taylor and Quesenberry 1996a). Table 7.1 lists the somatic chromosome number, life form, and geographic distribution of *Trifolium* species. The section *Lotoidea* Crantz. contains four species: *T. repens*, *T. hybridum*, *T. nigrescens*, and *T. ambiguum*, and section *Trifolium* also consists of four species: *T. pratense*, *T. incarnatum*, *T. alexandrinum*, and *T. hirtum* (Zohary and Heller 1984). Thus, these two taxonomic sections of the genus contain the two most important cultivated clovers and most of the other species of lesser importance.

These sections of the genus are probably the most variable for morphological traits, and they show the complete range of chromosome numbers ($2n = 10$ to 120) found in the genus. The section *Lotoidea* is suggested to be the ancestral section of all other sections (*Paramesus*, *Mistyllus*, *Vesicaria*, *Chronosemium*, *Trifolium*, *Trichocephalum*, and *Involucrarium*) of the genus *Trifolium* (Zohary and Heller 1984). Section *Lotoidea* in the *Trifolium* genus contains the agriculturally important annual species (*T. nigrescens*), followed by the biannual species, weakly perennial species (*T. ambiguum*), and long-lived perennial species (*T. repens*). Section *Trifolium* contains the important annual species (*T. hirtum*), biannual species (*T. alexandrinum* L.), weakly perennial species (*T. incarnatum*), and the long-lived perennial species (*T. pratense*). Species in these sections of the genus exhibit some of the most diverse seed dispersal mechanisms known in the genus (e.g., *T. polymorphum*, *T. stellatum*, and *T. dasyurum*).

Table 7.1 Taxonomic Classification for *Trifolium* Species

Species Botanical Name	Chromosome Number	Life Form	Distribution
T. acaule A. Rich.	16	P	Ethiopia, Uganda, Kenya
T. affine C. Presl.	16	A	Turkey, Bulgaria
T. africanum Ser.	32	P	South Africa
T. aintabense Boiss. & Hausskn.		A	Southeastern Turkey
T. albopurpureum Torr. & A. Gray	16	A	Western U.S.
T. alexandrinum L.	16	A	Only in cultivation, eastern Mediterranean
T. alpestre L.	16	P	Europe, Azerbaijan, Armenia, Ciscaucasia, Russian Federation
T. alpinum L.	16	P	Northern Spain, Austria, central France, Italy, Switzerland
T. amabile Kunth	16	P	Arizona, U.S.; Argentina, Bolivia, Chile, Colombia, Costa Rica, Ecuador, Guatemala, Mexico
T. ambiguum M. Bieb.	16, 32, 48	P	Iran, Iraq, Romania, Moldova, Turkey, Ukraine, Georgia, Azerbaijan, Armenia, Caucasus, Russian Federation
T. amoenum Greene	32	A	California, Washington, U.S.; Vancouver Island
T. andersonii A. Gray	16	P	California, Nevada, Oregon
T. andinum Nutt.	16	P	Arizona, Utah, Nevada, Wyoming
T. andricum P. Lassen	16	A	Greece
T. angulatum Waldst. & Kit.	16	A	Eastern Europe, Russian Federation
T. angustifolium L.	16	A	Mediterranean, Austria, Bulgaria, Iran, Iraq, Yugoslavia, Azerbaijan, Georgia, Dagestan, Russian Federation
T. apertum Bobrov	16	A	Turkey, former Soviet Union west, Greece, Italy
T. argutum Banks & Sol.	16	A	Eastern Mediterranean
T. arvense L.	14	A	Europe to former Soviet Union southward to Mediterranean; U.S.
T. aureum Pollich	14	A	Europe eastward to former Soviet Union and southward to Turkey and North Africa
T. baccarinii Chiov.	16	A	Ethiopia, Tanzania, Burundi, Cameroon, Zaire, Kenya, Nigeria, Rwanda, Uganda
T. badium Schreb.	14	P	Much of Europe, Iran
T. barbigerum Torr.	16	A	California
T. batmanicum Katzn.	16	A	Turkey
T. beckwithii. Brewer ex Watson	48	P	California, Oregon, Idaho, Montana, Nevada, South Dakota, Utah
T. bejariense Moric.	16	A	Mexico northeast to eastern Texas, southern Arkansas, and western Louisiana
T. berytheum Boiss. & Blanche	16	A	Turkey, Lebanon, Palestine
T. bifidum A. Gray	16	A	California
T. bilineatum Fresen.	16	A	Ethiopia
T. billardieri Spreng.	16	A	Coastal Lebanon and Israel
T. bivonae Guss.	32	P	Sicily
T. blancheanum Boiss.		A	Lebanon, Palestine

Table 7.1 Taxonomic Classification for *Trifolium* Species (*Continued*)

Species Botanical Name	Chromosome Number	Life Form	Distribution
T. bocconei Savi	12, 14	A	Southern Italy, Sicily, Balkan Peninsula, Crete
T. boissieri Guss. ex Soy.-Will. & Godr.	16	A	Mediterranean
T. bolanderi A. Gray		P	California
T. brandegei S. Wats.	16	P	Colorado, New Mexico
T. breweri S. Wats.	16	P	California, Oregon
T. brutium Ten.		A	Italy, Albania, Greece
T. buckwestiorum Isely		A	California
T. bullatum Boiss. & Hausskn.	16	A	Eastern Mediterranean
T. burchellianum Ser.		P	Angola, Ethiopia, Kenya, Lesotho, S. Africa, Tanzania, Uganda
T. calcaricum J. L. Collins & Wieboldt	16	P	Tennessee, Virginia
T. calocephalum Fresen.		A	Ethiopia
T. campestre Schreb.	14	A	Europe, Mediterranean
T. canescens Willd.	48	P	Caucasus, former Soviet Union, Iran, Turkey
T. carolinianum Michx.	16	A	Southern U.S., eastern Kansas, U.S.; Mexico
T. caucasicum Tausch.		P	Caucasus, former Soviet Union, northern Iran, Syria, Turkey
T. caudatum Boiss.		P	Turkey (endemic)
T. cernuum Brot.	16	A	Southwest Europe, Corsica, Belgium, Azores, Morocco
T. cheranganiense J. B. Gillett	16	P	Kenya, Uganda
T. cherleri L.	10	A	Europe, Mediterranean
T. chilaloense Thulin	16	A	Ethiopia (endemic)
T. chilense Hook. & Arn.	16	A	Chile
T. ciliolatum Benth.	16	A	Western U.S.
T. clusii Godr. & Gren.	16	A	France, Spain, Mediterranean, Iran, Iraq
T. clypeatum L.	16	A	Northeast Mediterranean
T. constantinopolitanum Ser.	16	A	Algeria, France, Switzerland, northeastern Mediterranean
T. cryptopodium. ex A. Rich.	16	P	Ethiopia, Kenya, Tanzania, Uganda
T. cyathiferum Tackh.	16	A	Western Baja California, U.S.; Vancouver Island
T. dalmaticum Vis.	10	A	Albania, Bulgaria, Greece, Yugoslavia
T. dasyphyllum Torr. & A. Gray	16, 24	P	Western Colorado, southwest Montana, eastern Utah, south central Wyoming
T. dasyurum C. Presl.	16	A	Northeastern Mediterranean, Iran, Iraq, Libya, Egypt
T. decorum Chiov.		A	Ethiopia
T. depauperatum Desv.	16	A	California, Washington, Oregon, U.S.; Chile, Peru
T. dichotomum Hook & Arn.	32	A	California, Washington, U.S.; Vancouver Island, Canada
T. dichroanthum Boiss.	16	A	Cyprus, Israel, Lebanon, Turkey

Continued

Table 7.1   Taxonomic Classification for *Trifolium* Species (*Continued*)

Species Botanical Name	Chromosome Number	Life Form	Distribution
T. diffusum Ehrh.	16	A	Europe, Ukraine, Armenia, Azerbaijan, Georgia, Turkey, Caucasus
T. douglasii House	16	P	Southeast Washington, northeast Oregon, western Idaho
T. dubium Sibth.	28	A	Europe, Russian Federation, Caucasus, Mediterranean
T. echinatum M. Bieb.	16	A	Europe, Mediterranean, Iraq, Iran, Transcaucasia, Caucasus
T. elgonense Gillett	16	A	Uganda, Kenya, Ethiopia
T. eriocephalum Nutt.	16	P	Montana, Idaho, Washington, Oregon, Utah, Nevada, California
T. eriosphaerum Boiss.	14	A	Southern Syria, Lebanon, Israel
T. erubescens Fenzl.	16	A	Turkey, eastern Mediterranean, Ethiopia
T. fragiferum L.	16	P	Europe, western Asia, Mediterranean countries, Iraq, Caucasus, Iran, Afghanistan, Pakistan, middle Asia, Ethiopia
T. fucatum Lindl.	16	A	California, Oregon
T. gemellum Pourr. ex Willd.	14	A	Portugal, Spain, Algeria, Morocco
T. glanduliferum Boiss.	16	A	Albania, Chios, eastern Mediterranean
T. globosum L.	10, 16	A	Greece, Bulgaria, Turkey, Cyprus
T. glomeratum L.	14, 16	A	Europe, South Africa, Mediterranean
T. gracilentum Torr. & Gray	16	A	Arizona, Nevada, Pacific coastal states, U.S.; Mexico
T. grandiflorum Schreb.	16	A	Mediterranean, Bulgaria, Yugoslavia, Iraq, northwest Iran
T. gymnocarpon Nutt. in Torr. & Gray		P	Oregon, California, Nevada, Idaho, Colorado, Wyoming, Arizona, New Mexico
T. haussknechtii Boiss.	16	A	Turkey, Syria, Iraq
T. haydenii Porter	16	P	Southeast Idaho, southwest Montana, northwest Wyoming
T. heldreichianum (Gibelli & Belli) Hausskn.	16	P	Albania, Bulgaria, northern Greece, Turkey
T. hirtum All.	10	A	Much of central Europe, Mediterranean, Iraq, Iran, Tunisia, Algeria, Morocco, Armenia, Azerbaijan
T. howellii S. Watson		P	Oregon, California
T. hybridum L.	16	P	Mediterranean, Finland, Caucasus, Bulgaria, Iraq, Iran
T. incarnatum L.	14	A	British Isles, Spain, Portugal, France, Italy, Balkan Peninsula, Turkey, Albania, Bulgaria, Greece, Italy, Sicily, Romania, Yugoslavia
T. israeliticum Zohary & Katzn.	12	A	Israel
T. isthmocarpum Brot.	16	A	Mediterranean
T jokerstii. Vincent & R. Morgan		A	California
T. kingii S. Watson	16	P	Utah, Nevada, Arizona, Oregon, California
T. lanceolatum J. B. Gillett	16	A	Kenya, Tanzania, Ethiopia
T. lappaceum L.	16	A	Canary Islands, Iran, Iraq, Mediterranean

Table 7.1   Taxonomic Classification for *Trifolium* Species (*Continued*)

Species Botanical Name	Chromosome Number	Life Form	Distribution
T. latifolium (Hook.) Greene	16, 32	P	Idaho, Montana, Oregon, Washington
T. latinum Sebast.	16	A	France, Italy, Greece, Bulgaria, Turkey
T. leibergii A. Nelson & Macbr.		P	Northeast Oregon, Nevada (Elko Co.)
T. lemmonii S. Watson		P	California, southern Nevada
T. leucanthum M. Bieb.	14, 16	A	Mediterranean, Germany, Bulgaria, France, Iraq, Iran, Ukraine, Yugoslavia
T. ligusticum Balb. ex Loisel.	12, 14	A	Mediterranean, Canary Islands
T. longidentatum Nabelek		P	Turkey
T. longipes Nutt.	16, 32, 48	P	Much of western U.S.
T. lugardii Bullock	16	A	Kenya, Uganda
T. lupinaster L.	16, 32, 40, 48	P	Korea, Kazakhstan, Lithuania, Estonia, Belarus, Kyrgyzstan, Moldova, Russian Federation, northern China, Czechoslovakia, Japan, Mongolia, Poland, Romania, Latvia, Ukraine
T. macilentum Greene	16	P	California, Nevada, Arizona, Utah
T. macraei Hook & Arn.	16	A	Califonia, Oregon, U.S.; Chile
T. macrocephalum (Pursh) Poir	32, 48	P	Northeast California, southern Idaho, northern Nevada, Oregon, Washington
T. masaiense J. B. Gillett	16	A	Tanzania, Uganda
T. mattirolianum Chiov.	16	A	Ethiopia (endemic)
T. medium L.	48–80	P	Most of Europe, Iran, Latvia, Turkey, Estonia, Belarus, Ukraine, Moldova, Armenia, Azerbaijan, Georgia, Ciscaucasia, Russian Federation, western Siberia
T. meduseum Blanche ex Boiss.	14	A	Syria, Lebanon, Caucasus, Armenia
T meironense Zoh. & Lern.	16	A	Syria, Iraq, Lebanon, Israel
T. michelianum Savi.	16	A	Mediterranean, Portugal, Italy, Romania
T. micranthum Viv.	14, 16	A	North Africa northward to the Mediterranean and Europe
T. microcephalum Pursh.	16	A	Western U.S., South America, British Colombia, Vancouver Island
T. microdon Hook & Arn.	16	A	California, Oregon, U.S.; British Columbia, Chile
T. miegeanum Maire	16	A	Morocco
T. monanthum A. Gray	16	A	North America, California, Nevada
T. montanum L.	16	P	Much of Europe, Caucasus, and Turkey
T. mucronatum Willd. ex Spreng	16	P	Arizona, southern California, Colorado, New Mexico, western Texas, Utah (Kane Co.), U.S.; northern and central Mexico
T. multinerve A. Rich.	16	A	Uganda, Kenya, eastern Congo, Sudan, Eritrea, Ethiopia
T. mutabile Port.	16	A	Italy, Sicily, Albania, Yugoslavia, Greece, Turkey
T. nanum Torr.	16	P	Colorado, southwest Montana, northern New Mexico, eastern Utah, Wyoming

Continued

Table 7.1 Taxonomic Classification for *Trifolium* Species (*Continued*)

Species Botanical Name	Chromosome Number	Life Form	Distribution
T. nigrescens Viv.	16	A	Mediterranean, Iraq, Iran, Yugoslavia, Caucasus, Armenia, Czechoslovakia, France, Greece,Hungary, Iran, Italy, Morocco, Poland, Portugal, Romania, Ukraine, Switzerland, Turkey, Yugoslavia, Moldova, U.S., Argentina, Brazil, Chile
T. noricum Wulfen	16	P	Albania, Austria, Greece, Italy, Switzerland, Yugoslavia
T. obscurum Savi	16	A	Italy, Turkey, Algeria, Morocco
T. obtusiflorum Hook. & Arn.	16	A	California, Oregon
T. occidentale Coombe		P	England, France, Portugal, Spain, Greece, Yugoslavia, Albania, Turkey
T. ochroleucum Huds.	16	P	Albania, eastern Spain, Algeria, U.K., Germany, Austria, Belgium, Bulgaria, Czechoslovakia, France, Greece, Hungary, Iran, Italy, Morocco, Poland, Portugal, Romania, Ukraine, Switzerland, Turkey, Yugoslavia, Moldova
T. oliganthum Steud.	16	A	California, Oregon, Washington, New Mexico, U.S.; Vancouver Island
T. ornithopodioides L.	16	A	Spain, U.K., Germany, France, Morocco, Algeria, Hungary, Ireland, Italy, Netherlands, Portugal, Romania, Yugoslavia
T. owyheense Gilkey		P	Oregon, Idaho
T. pachycalyx Zoh.		A	Only in Turkey
T. palaestinum Boiss.	16	A	Lebanon, Palestine, Israel
T. pallescens Schreb.	16	P	Albania, Spain, Germany, Austria, Bulgaria, France, Italy, Romania, Switzerland, Yugoslavia
T. pallidum Waldst. & Kit.	16	A	North Africa northward to Turkey, Italy, and south central Europe
T. pannonicum Jacq.	96, 98–126, 128, 130–180	P	Much of central and southern Europe, Ukraine, Turkey
T. parnassii Boiss. and Sprun.	16	P	Only in Greece
T. parryi A. Gray	16	P	Colorado, southern Idaho, southwest Montana, northern New Mexico, eastern Utah, Wyoming
T. patens Schreb. in Sturm.	16	A	Spain to Poland, east to Turkey, Syria, and Israel
T. patulum Tausch		P	Albania, Greece, Italy, Yugoslavia
T. pauciflorum D'Urv.	16	A	Greece, Aegean Islands, Turkey, Lebanon, Syria, Israel, Iraq
T. philistaeum Zoh.	16	A	Israel, northern Sinai
T. phleoides Pourr.	14	A	Mediterranean, Germany, Bulgaria, northern Iran
T. physanthum Hook. & Arn. in Hook.		A	Chile
T. physodes Steven ex M. Bieb.	16	P	Albania, Spain, Algeria, Caucasus, former Soviet Union, Cyprus, Greece, Iran, Iraq, Israel, Italy, Jordan, Morocco, Portugal, Syria, Turkey, Yugoslavia

Table 7.1   Taxonomic Classification for _Trifolium_ Species (_Continued_)

Species Botanical Name	Chromosome Number	Life Form	Distribution
T. pichisermollii Gillett	16	A	Central Ethiopia (endemic)
T. pignantii Brogn. & Bory	16	P	Yugoslavia, Albania, Greece, Bulgaria
T. pilulare Boiss.	14	A	Aegean Islands, Turkey, Cyprus, Syria, Lebanon, Israel, western Iran, northern Iraq
T. pinetorum Greene	16	P	Arizona, New Mexico, Utah, U.S.; Mexico
T. plebeium Boiss.	16	A	Turkey, Syria, Lebanon, Palestine
T. plumosum Douglas ex Hook.	32	P	Southern Washington, northeast Oregon, Idaho
T. polymorphum Poir.	16, 32	P	Western Louisiana, east Texas, U.S.; Argentina, Brazil, Chile, Paraguay, Peru, Uruguay
T. polyodon Greene		A	California (endemic)
T. polyphyllum C. A. Mey	16	P	Northeast Turkey, Georgia, Russian Federation
T. polystachyum Fresen.	16	P	Angola, Zaire, Ethiopia, Kenya, Sudan, Uganda, Zambia
T. pratense L.	14	P	Most of Europe, Asia, middle eastern countries, Russian Federation, Azerbaijan, Georgia, Siberia, Turkmenistan, Kazakhstan, Kyrgyzstan, Tajikistan
T prophetarum, Hossain.		A	Only in Israel
T. pseudostriatum, Baker	16	A	Uganda, Tanganyika, eastern Congo, north Malawi, Rwanda, Burundi
T. purpureum Loisel.	14	A	Mediterranean, Romania, Syria, Iran, Iraq, Caucasus, Canary Islands
T. purseglovei J. B. Gillett		P	Zaire, Rwanda, Uganda
T. quartinianum A. Rich.		A	Eritrea, northern Ethiopia, Uganda
T. radicosum Boiss. & Hohen.		P	Iran, Iraq
T. reflexum L.	16	A	South, midwestern Pennsylvania, U.S.; Mexico
T. repens L.	32	P	Most of Europe, middle eastern countries, Russian Federation, Siberia, Armenia, Azerbaijan, Georgia, Ciscaucasia, Tajikistan, Uzbekistan, Kazakhstan, Turkmenistan, Kyrgyzstan, Afghanistan, Pakistan, Iran, Iraq, Mediterranean countries
T. resupinatum L.	14, 16, 32	A	Central and southern Europe, Mediterranean, southwest Asia
T. retusum L.	16	A	Europe, Turkey, Iraq, Caucasus, Crimea, Russia, north Africa
T. riograndense Burkart	16	P	Argentina, Brazil, Uruguay
T. rubens L.	16	P	Albania, northern Spain, Germany, Austria, Czechoslovakia, France, Hungary, Italy, central and northern Poland, Romania, former Soviet Union, Switzerland, Turkey, Yugoslavia
T. rueppellianum Fresen.	16	A	Uganda, Kenya, Sudan, Tanganyika, Fernando Po, eastern Congo, Eritrea, Ethiopia, Cameroon

Continued

Table 7.1   Taxonomic Classification for *Trifolium* Species (*Continued*)

Species Botanical Name	Chromosome Number	Life Form	Distribution
T. salmoneum Mout.	16	A	Syria, Palestine
T. saxatile All.		A	Austrian, French, Italian, and Swiss Alps
T. scabrum L.	10,16	A	Europe, Mediterranean
T. schimperi A. Rich.		A	Ethiopia (endemic)
T. scutatum Boiss.	16	A	Turkey, Cyprus, Syria, Lebanon, Palestine, Iraq, Libya
T. sebastianii Savi		A	Italy, Greece,Bulgaria, Yugoslavia, Turkey, Georgia, Armenia, Azerbaijan, Iran
T. semipilosum Fresen.	16	P	Yemen, Ethiopia, Kenya, Tanzania, Uganda
T. setiferum Boiss.		A	Southern Italy, Bulgaria, eastern and southern Greece, Albania, Serbia, western Turkey
T. simense Fresen.		P	Burundi, Cameroon, Zaire, Equatorial Guinea, Ethiopia, Kenya, Malawi, Nigeria, Rwanda, Sudan, Tanzania, Uganda, Zambia
T. siskiyouense J. M. Gillett		P	California, Oregon
T. somalense Taub. ex Harms		P	Southern Ethiopia
T. spadiceum L.	16	A	Europe, Turkey, Iran
T. spananthum Thulin	16	P	Ethiopia
T. spumosum L.	16	A	England, Spain, southern France, Corsica, Sardinia, Sicily, Greece, Aegean Islands, Cyprus, Turkey, Syria, Lebanon, Israel, Iraq, Morocco
T. squamosum L.	16	A	Europe, northern Africa, eastern Mediterranean
T. squarrosum L.	14, 16	A	Germany, southeast Europe, Balkan Peninsula, Crimea, Algeria, Morocco, Mauritania, Canary Islands
T. stellatum L.	12	A	British Isles, Iran, Mediterranean, Iraq, Caucasus, Transcaucasia
T. steudneri Schweinf.	16	A	Uganda, Kenya, Ethiopia, Eritrea
T. stoloniferum Muhl.	16	P	Illinois, eastern Kansas, Kentucky, Missouri, New Jersey, Ohio, Virginia, West Virginia
T. stolzii Harms		P	Tanzania, Ethiopia
T. striatum L.	14	A	Europe, southern Russia, Balkan peninsula, Crimea, Turkey, Iraq, Iran, Transcaucasia, Tunisia, Algeria, Morocco, Madeira, Canary Islands
T. strictum L.	16	A	Central and southern Europe, Turkey, Morocco, Algeria
T. subterraneum L.	16	A	Mediterranean, England, Bulgaria, Romania, southern Russia, Iran, Canary Islands, Caucasus, Albania, Serbia
T. suffocatum L.	16	A	England, Mediterranean, Canary Islands
T. sylvaticum Gerard ex Loisel.	14, 16	A	Belgium, Portugal, Spain, France, Balkan Peninsula, Turkey, Syria, Iraq
T. tembense Fresen.	16	A	Uganda, Kenya, Eritrea, Tanganyika, eastern Congo, Ethiopia

Table 7.1   Taxonomic Classification for *Trifolium* Species (*Continued*)

Species Botanical Name	Chromosome Number	Life Form	Distribution
T. thalii Vill.	16	P	Northern Spain, German Alps, Austria, French Alps and Pyrenees, Italian mountains, Morocco, Switzerland
T. thompsonii C. V. Morton	16	P	Washington
T. tomentosum L.	16	A	Mediterranean
T. triaristatum Bert. ex Colla		A	Chile
T. trichocalyx Heller		A	California (endemic)
T. trichocephalum M. Bieb.	14–48	P	Caucasus, former Soviet Union, northern Iran, Turkey
T. trichopterum Panc.	14	A	Yugoslavia, Albania, Greece, Bulgaria, Turkey
T. tumens Steven ex M. Bieb.	16	P	Caucasus, former Soviet Union, Iran, Turkey
T. ukingense Harms.		P	Tanzania
T. uniflorum L.	32	P	Southern France, Greece, southern Italy, northeast Libya, Turkey
T. usambarense Taub.	16	A	Burundi, Cameroon, eastern Zaire, equatorial Guinea, southern Ethiopia, Kenya, Malawi, Nigeria, Rwanda, Tanzania, Uganda, Zambia
T. variegatum Nutt.	16	A	Vancouver Island, western U.S., Mexico
T. vavilovii Eig.	16	A	Syria, Palestine, Iraq
T. velebiticum Degen		P	Albania, northwest Yugoslavia
T. velenovskyi Vandas.	16	A	Bulgaria, Albania, Yugoslavia, Greece
T. vernum Phil.	16	A	Chile
T. vesiculosum Savi.	16	A	Albania, Bulgaria, Hungary, Yugoslavia, Romania, Greece, Italy, Sicily, Corsica, Crimea, Russia, Serbia, western Turkey
T. vestitum Zoh		P	Only in Chile
T. virginicum Small	16	P	Northwest Maryland, south central Pennsylvania, western Virginia, West Virginia
T. wentzelianum Harms.		P	Tanganyika, Ethiopia
T. wettsteinii Dorfl. & Hayek		P	Northern Albania
T. wigginsii Gillett		P	Mexico
T. willdenovii Spreng.	16	A	Western U.S.
T. wormskioldii Lehm.	16, 32	P	Much of western U.S., British Columbia, Canada, Mexico

Normal clover flowers have five petals with a keel enclosed by two wing petals. Both keel and wing petals consist of a claw. At the base of the keel and wing is a swollen portion serving as a mechanism to return flower parts to their normal position following pollinator visits. Nine stamens are united in an adaxially open U-shaped column, and both the ovary and style are enclosed within this column. The style in cross-pollinated species is curved upwards near the tip and the stigma projects beyond the circle of anthers. In self-pollinated annuals, the style is usually short or curved backwards, allowing the stamens to lie close to or touching the stigma. The nectar source is at the base of the ovary in a swollen receptacle. The ovary contains one to eight ovules, and many abort. The perennial species usually develop six ovules; annuals develop only one or two ovules.

A mechanism triggered by pollinators or physical disturbance exists in clover; this allows, in normally cross-pollinated species such as *T. wormskioldii*, a method ensuring self-pollination if cross-pollination fails to occur. The calyx is essential for propagation. All sections have standards which turn upwards at anthesis except for the section *Chronosemium,* where the standards turn down, forming a reversed, spoon-shaped blade. Elongated or short petal claws exist, and the elongation appears to be an adaptation to pollinators that enable insects with long tongues to reach nectaries of long-corolla plants while insects with short tongues are unable to do so. Most annual species are selfers and have short corollas, thus eliminating the need for pollinators. The legume or pod in clover is enclosed in the calyx and the withered corolla. The margin is thick with membranous pericarp tissue, and dehiscence may be along the suture or not at all (Gillett 1985). Floral characteristics vary from large, open corollas with evident bee attractants to very small, cleistogomous flowers.

Chromosome numbers (basic chromosome number) in *Trifolium* are among the most diverse of any closely related genera, varying from $x = 8$ (80% of species in the genus) to $x = 7$ (15%), $x = 6$ (2%), and $x = 5$ (3%) (Zohary and Heller 1984). Additionally, polyploidy exists at both the $x = 8$ and $x = 7$ levels such that the reported chromosome numbers for species in the genus include $2n = 10, 12, 14, 16, 28, 32, 48, 64, 72, 80,$ and 120 (Cleveland 1985). The preceding two sections of the genera contain species encompassing the range of these chromosome numbers.

It is evident that changes in chromosomal number in *Trifolium* have proceeded by various chromosomal evolution mechanisms, including translocations, inversions, duplications, and deletions. If one accepts that $x = 8$ is the original base chromosome number for the genus, then it would appear that Robertsonian chromosome number changes via centric fusion have been a major force in the genus. Chromosome pairing results in an interspecific hybrid between $x = 8$ and $x = 7$ species; *T. pratense* and *T. diffusum,* respectively, support these conclusions. The $2n = 2x = 15$ hybrid obtained showed a range of multivalent configurations (trivalents, quadrivalents, and up to septivalents) indicative of multiple chromosome rearrangements among the chromosomes of these two species that produce an interspecific hybrid easily (Schwer and Cleveland 1972a; Taylor et al. 1963). The amphidiploid $2n = 4x = 30$ hybrid of these same species produced by crossing autotetraploid versions of each parent showed primarily bivalent configurations, with a low level of quadrivalents. This would be expected from species that differ by a series of complex structural chromosomal rearrangements (Schwer and Cleveland 1972b; Taylor and Quesenberry 1996b). Similar findings have been reported in hybrids of *T. pratense* and *T. pallidum,* another $2n = 2x = 16$ annual species (Armstrong and Cleveland 1970). Other studies of interspecific hybrids in section *Trifolium* (Katznelson 1971; Putiyevsky and Katznelson 1973, 1974) have shown that, even among $2n = 16$ species related to *T. alexandrinum* (also $2n = 16$), the chromosomes differed by from one to three translocations.

Polyploidy evidently has also been a major factor in species evolution in *Trifolium.* Various studies of interspecific hybrids between the polyploid white clover and related diploid species show that several diploid annual and perennial species have chromosome relationships to white clover (Chen and Gibson 1970a, 1970b; Kazimierski and Kazimierska 1968, 1970, 1972). Recent research has suggested that heritable variation for seed production traits useful in breeding white clover (*T. repens* $2n = 4x = 32$) could be obtained from crosses with *T. nigrescens* ($2n = 2x = 16$) (Marshall et al. 1995). Molecular fluorescent *in situ* hybridization (FISH) studies of the 5S and 18S-26S r DNA loci in white clover and *T. nigrescens* strongly support the allotetraploid origin of white clover with *T. nigrescens* spp. *petrisavii* as one of the diploid ancestors (Ansari et al. 1999). Other research has shown that a polyploid series bridge can be developed between white clover and kura clover (*T. ambiguum* M. Bieb. $2n = 6x = 48$), further supporting the role of polyploidy in perennial *Trifolium* species evolution (Hussain and Williams 1997a). It appears that functional unreduced gametes ($2n$ gametes) likely are a mechanism of increased ploidy and movement of genetic material among ploidy levels (Hussain and Williams 1997b).

In section *Trifolium,* a polyploid series exists in the species *T. medium* var. *sarosiense* and *T. medium,* where chromosome numbers of $2n = 48, 64, 72,$ and 80 have been reported. Hybrids between these ploidy levels ($2n = 58$–60) were mostly fertile (Quesenberry and Taylor 1977). Additionally, partially fertile hybrids between the polyploid *T. medium* var. *sarosiense* and autotetraploid $4x$ *T.*

alpestre were produced (Quesenberry and Taylor 1978), supporting polyploidy as a mechanism of evolution in these perennial species.

It is of interest that species of the genus *Trifolium* native to the Americas only have basic chromosome numbers of $x = 8$, although polyploids of $2n = 32$ and 48 have been reported (Gillett 1985). Species in the Americas are classified as members of either section *Lotoidea* or section *Involucrarium* Hook. The latter section of the genus is only found in Western and North America as well as South America, representing a unique evolutionary branch of the genus (Zohary and Heller 1984). Apparently, chromosome number rearrangements have not been an important evolutionary mechanism among species in the Americas.

Interspecific hybridization barriers are generally strong in *Trifolium*, with hybrids only generally achieved among species classified as closely related members of a subsection of the genus (Taylor and Quesenberry 1996a). The implications of these findings for germplasm managers are that, for the important cultivated species, the primary gene pool (Harlan and de Wet 1971) may only consist of diversity within that species. Additionally, the secondary gene pool is likely a very small number of species. Clover collections should strive to achieve complete representation of the scope of known *Trifolium* species (perhaps the tertiary gene pool for cultivated species). However, when resources for collection management and maintenance are scarce, limited numbers of accessions of these species from diverse areas should be adequate.

7.7 APPLICATION OF MOLECULAR TECHNIQUES TO CHARACTERIZE GENETIC DIVERSITY OF CLOVER SPECIES

Variability in molecular marker patterns random amplified polymorphic DNA (RAPD), restriction fragment length polymorphism (RFLP), AFLP, and simple sequence repeats (SSRs) among and within accessions of species has been suggested as a useful tool for germplasm managers to access the existing variability in collections and the need for collection and maintenance of additional accessions of a species. These techniques have been applied to various clover species. The *Trifolium* is one of the largest genera in the legume family. The genus comprises annual to perennial species and outbreeders to self-pollinators, with a wide range of chromosome numbers (from 10 to 180) and ploidy levels from $2x$ to at least $8x$. From the literature, the basic chromosome number (x) varies from $x = 5$ to $x = 6$, 7, or 8. Furthermore, the rate of misidentification within the genus could exceed 5–10% (Ellison et al. 2006). The mean of pollination and annuity, large number of species, level of ploidy, and chromosome number—plus a high rate of misidentification—make characterization, classification, and evaluation of clover germplasm very complicated. The genetic diversity of some clover species has been evaluated with early developed molecular markers including RAPD, RFLP, and AFLP (Ulloa, Ortega, and Campos 2003; Isobe et al. 2003; Herrmann et al. 2005).

In red clover, RAPD band patterns have generally shown high levels of genetic variation both within and among accessions (Kongkiatngam et al. 1995; Campos-de Quiroz and Ortega-Klose 2001; Ulloa et al. 2003). Molecular phylogenetic relationships among clover species were studied using sequences from nuclear ribosomal DNA internal transcribed spacer (nrDNA ITS) and cpDNA *trnL* (Ellison et al. 2006). Simple sequence repeats have been demonstrated as highly polymorphic DNA markers for characterization and evaluation of germplasm within and among accessions in many plant species. Recently, a total of 7,159 SSR markers has been developed and used for genetic map construction in red clover (Sato et al. 2005). In addition, synteny (referring to some degree of conservation of gene content, order, and orientation between chromosomes of different species) was identified among red clover, *Medicago truncatula* Gaertner, and *Lotus japonicus* (Regel) K. Larsen when the red clover sequences from mapped adjacent SSR markers were blasted against the genome database of these two model legumes. This finding opens a new avenue for clover researchers.

They can explore and use *M. truncatula* and *L. japonicus* genomic resources (DNA markers, DNA sequences, and gene orders) to study clover species.

Several advanced molecular techniques, including flow cytometry (flow cytometer equipped with laser detectors), FISH (equipped with sensitive cooled CCD cameras), PCR (polymerase chain reaction), and automated DNA sequencing, are available and allow study of clover species thoroughly from different levels using different techniques. The ploidy level of clover species is closely related to nuclear DNA content. At the DNA content level, nuclei from different clover species can be prepared, stained with fluorescence, and run through a flow cytometer. The ploidy level of a species can be estimated by comparison of nuclear DNA content. To trace the progenitors of a polyploidy species at chromosome or chromosome fragment level, GISH (genomic *in situ* hybridization) and/or FISH techniques can be employed. Whole genomic DNA, whole chromosome or chromosome arms from a potential progenitor can be labeled and used as probes to hybridize with metaphase chromosomes from a polyploidy species. The base chromosome set, chromosomes or chromosome fragments in the polyploidy species may be traced to the potential progenitors by examination of the hybridization signals.

At the DNA fragment level or DNA marker level, a group of mapped adjacent markers can be amplified, mapped, or checked by BLAST search to detect macrosynteny among species and genera. Macrosynteny (usually by marker orders) from many chromosome regions has been identified in the grass and legume families. At the DNA sequence level, microsynteny (referring to two or more homologous DNA sequences physically linked to one another in different genomes) may be identified by comparison of sequences from homologous chromosome regions in different species (e.g., comparison of homologous BAC sequences). If some sequence (i.e., a gene) contributes to an agriculturally important trait, the sequence can be amplified from different accessions within the same species to mine useful alleles in the clover germplasm. A decade ago, not much genomic information was available for the clover species. Now, in the genomics era, a lot of genomic resources are being generated by advanced molecular techniques from several model legume species (*M. truncatula, L. japonicus,* soybean, and other legume species) and from clover. Application of molecular techniques to exploring existed genomic resources, clover species can be well classified to avoid misidentification and get a clue as to how clover species evolved or diverged within the *Trifolium* genus by molecular phylogenetics. The clover germplasm can be well characterized by using many DNA markers that are evenly distributed in the genome. The genetic diversity of the clover species can be well evaluated at the DNA sequence level to identify point mutations, small insertions and deletions (InDels). Eventually, different alleles are conserved in the germplasm management program and utilized in the clover improvement program.

Applications of molecular techniques were used in the production of an SSR and AFLP molecular marker-based framework genetic map of white clover (Jones et al. 2003). This map was constructed using an F_2 progeny set derived from the intercross of fourth- and fifth-generation inbred genotypes carrying a self-fertile allele (S_f). Other recent research reports the development of a functional, high-throughput, genetic linkage map of white clover based on SSRs from both genomic and EST sequences (Sawbridge et al. 2003). This map was based on 92 F_1 progeny from a cross between two highly heterozygous genotypes exhibiting contrasting phenotypes for seed production, forage performance, and pathogen response. The map is estimated to cover >120 cM and includes all 16 linkage groups (Barrett et al. 2003). Development of these genetic maps for important clover species should enhance the efforts of clover germplasm managers to identify and maintain variability for important genetic traits.

7.8 GENE POOLS

7.8.1 Primary Gene Pool

We will focus on the gene pools for the cultivated major and minor clover species and their relatives (Table 7.2). The clover primary gene pool is considered important because of its usefulness for

Table 7.2 Primary, Secondary, and Tertiary Gene Pools for the Cultivated Major and Minor *Trifolium* Species

Cultivated Species	Primary Gene Pool	Secondary Gene Pool	Tertiary Gene Pool
	Major Species		
T. pratense	Cultivars, landraces, wild/ naturalized, botanical varieties	*T. diffusum, T. pallidum*	*T. alpestre, T. heldreichianum, T. medium, T. noricum, T. rubens*
T. repens	Cultivars, landraces, wild/ naturalized, botanical varieties	*T. argutum, T. nigrescens, T. uniflorum, T. ambiguum, T. isthmocarpum*	
	Minor Species		
T. alexandrinum	Cultivars, landraces, wild/ naturalized botanical varieties	*T. berytheum, T. salmoneum, T. apertum, T. meironense*	
T. ambiguum	Cultivars, landraces, wild/ naturalized		*T. montanum, T. repens*
T. dubium	Cultivars, landraces, wild/ naturalized		
T. fragiferum	Cultivars, landraces, wild/ naturalized	*T. neglectum*	*T. resupinatum, T. physodes*
T. glomeratum	Landraces, wild/ naturalized		
T. hirtum	Cultivars, landraces, wild/ naturalized		
T. hybridum	Cultivars, landraces, wild/naturalized		
T. incarnatum	Cultivars, landraces, wild/ naturalized		
T. medium	Cultivars, landraces, wild/ naturalized	*T. sarosiense*	*T. pratense*
T. nigrescens	Landraces, wild/ naturalized	*T. repens*	
T. resupinatum	Cultivars, landraces, wild/ naturalized	*T. clusii*	*T. neglectum, T. fragiferum*
T. subterraneum	Cultivars, landraces, wild/ naturalized		
T. vesiculosum	Cultivars, landraces, wild/ naturalized	*T. leiocalycinum, T. mutabile*	

cultivar development. It can be separated into several subpools, which differ in their relative value to humans (Morris and Greene 2001). The primary gene pool can be divided into four subpools ranging from current and obsolete cultivars, selection-based populations such as regional strains and breeding populations, landraces, noncultivated germplasm, or naturally occurring endemic wild populations (Taylor and Quesenberry 1996a) for the major cultivated types including both *T. pratense* and *T. repens*. The primary gene pool for the minor cultivated species including *T. ambiguum*, *T. fragiferum, T. hybridum, T. medium* (Townsend 1985), *T. dubium, T. incarnatum, T. resupinatum* (Knight 1985), *T. hirtum* (Love 1985), *T. subterraneum* (McGuire 1985), and *T. vesiculosum* (Miller and Wells 1985) consists primarily of cultivars and landraces as well as wild and naturalized forms. The primary gene pool for *T. alexandrinum* consists of cultivars, landraces, wild and naturalized forms, and botanical varieties. Both *T. glomeratum* and *T. nigrescens* consist of landraces as well

as wild and naturalized forms (Knight 1985). These subpools include current and obsolete cultivars, selection-based populations such as regional strains and breeding populations, landraces, noncultivated germplasm, or naturally occurring endemic wild populations (Taylor and Quesenberry
1996a). The majority of forage legume cultivars have been derived from the primary gene pool with
cultivars, selected strains, and landraces representing the most important subpool for cultivar development in the United States (Rumbaugh 1990, 1991).

The first, third, and fourth components include cultivars, landraces, and cultivated species that
have escaped to form naturalized populations across a large geographic range, respectively. We
would consider the first and third components to be a good genetic resource for forage and pasture
improvement. The fourth component (naturalized populations) contains a tremendous amount of
diversity that may have new adaptive traits particularly useful to breeders (Pederson and Brink
1998; Brink et al. 1999). These populations could provide a rich source of alleles that could be
used to improve germplasm used in phytoremediation or as forage on acid soils. The fifth component would be particularly important to botanical research and would be a source of variation for
improving crops for both traditional and nontraditional uses.

7.8.2 Secondary and Tertiary Gene Pools

Generally, plant breeders use the secondary gene pool if they cannot find desirable alleles in the
primary gene pool and then turn to the tertiary gene pool only after exhausting the secondary pool
(Fehr 1987). Based on their ability to hybridize with the cultivated *Trifolium* species as reported in
the literature, the related species in the secondary and tertiary gene pools are as follows. *T. pretense,
T. diffusum* (Taylor et al. 1963) and *T. pallidum* (Armstrong and Cleveland 1970) qualify as members
of the secondary gene pool; *T. alpestre* (Evans 1962), *T. heldreichianum* (Taylor and Quesenberry
1996a), *T. medium* (Sawai et al. 1990), *T. noricum* (Taylor and Quesenberry 1996a), and *T. rubens*
(Evans 1962) qualify as members of the tertiary gene pool. *Trifolium argutum* (Kazimierski,
Kazimierska, and Strzyzewska 1972), *T. nigrescens* (Keim 1953), and *T. uniflorum* (Williams and
Williams 1981) qualify as secondary gene pool members, and *T. ambiguum* (Williams and Verry
1981) and *T. isthmocarpum* (Kazimierski et al. 1972) qualify as tertiary gene pool members for
T. repens. Trifolium berytheum, T. salmoneum, T. apertum, and *T. meironense* (Putiyevksy and
Katznelson 1973) qualify as secondary gene pool members for *T. alexandrinum. Trifolium mon-
tanum* (Rupert and Evans 1980) and *T. repens* (Williams and Verry 1981) are tertiary gene pool
members of *T. ambiguum. T. fragiferum,* and *T. neglectum* (Taylor and Gillett 1988) qualify as a
secondary gene pool member and *T. resupinatum* and *T. physodes* (Taylor and Gillett 1988) are
tertiary gene pool members. *Trifolium sarosiense* (Quesenberry and Taylor 1977) and *T. pretense*
(Merker 1984) qualify as secondary gene pool and tertiary gene pool members, respectively, for
T. medium. Trifolium repens (Brewbaker and Keim 1953) is a secondary gene pool member of *T.
nigrescens.* For *T. resupinatum, T. clusii* (Taylor and Gillett 1988) qualifies as a secondary gene
pool member. While both *T. neglectum* and *T. fragiferum* (Taylor and Gillett 1988) are tertiary
members. Both *T. leiocalycinum* and *T. mutabile* (Taylor and Giri 1984) are secondary members
for *T. vesiculosum.*

As curators review the literature on trends in crop use and as they consult with crop germplasm
committees (CGCs), botanists, taxonomists, conservationists, and breeders, additional species could
be identified and placed in the tertiary class based upon transgenic techniques used to move any
gene into *Trifolium.* The interest in isoflavones may promote some clover species into the tertiary
class based on their high production of isoflavones and the potential for transferring these genes into
cultivated species using genetic transformation techniques.

The quaternary gene pool would include all other nonrelated and noncultivated species in the
genus *Trifolium* (Greene and Morris 2001). At present this would include approximately 200 species. In terms of biological diversity, the quaternary pool would contain the greatest portion of

variation relative to the other pools. The actual number of accessions representing this pool will be a function of institute objectives and available resources. For example, in the U.S. *Trifolium* collection, high priority was given to obtaining a complete collection of all nonrelated, noncultivated species of the genus (Taylor et al. 1977). Based on a global survey, the U.S. *Trifolium* collection has the most extensive representation of wild species and, as such, makes a significant global contribution to botanical research and conservation (Greene 1998). Additionally, the Clover and Special-Purpose Legume Crop Germplasm Committee (CSPL-CGC) has placed a priority on obtaining accessions that represent rare or endangered species endemic to the United States.

7.9 GERMPLASM ENHANCEMENT

7.9.1 Conventional Breeding

Breeding procedures utilized for clover species vary for self-pollinated versus cross-pollinated species and annual versus perennial species (Cope and Taylor 1985; Williams 1987; Taylor and Quesenberry 1996a). Cultivars of self-pollinated clovers have been produced by phenotypic selection within a range of adapted ecotypes evaluated under specific environmental conditions or for specific traits. Other clover breeders have conducted hybridization between existing adapted ecotypes or cultivars followed by selection over a number of generations to produce pure lines or populations. Cultivars of perennial cross-pollinated species are often bred as synthetic varieties from controlled pollination between elite clones or lines. These clones are selected from local ecotypes, adapted populations, improved germplasms, or cultivars. Polycross progeny testing of clones or families for combining ability, productivity, and pest resistance, including disease resistance, is conducted over a range of environments to produce prospective cultivars with adaptation under the conditions in which the cultivar will be grown. Cultivars of cross-pollinated species have also been produced by mass selection, phenotypic recurrent selection, and backcrossing.

7.9.2 Introgression of Economically Valuable Traits through Wide Crosses

Recent interest in introgression of valuable traits through wide crosses in *Trifolium* species has centered on the interspecific hybrids *T. ambiguum* × *T. repens* and *T. repens* × *T. nigrescens*. *Trifolium repens* × *T. nigrescens* hybrids backcrossed to white clover have greater flower production and seed yield potential than white clover (Marshall et al. 2002). Few differences in yield, persistence, and forage quality have been noted between *T. repens* × *T. nigrescens* hybrids backcrossed to white clover (Marshall, Abberton, et al. 2003). Further backcross generations are being developed over a range of white clover leaf sizes to produce new cultivars with improved seed production. Numerous studies have been conducted to attempt to introgress the rhizomotous trait, improved perenniality, and drought tolerance of *T. ambiguum* into the stoloniferous white clover. Backcross hybrids between *T. ambiguum* and *T. repens* have been shown to have greater drought tolerance, lower crude protein, comparable *in vitro* dry matter digestibility, and comparable dry matter yields over time compared to white clover (Marshall et al. 2001; Marshall, Abberton, et al. 2003; Marshall, Williams, et al. 2003; Marshall et al. 2004). A fertile bridge at the hexaploid level of crosses between *T. ambiguum* and *T. repens* should facilitate further hybridization between these two species (Hussain and Williams 1997a). Discovery of a chromosome-specific marker linked to the locus controlling rhizome development in *T. ambiguum* and backcross hybrids with white clover should facilitate introgression of rhizomes into white clover (Abberton et al. 2003). Nitrogen fixation capability may be a problem in clover interspecific hybrids, though *T. ambiguum* × *T. repens* hybrids have been reported to have effective nitrogen fixation utilizing rhizobia from white clover (Lowther et al. 2002).

7.9.3 Somaclonal Variation

Little successful use of somaclonal variation as a source of genetic variation has been reported in clover species. Somaclonal variation for somatic embryogenesis was observed in red clover and was closely linked to a gene expressing enhanced cold tolerance (Nelke et al. 1999). Red clover somaclones selected under low phosphorus concentration did not have improved phosphorus uptake as whole plants (Bagley and Taylor 1987). At present, no variety has been released through somaclonal variation.

7.9.4 Genetic Transformation

Development of efficient plant regeneration protocols is the initial step in utilization of genetic transformation for clover improvement. Successful regeneration protocols and documented transformation have now been reported in a number of clover species (Quesenberry et al. 1996; Ding et al. 2003; Moriuchi et al. 2004). Transformation has been used in *Trifolium* species to produce subterranean clover with virus resistance (Chu et al. 1999), white clover with increased anthocyanin production (De Majnik et al. 2000), white clover with *Arabidopsis thaliana* promoter genes (Lin et al. 2003), subterranean clover with herbicide resistance (Dear et al. 2003), and red clover (Sullivan and Quesenberry 2006). Transgenic white clover was used to express an antigen of *Mannheimia haemolytica,* which causes shipping fever in cattle, with the intent of developing an edible vaccine against shipping fever (Lee et al. 2001). This range of studies indicates that genetic transformation is being utilized as more than merely a mechanism for clover improvement alone within *Trifolium* species.

7.10 CONCLUSION

Over the last several decades, clover research has focused primarily on breeding for forage quality, disease resistance, ploidy levels, and taxonomic information. Clover research has made significant strides in the areas of phytoremediation, cover cropping, genetic diversity for forage traits, gene pool evaluation, use of molecular markers to detect genetic variability, cytogenetics, and germplasm enhancement. However, several clover traits of importance are currently being realized, with more discoveries expected on the horizon. Some notable phytochemical traits in red clover are various flavones such as daidzein and genistein, which may be antioxidant as well as anticancer agents in humans.

ACKNOWLEDGMENTS

The authors thank the USDA, ARS, Plant Genetic Resources Conservation Unit, Southern Region Plant Introduction Station (S9), University of Florida, and University of Georgia for partial support of this research. We appreciated the constructive comments from Norman Taylor and Rob Dean.

REFERENCES

Abberton, M. T., T. P. T. Michaelson-Yeates, C. Bowen, et al. 2003. Bulked segregant AFLP analysis to identify markers for the introduction of the rhizomatous habit from *Trifolium ambiguum* into *T. repens* (white clover). *Euphytica* 134:217–222.

Ansari, H. A., N. W. Ellison, S. M. Reader, et al. 1999. Molecular cytogenetic organization of 5S and 18S-26S r DNA loci in white clover (*Trifolium repens* L.) and related species. *Annals of Botany* 83:199–206.

Armstrong, K. C., and R. W. Cleveland. 1970. Hybrids of *Trifolium pratense* × *Trifolium pallidum*. *Crop Science* 10:354–357.

Association of Official Seed Analysts. 2000. Rules for testing seeds. AOSA.

Bagley, P. C., and N. L. Taylor. 1987. Evaluation of phosphorus efficiency in somaclones of red clover. *Iowa State Journal of Research* 61:459–480.

Barrett, B. A. Griffiths, G. Bryan, et al. 2003. A functional high-throughput genetic linkage map of white clover. In *Molecular Breeding of Forage and Turf, Third International Symposium*, p. 84, May 18–22, 2003, Dallas, TX.

Brewbaker, J. L., and W. F. Keim. 1953. A fertile interspecific hybrid in *Trifolium* (4n *T. repens* L. x 4n *T. nigrescens* Viv.). *The American Naturalist* 87:323–326.

Brink, G. E., G. A. Pederson, M. W. Alison, et al. 1999. Growth of white clover ecotypes, cultivars, and germplasms in the southeastern USA. *Crop Science* 39:1809–1814.

Campos-de Quiroz, H., and F. Ortega-Klose. 2001. Genetic variability among elite red clover (*Trifolium pratense* L.) parents used in Chile as revealed by RAPD markers. *Euphytica* 122:61–67.

Chen, C.-C., and P. Gibson. 1970a. Meiosis in two species of *Trifolium* and their hybrids. *Crop Science* 10:188–189.

———. 1970b. Chromosome pairing in two interspecific hybrids of *Trifolium*. *Canadian Journal of Genetics and Cytology* 12:790–794.

Chu, P. W. G., B. J. Anderson, M. R. I. Khan, et al. 1999. Production of *bean yellow mosaic virus* resistant subterranean clover (*Trifolium subterraneum*) plants by transformation with the virus coat protein gene. *Annals of Applied Biology* 135:469–480.

Cleveland, R. W. 1985. Reproductive cycle and cytogenetics. In *Clover science and technology*, ed. N. L. Taylor, 71–110. Madison, WI: ASA, CSSA, and SSSA.

Cope, W. A., and N. L. Taylor. 1985. Breeding and genetics. In *Clover science and technology*, ed. N. L. Taylor, 384–385. Madison, WI: ASA, CSSA, and SSSA.

Dear, B. S., G. A. Sandral, D. Spencer, et al. 2003. The tolerance of three transgenic subterranean clover (*Trifolium subterraneum* L.) lines with the *bxn* gene to herbicides containing bromoxynil. *Australian Journal of Agricultural Research* 54:203–210.

De Majnik, J., J. J. Weinman, M. A. Djordjevic, et al. 2000. Anthocyanin regulatory gene expression in transgenic white clover can result in an altered pattern of pigmentation. *Australian Journal of Plant Physiology* 27:659–667.

Dillard, H. R., R. R. Bellinder, and D. A. Shah. 2004. Integrated management of weeds and diseases in a cabbage cropping system. *Crop Protection* 23:163–168.

Ding, Y. L., G. Aldao-Humble, E. Ludlow, et al. 2003. Efficient plant regeneration and *Agrobacterium*-mediated transformation in *Medicago* and *Trifolium* species. *Plant Science* 165:1419–1427.

Duke, J. A. 1981. *Handbook of legumes of world economic importance*. New York: Plenum Press.

Ellison, N. W., A. Liston, J. J. Steiner, W. M. Williams, and N. L. Taylor. 2006. Molecular phylogenetics of the clover genus (*Trifolium*-Leguminosae). *Molecular Phylogenetics and Evolution* 39:688–705.

Evans, A. M. 1962. Species hybridization in *Trifolium*. I. Methods of overcoming species incompatibility. *Euphytica* 11:164–176.

Fehr, W. R. 1987. *Principles of cultivar development*, vol. 1. *Theory and technique*. New York: McGraw–Hill Inc.

Fernald, M. L. 1950. Leguminosae: *Trifolium*. In *Gray's manual of botany*, 8th ed., 891. New York: American Book Co.

Gillett, J. M. 1985. Taxonomy and morphology. In *Clover science and technology*, ed. N. L. Taylor, 7–69. Madison, WI: ASA, CSSA, and SSSA.

Glob, P. V. 1970. *The bog people. Iron Age man preserved*. Ithaca, NY: Cornell Univ. Press.

Greene, S. L. 1998. U.S. *Trifolium* germplasm collection: Celebrating a century of plant collecting, introduction and conservation. p. 42. In *Proceedings of 15th* Trifolium *Conference*, Madison, WI, June 10–12, 1998.

Greene, S. L., and J. B. Morris. 2001. The case for multiple-use plant germplasm collections and a strategy for implementation. *Crop Science* 41:886–892.

Hanson, A. A. 1974. Importance of forages in agriculture. In *Forage fertilization*, ed. D. A. Mays. Madison, WI: ASA, CSSA, and SSSA.

Harlan, J. R., and J. M. J. de Wet. 1971. Toward a rational classification of cultivated plants. *Taxon* 20:509–517.

Herrmann, D., B. Boller, F. Widmer, and R. Kölliker 2005. Optimization of bulked AFLP analysis and its application for exploring diversity of natural and cultivated populations of red clover. *Genome* 48:474–486.

Hussain, S. W., and W. M. Williams. 1997a. Development of a fertile genetic bridge between *Trifolium amgbiguum* M. Bieb. and *T. repens* L. *Theoretical and Applied Genetics* 95:678–690.

———. 1997b. Evidence of functional unreduced gametes in *Trifolium repens* L. *Euphytica* 97:21–24.

Isobe, S., I. Klimenko, S. Ivashuta, M. Gau, and N. Kozlov. 2003. First RFLP linkage map of red clover (*Trifolium pratense* L.) based on cDNA probes and its transferability to other red clover germplasm. *Theoretical and Applied Genetics* 108:105–112.

Jacobson, L. H., T. A. Nagle, N. G. Gregory, R. G. Bell, G. Le Roux, and J. M. Haines. 2002. Effect of feeding pasture-finished cattle different conserved forages on *Escherichia coli* in the rumen and faeces. *Meat Science* 62:93–106.

Jones, E. S., L. J. Hughes, M. C. Drayton, et al. 2003. An SSR and AFLP molecular marker-based genetic map of white clover (*Trifolium repens* L.). *Plant Science* 165:531–539.

Katznelson, J. 1971. Semi-natural interspecific hybridization in plants. *Euphytica* 20:266–269.

Kazimierski, T., and E. M. Kazimierska. 1968. Investigations of the hybrids of the genus *Trifolium*. I. Sterile hybrid *Trifolium repens* L. × *T. xerocephalum* Fenzl. *Acta Societatis Botanicorum Poloniae* 37:549–560.

———. 1970. Investigations of the hybrids of the genus *Trifolium*. III. Morphological tratis and cytogenetics of the hybrid *Trifolium repens* × *T. nigrescens* Viv. *Acta Societatis Botanicorum Poloniae* 39:569–592.

———. 1972. Investigations of the hybrids of the genus *Trifolium*. IV. Cytogenetics of the cross *Trifolium repens* × *T. isthmocarpum* Brot. *Acta Societatis Botanicorum Poloniae* 41:129–147.

Kazimierski, T., E. M. Kazimierska, and C. Strzyzewska. 1972. Species crossing in the genus *Trifolium* L. *Genetica Polonica* 13:11–31.

Keim, W. F. 1953. Interspecific hybridization in *Trifolium* utilizing embryo culture techniques. *Agronomy Journal* 45:601–606.

Knight, W. E. 1985. Miscellaneous annual clovers. In *Clover science and technology,* ed. N. L. Taylor, 547–562. Madison, WI: ASA, CSSA, and SSSA.

Kongkiatngam, P. et al. 1995. Genetic variation within and between two cultivars of red clover (*Trifolium pratense* L.): Comparisons of morphological, isozyme, and RAPD markers. *Euphytica* 84:237–246.

Lam, A. N. C., M. Demasi, M. J. James, A. J. Husband, and C. Walker. 2004. Effect of red clover isoflavones on cox-2 activity in murine and human monocyte/macrophage cells. *Nutrition and Cancer* 49:89–93.

Lee, R. W. H., A. N. Pool, A. Ziauddin, R. Y. C. Lo, P. E. Shewen, and J. N. Strommer. 2003. Edible vaccine development: stability of *Mannheimia haemolytica* A1 leukotoxin 50 during post-harvest processing and storage of field-grown transgenic white clover. *Molecular Breeding* 11:259–266.

Lee, R. W. H., J. Strommer, D. Hodgins, et al. 2001. Towards development of an edible vaccine against bovine pneumonic pasteurellosis using transgenic white clover expressing a *Mannheimia haemolytica* A1 leukotoxin 50 fusion protein. *Infection and Immunity* 69:5786–5793.

Lin, Y. H., E. Ludlow, R. Kalla, et al. 2003. Organ-specific, developmentally-regulated and abiotic stress-induced activities of four *Arabidopsis thaliana* promoters in transgenic white clover (*Trifolium repens* L.). *Plant Science* 165:1437–1444.

Love, R. M. 1985. Rose clover. In *Clover science and technology,* ed. N. L. Taylor, 535–546. Madison, WI: ASA, CSSA, and SSSA.

Lowther, W. L., H. N. Pryor, R. P. Littlejohn, et al. 2002. *Rhizobium* specificity of *Trifolium ambiguum* M.Bieb. × *T. repens* L. hybrid clovers. *Euphytica* 127:309–315.

Marshall, A. H., M. T. Abberton, T. A. Williams, et al. 2003. Forage quality of *Trifolium repens* L. × *T. nigrescens* Viv. Hybrids. *Grass Forage Science* 58:295–301.

Marshall, A. H., T. P. T. Michaelson-Yeates, M. T. Abberton, et al. 2002. Variation for reproductive and agronomic traits among *T. repens* × *T. nigrescens* third generation backcross hybrids in the field. *Euphytica* 126:195–201.

Marshall, A. H., C. Rascle, M. T. Abberton, et al. 2001. Introgression as a route to improved drought tolerance in white clover (*Trifolium repens* L.). *Journal of Agronomy and Crop Science* 187:11–18.

Marshall, A. H., A. Williams, M. T. Abberton, et al. 2003. Dry matter production of white clover (*Trifolium repens* L.), Caucasian clover (*T. ambiguum* M. Bieb.), and their associated hybrids when grown with a grass companion over 3 harvest years. *Grass Forage Science* 58:63–69.

Marshall, A. H., T. A. Williams, M. T. Abberton, et al. 2004. Forage quality of white clover (*Trifolium repens* L.) × Caucasian clover (*T. ambiguum* M. Bieb.) hybrids and their grass companion when grown over three harvest years. *Grass Forage Science* 59:91–99.

Marshall, A. H., T. P. T. Michaelson-Yeates, P. Aluka, and M. Meredith. 1995. Reproductive characters of interspecific hybrids between *Trifolium repens* L. and *T. nigrescens* Viv. *Heredity* 74:136–145.

McGuire, W. S. 1985. Subterranean clover. In *Clover science and technology,* ed. N. L. Taylor, 515–534. Madison, WI: ASA, CSSA, and SSSA.

Merker, A. 1984. Hybrids between *Trifolium medium* and *Trifolium pratense. Hereditas* 101:267–268.

Miller, J. D., and H. D. Wells. 1985. Arrowleaf clover. In *Clover science and technology,* ed. N. L. Taylor, 503–514. Madison, WI: ASA, CSSA, and SSSA.

Moriuchi, H., Y. Fujikawa, M. A. M. Aly, et al. 2004. Hairy root-mediated transgenic-plant regeneration in Egyptian clover (*Trifolium alexandrinum* L.). *Plant Biotechnology* 21:165–168.

Morris, J. B., and S. L. Greene. 2001. Defining a multiple-use germplasm collection for the genus *Trifolium. Crop Science* 41:893–901

NPGS (National Plant Germplasm System). 2003. Germplasm resources information network (GRIN). Database management unit (DBMU), National Plant Germplasm System, U.S. Dep. Agric., Beltsville, MD.

Nelke, M., J. Nowak, J. M. Wright, et al. 1999. Enhanced expression of a cold-induced gene coding for a glycine-rich protein in regenerative somaclonal variants of red clover (*Trifolium pratense* L.). *Euphytica* 105:211–217.

Palmroth, M. R. T., J. Pichtel, and J. A. Puhakka. 2002. Phytoremediation of subarctic soil contaminated with diesel fuel. *Bioresource Technology* 84:221–228.

Pederson, G. A., and G. E. Brink. 1998. Cyanogenesis effect on insect damage to seedling white clover in a bermudagrass sod. *Agronomy Journal* 90:208–210.

Putiyevsky, E., and J. Katznelson. 1973. Cytogenetic studies in *Trifolium* spp. related to berseem. I. Intra- and interspecific hybrid seed formation. *Theoretical and Applied Genetics* 43:351–358.

———. 1974. Cytogenetic studies in *Trifolium* spp. related to berseem III. The relationships between the *T. scutatum, T. plebeium,* and the *echinata* group. *Theoretical and Applied Genetics* 44:184–190.

Quesenberry, K. H., and N. L. Taylor. 1977. Interspecific hybridization in *Trifolium* L. Sect. *Trifolium* Zoh. II. Fertile polyploid hybrids between *T. medium* and *T. sarosiense* Hazsl. *Crop Science* 17:141–145.

———. 1978. Interspecific hybridization in *Trifolium* L. Section *Trifolium* Zoh. III. Partially fertile hybrids of *T. sarosiense* Hazsl. and 4*x T. alpestre* L. *Crop Science* 18:536–540.

Quesenberry, K. H., D. S. Wofford, R. L. Smith, et al. 1996. Production of red clover transgenic for neomycin phosphotransferase II using *Agrobacterium. Crop Science* 36:1045–1048.

Rangappa, M., A. A. Hamama, and H. L. Bhardwaj. 2002. Legume and grass cover crops for seedless watermelon production. *HortTechnology* 12:245–249.

Rumbaugh, M. D. 1990. Special purpose forage legumes. p. In *Advances in new crops,* ed. J. Janick and J. E. Simon, 183–190. Portland, OR: Timberline Press.

———. 1991. Plant introductions: The foundation of North American forage legume cultivar development. In *Use of plant introductions in cultivar development,* part 1, ed. H. L. Shands and L. E. Wiesner, 103–114. Madison, WI: SSA Spec. Publ. 17.

Rupert, E. A., and P. T. Evans. 1980. Embryo development after interspecific cross-pollinations among species of *Trifolium,* section *Lotoidea. American Society of Agronomy Abstracts* 68.

Rutter, S. M. et al. 2004. Dietary preference of dairy cows grazing ryegrass and white clover. *Journal of Dairy Science* 87:1317–1324.

Sato, S. et al. 2005. Comprehensive structural analysis of the genome of red clover (*Trifolium pratense* L.). *DNA Research* 12:301–364.

Sawai, A. et al. 1990. Interspecific hybrids of *Trifolium medium* L. × 4*x Trifolium pratense* L. obtained through embryo culture. *Journal of the Japanese Society of Grassland Science* 33:157–162.

Sawbridge, T., E. Ong, C. Binnion, et al. 2003. Generation and analysis of expressed sequence tags in white clover (*Trifolium repens* L). *Plant Science* 165:1077–1087.

Schwer, J. F., and R. W. Cleveland. 1972a. Diploid interspecific hybrids of *Trifolium pratense* L., *T. diffisum* Ehrh., and some related species. *Crop Science* 12:321–324.

———. 1972b. Tetraploid and triploid interspecific hybrids of *Trifolium pratense* L., *T. diffisum* Ehrh., and some related species. *Crop Science* 12:419–422.

Sullivan, M. L., and K. H. Quesenberry. 2006. Red clover (*Trifolium pretense*). In *Methods in molecular biology,* 2nd ed., ed. K. Wang, 369–384. Totowa, NJ: Humana Press.

Taylor, N. L., et al. 1963. Interspecific hybridization of red clover (*Trifolium pratense* L.). *Crop Science* 3:549–552.

Taylor, N. L., P. B. Gibson, and W. E. Knight. 1977. Genetic vulnerability and germplasm resources of the true clovers. *Crop Science* 17:632–634.

Taylor, N. L., and J. M. Gillett. 1988. Crossing and morphological relationships among *Trifolium* species closely related to strawberry and Persian clover. *Crop Science* 28:636–639.

Taylor, N. L., and N. Giri. 1984. Crossing and morphological relationships among species closely related to arrowleaf clover. *Crop Science* 24:373–375.

Taylor, N. L., and K. H. Quesenberry. 1996a. Biosytematics and interspecific hybridization. In *Red clover science,* 11–24. Dordrecht, the Netherlands: Kluwer Academic Publishers.

———. 1996b. *Red clover science. Current plant science and biotechnology in agriculture.* Dordrecht, the Netherlands: Kluwer Academic Publishers.

Townsend, C. E. 1985. Miscellaneous perennial clovers. In *Clover science and technology,* ed. N. L. Taylor, 563–578. Madison, WI: ASA, CSSA, and SSSA.

Ulloa, O., F. Ortega, and H. Campos. 2003. Analysis of genetic diversity in red clover (*Trifolium pratense* L.) breeding populations as revealed by RAPD genetic markers. *Genome* 46:529–535.

Williams, E., and I. M. Verry. 1981. A partially fertile hybrid between *Trifolium repens* and *T. ambiguum*. *New Zealand Journal of Botany* 19:1–7.

Williams, W. M. 1987. Genetics and breeding. In *White clover,* ed. M. J. Baker and W. M. Williams, 343–419. Wallingford, U.K.: CAB International.

Williams, W. M., and E. G. Williams. 1981. Use of embryo culture with nurse endosperm for interspecific hybridization in pasture legumes. p. In *Proceedings of the XIV International Grassland Congress,* ed. J. A. Smith and V. W. Hays, 163–165, Lexington, KY, June 15–24. Boulder, CO: Westview Press.

Zohary, M., and D. Heller. 1984. Taxonomic part. In *The genus* Trifolium, 33–587. Jerusalem: Israel Academy of Sciences and Humanities.

CHAPTER 8

Bermudagrass

Yanqi Wu and Charles M. Taliaferro

CONTENTS

8.1 INTRODUCTION

Among the most important forage crops, bermudagrass is widely used in tropical and warmer regions of the world (Burton 1965; Harlan 1970). The warm-season, sod-forming, long-lived perennial grass is also important for turf use, soil stabilization, and remediation of contaminated soils (Beard 1973; Taliaferro 2003). "Bermuda grass" or "bermudagrass" is the most widely used common name applied to plants in genus *Cynodon* L. C. Rich. of grass family (Gramineae or Poaceae) (Burton and Hanna 1995). The common name now widely used has an American origin. Kneebone (1966a) indicated the name is probably because of the early introductions into Georgia and the Carolinas from the West Indies. Although there is no definitive published information concerning the name source of bermudagrass, Tracy (1917) indicated that the name itself suggested it may have been introduced to America by way of the island of Bermuda.

The use of bermudagrass as a common name has increased beyond North America over the past century. This is probably because many of the commercially successful cultivars were developed in the United States and exported to many other countries in Asia, Australia, South America, Africa, and Europe. However, several other common names exist for *Cynodon;* some probably greatly predate the inception of bermudagrass. Ho et al. (1997) indicated bermudagrass plants are called "couch" or "couch grass" in Australia; Kneebone (1966a) recorded "doub," "dhoub," "durva," "dwiva," and "durbha" grass as common names used in India. In local languages in other countries, like Germany, France, and China, "dog tooth grass"—because *Cynodon* in Greek means "dog's tooth"—or "wire grass"—because of its abundant and thin stolons and creeping growth habits—is used. In South Africa, more distinctive names are used for different specific forms. "Kweek," "coarse kweek," "quick grass," "couch," and "kweekgrass" are used to denote plants of *C. dactylon* (L.) Pers., "Florida grass" for *C. transvaalensis* Burtt-Davy, and "Magennis grass" for *C. magennisii* Hurcombe, as documented by Chippindall (1955) and Beard and Watson (1982). *Cynodon dactylon* is known as "chiendent" in Mauritius (Rochecouste 1962).

Cynodon plants can be divided into the rhizomatous and the nonrhizomatous on the basis of presence or absence of rhizomes in mature plants. The robust, nonrhizomatous forms are referred to as "star grass" and the rhizomatous taxa as "bermudagrass" by Harlan (1970). In the United States, *C. dactylon* (L.) Pers. var. *dactylon* is widely referred to as common bermudagrass; var. *aridus* Harlan et de Wet is often termed "giant bermudagrass" and *C. transvaalensis* African bermudagrass (Burton and Hanna 1995; Kneebone 1966a; Taliaferro 1992).

Star grass includes three East African species containing plants that generally are robust and not cold hardy (Harlan 1970). Before the publication of a revised taxonomic classification system for the genus *Cynodon* by Harlan, de Wet, Huffine, et al. (1970), there was much confusion over the species classification of *Cynodon* plants from East Africa. Harlan, de Wet, and Rawal (1970a) indicated at least four *Cynodon* species in East Africa on the basis of morphology, distribution, ecological behavior, cytology, and ease of crossing under experimental conditions. Recently, Taliaferro, Rouquette, and Mislevy (2004) used stargrass as a common name for the East African species *Cynodon aethiopicus* Clayton et Harlan, *Cynodon nlemfuensis* Vanderyst, and *Cynodon plectostachyus* (K. Schum.) Pilg.; they used bermudagrass as a common name of all other *Cynodon* taxa. In this chapter, bermudagrass is used as a general name for *Cynodon* plants, and, if needed, specific names are applied.

8.2 DESCRIPTION AND CROP USE

8.2.1 Morphology

Morphology for bermudagrass can be characterized by its typical characteristics of root systems, stems, leaves, inflorescences, and seeds, although it is very diverse. Bermudagrass has an extensive root system, which is fibrous and can be quite deep (Beard 1973). Burton, DeVane, and Carter (1954) demonstrated that roots of certain common bermudagrass strains can reach about 2.4 m in deep, sandy soil. However, research data on root systems for other *Cynodon* species are limited. At the nodes of bermudagrass stolons, fibrous root systems can develop from the adventitious roots growing from the node side touching the soil surface. Bermudagrass has three forms of stems: two aboveground types (i.e., stolons and shoots) and one underground form (i.e., rhizomes). Zheng, Liu, and Chen (2003) observed rhizome characters (i.e., depth, density, and diameter) of 50 native Chinese bermudagrass accessions originated from tropical (22.5° N latitude) to cold temperate (43.6° N latitude) regions. They found that the rhizome trait descriptors were highly correlated with latitudes of their origins and that in a uniform nursery, the rhizomes of bermudagrasses grew deeper, denser, and thicker with increasing north latitude of origin.

Stems consist of several nodes and internodes. Stem nodes are glabrous, containing meristem tissues capable of producing new roots and shoots. Duell (1961) reported that some bermudagrass plants have multiple (two to three) leaved nodes, although one leaf at each node is common (Beard 1973). Clayton and Renvoize (1986) described leaf blades as basically flat and linear, with a ligule that is a short membrane with ciliate margin. Foliage color varies from light green to darker green. Leaves are rolled or folded in the bud (Watson and Dallwitz 1992). Leaf sheaths are overlapped at the base and split toward the blade. Most spikelets of *Cynodon* plants contain one floret. In rare cases, plants of *C. dactylon* var. *dactylon* may have spikelets containing two fertile florets. For example, two germplasm accessions collected from Fujian and Sichuan provinces in China by the senior author had two florets in some spikelets, but only one floret in other spikelets (Y. Wu, unpublished results).

Inflorescences consist of two or more unilateral racemes forming one or more whorls. For each raceme, spikelets are biseriate and partially overlapped on one side of the rachis (Watson and Dallwitz 1992). The racemes, sometime referred to as digitates or digitate spikes, persist on the plant, but florets shatter from the plant at maturity. Florets are laterally compressed. Clayton and Renvoize (1986) indicated that *Cynodon* L. C. Rich. plants are clearly characterized morphologically from the plants of other genera of the tribe Cynodonteae Dumort. by awnless lemmas and acute glumes. Bermudagrass seeds are relatively small, consisting of a caryopsis enclosed in the lemma and palea. Bermudagrass seed is commercially sold in two forms: hulled (i.e., a caryopsis enclosed by a lemma and a palea) or unhulled (i.e., a caryopsis without enclosing bracts). Ahring and Todd (1978) reported seed numbers per gram ranging from 3,080 to 4,629 for caryopses and from 2,198 to 2,847 for hulled seeds.

8.2.2 Biology

Bermudagrass plants are warm-season, long-lived perennials. Distribution and growth of bermudagrass are influenced mainly by temperature, water availability, light, and edaphic conditions, and to lesser extents by disease and insect pests. In uncultivated conditions, plants typically follow a secondary succession pattern, invading disturbed areas. Bermudagrass has active growth when temperatures exceed 10°C. Compared to the commonly used warm- and cool-season grasses, bermudagrass has excellent heat tolerance but is relatively poor in low temperature hardiness (Beard 1973). Beard (1973) indicated that the optimum temperature range for bermudagrass growth is 25–35°C, similar to that of many other warm-season grasses. In tropical and subtropical climates, where temperatures never or seldom reach 0°C, bermudagrass plants remain in a nondormant state throughout the year. Sporadic light frost events that may occur in subtropical climates during the cool season temporarily slow the growth and cause discoloration of foliage due to tissue injury. Low temperatures may also induce anthocynanin pigmentation in foliage (Taliaferro 2003). In temperate climates, adapted bermudagrass forms become dormant in response to temperatures ≤ 0°C. New growth in the spring may be generated from buds of rhizomes, surviving stolons, or combinations thereof.

Along with precipitation, winter temperature is a key deciding factor for bermudagrass population geographic dispersion and adaptation and for individual plant persistence. Taliaferro et al. (2004) noted that the very limited freeze tolerance of stargrass mandates its use in tropical and subtropical regions. In contrast, large variation in freeze tolerance exists in common bermudagrass, *C. dactylon* (L.) Pers. var. *dactylon,* the cosmopolitan taxon of *Cynodon.* According to de Wet and Harlan (1971), this taxon is found on every continent and most islands between 45° N and S latitudes. Plants of tropical origin generally have little freeze tolerance, but those from temperate regions can tolerate freezing temperatures. In the continental United States, bermudagrass is predominantly used commercially south of about 38° N latitude, although winter-hardy forms are sparsely distributed further north (Taliaferro 2003). Taliaferro et al. (2004) noted that stargrass in the United States is mainly cultivated in the southern half of Florida because of its low freeze tolerance. The ability of bermudagrass plants to survive winters with frequent and/or sustained freezing events may be by escape or their ability to acclimate physiologically to freezing temperatures. Plants with rhizomes well below the soil surface may escape freezing temperatures. Anderson, Kenna, and Taliaferro (1988) studied levels of acclimation in field-grown bermudagrass plants in Oklahoma and reported that the greatest levels of freeze tolerance occurred during December and January. After January, the bermudagrass plants gradually lost freezing tolerance as the deacclimation process progressed. Bermudagrass winter kill becomes a more serious problem as its utilization extends into the transition zone and even colder regions.

Bermudagrass is a typical C_4 plant. A characteristic of the C_4 carbon fixation chemical pathway is the early production of carboxylation products, such as malate and aspartate, important in photosynthesis. Downton and Tregunna (1968) reported that *C. dactylon* had a photosynthetic rate of 60 mg CO_2/dm^2 h and CO_2 compensation value of 5 ppm. Photosynthetic rates of C_4 plants range from 50 to 80 mg CO_2/dm^2 h, while those of C_3 plants are from 30 to 45 mg CO_2/dm^2 h. Under normal physiological conditions, photosynthetic CO_2 compensation concentrations of C_4 plants are 5 ppm or less, and CO_2 compensation concentrations of C_3 plants are from 37 to 50 ppm (Black, Chen, and Brown 1969). Chen et al. (1969) confirmed the low value of CO_2 compensation concentration in bermudagrass. Dudeck and Peacock (1992) included *C. dactylon* among monocotyledons with high photosynthetic capacity. Krans, Beard, and Wilkinson (1979) demonstrated that Tifgreen, an interspecific hybrid of *C. dactylon* with *C. transvaalensis,* had Kranz leaf anatomy and a low value of CO_2 compensation concentration. C_4 plants typically have Kranz leaf anatomy, which is leaf vascular bundles having two concentric chlorophyllous layers, an inner parenchyma bundle sheath layer and an outer mesophyll layer (Krans et al. 1979).

Bermudagrass is best adapted to conditions of full sunlight and usually fails to persist under conditions where shading persistently occurs more than 50% of the time. Burton, Jackson, and Knox (1959) reported that forage yield, root and rhizome yields, and belowground available carbohydrates of cv. Coastal bermudagrass decreased as shade intensity increased. They additionally reported that at the level of 200 lb of N application per acre (220 kg N/ha), shade reduced available carbohydrates and increased moisture, lignin, crude protein, true protein, P, Ca, and Mg concentrations in the forage. At higher N level, shading caused more forage yield reduction but less change in chemical composition (Burton et al. 1959). The low shade tolerance of bermudagrass especially impacts its use for turf, where landscapes frequently include trees and buildings that cast shade. Reported effects of shading on turf bermudagrass include reduced sod density associated with a reduction in lateral growth and more upright, taller shoots with longer internodes (McBee and Holt 1966). Gaussoin, Baltensperger, and Coffey (1988) reported that a 90% low-light treatment significantly reduced forage biomass production in potted plants. Interestingly, Gaussion et al. (1988) reported that the low-light treatment reduced leaf length, shoot elongation, stem internode length, visual color, chlorophyll concentration, and dry weight. These morphological changes differ from those reported by Burton et al. (1959). Jiang, Duncan, and Carrow (2004) reported that low-light stress drastically reduced the turf quality of two hybrid bermudagrasses (*C. dactylon* by *C.*

transvaalensis cvs. TifEagle and Tifsport). They reported that, under severe shade conditions (70 and 90% low light levels), hybrid bermudagrasses have significantly lowered canopy photosynthetic rates compared to full-light conditions (Jiang et al. 2004).

Some variation among bermudagrass cultivars for shade tolerance has been reported. McBee and Holt (1966) reported variation in response to shading among four bermudagrass cultivars. Recently, Hanna and Maw (2007) reported the development of a turf bermudagrass cultivar with enhanced shade tolerance.

Bermudagrass is predominantly naturally distributed in areas receiving moderate to high annual precipitation (~≥500 mm), but it is generally considered to have excellent tolerance of sporadic droughty conditions (Beard 1973). In arid climates, it may persist in areas that accumulate soil moisture, such as along waterways or around ponds. Bermudagrass drought tolerance is attributed mainly to its ability to enter a dormant state when available soil moisture is low and to an extensive root system that allows extraction of water from low soil depths. Little is known of the genetic and physiological mechanisms by which bermudagrass enters dormancy in response to severe drought, but they are likely similar to the mechanisms inducing the state in response to low temperatures. As in the case of freezing temperatures, severe drought may cause aboveground parts of plants to die with subsequent new growth generated from buds of mainly crowns and rhizomes.

Burton et al. (1954) and Burton, Prine, and Jackson (1957) reported differences among bermudagrass cultivars in rooting characteristics. Common bermudagrass (bermudagrass naturally selected for adaptation to the southern United States) had greater root density than cvs. Coastal and Suwannee in the top 15-cm soil depth, but the roots of the hybrid cultivars penetrated to greater depths and produced higher yields of roots at the lower depths. Hybrid bermudagrasses had more than 30% of their roots below the depth of 60 cm, but the common form had only about 15%. Burton et al. (1954) reported that the deeper roots of hybrid cultivars correlated well with their greater drought tolerance. They also observed that Coastal bermudagrass developed a root system to a depth of about 2 m in a sandy field, and heavy nitrogen application increased the rate of root penetration. This deep root pattern of hybrids enabled them to draw more effectively on the moisture reserves in the soil below the 60-cm depth; consequently, they suffered less drought stress than common bermudagrass.

Doss, Asheley, and Bennett (1960) reported that the effective rooting depth of bermudagrass decreases as soil moisture increases. Adams, Elkins, and Beaty (1966) reported that fertilizer applications significantly increased the weight of bermudagrass roots to a depth of about 1 m for Coastal and only 0.3 m for a common strain. They also reported that roots as a percentage of the total plant biomass declined as the amount of fertilizer application increased. Holt and Fisher (1960) found a slight change in root weight distribution toward deeper roots with nitrogen applications in Coastal bermudagrass. Frequent defoliations may reduce bermudagrass tolerance of stresses, particularly drought, due to a decrease of total root production (Taliaferro, Rouquette, et al. 2004; Alcordo, Misley, and Rechcigl 1991).

Bermudagrass tolerance of waterlogging and inundation makes it a good choice for use in frequently flooded areas compared to other grasses (Beard 1973). However, its tolerance varies with duration and depth of submergence, physiological condition of the plant tissue, and temperature. Gamble (1958) reported that bermudagrass demonstrates remarkable tolerance of periodic inundations in areas above flooding sediment pools during the spring and early summer. De Gruchy (1952) observed that bermudagrass plants can withstand inundations for 2–3 months in shallow water areas. Juska and Hanson (1964) observed, however, that bermudagrass cannot thrive on persistently waterlogged soils.

Bermudagrass collectively is adapted to a wide range of soils in terms of texture, pH value, and fertility. The widely adaptive capability of bermudagrass to soils is demonstrated by its widely geographic dispersion on various soils between 45° N and S latitudes in the world. Bermudagrass grows better on fine-textured, well-drained soils than on light sandy soils because of the higher fertility level and moisture retention associated with the fine-textured soils (Beard 1973). However, high forage yields can be harvested on bermudagrass grown on deep sands if moisture and fertilizer are adequate (Burton et al. 1954). Soil nutrients, especially nitrogen, are critical in maintaining healthy, vigorous bermudagrass stands. In addition to soil fertility, many other factors, including climate, timing of fertilizer applications, cultivar, plant development stage, and their interactions, influence the quantity and quality of bermudagrass forage at different degrees. A large amount of research data accumulated over the past 60 years on fertilization of bermudagrass was recently reviewed by Taliaferro, Rouquette, et al. (2004).

Bermudagrass generally tolerates a wide range of soil pH, though substantial variation exists within the species for tolerance of low- and high-pH conditions and salinity (Marcum 2000). Lundberg, Bennett, and Mathias (1976) tested three commercial and five experimental bermudagrass cultivars grown on soils with pH values ranging from 3.4 to 6.4. Their results indicated that all experimental bermudagrasses survived on low pH soils, but with varying reductions in forage and root yields. Increasing the pH to ~6.0 via application of lime dramatically increased root and forage yields. However, the cultivars responded differently to various lime levels (Lundberg et al. 1976). The tolerance of bermudagrass to salinity as indicated by 50% reduction in shoot growth has been reported to range from moderate (17 dS m^{-1}) to high (34 dS m^{-1}) (Marcum 2000; Dudeck and Peacock 1993; Dudeck et al. 1983; Francois 1988). Marcum and Pessarakli (2006) recently reported data indicating that up to 40 dS m^{-1} salinity was required to reduce shoot growth by 50% in some bermudagrass cultivars. The level of salinity tolerance of bermudagrass plants is strongly associated with their ability to excrete and efficiency in excreting salt via two-cell salt glands embedded in the epidermis of leaf blades on both abaxial and adaxial surfaces (Marcum 2000; Marcum and Pessarakli 2006).

Bermudagrass has relatively few serious insects and diseases, though some fungi, insects, nematodes, bacteria, and viruses may damage it (Taliaferro, Rouquette, et al. 2004). At or above threshold infestations, these biotic pests injure bermudagrass with resultant reductions in biomass production, forage quality, and stand density. Taliaferro (1995) noted that the widespread use of only a few industry-standard clonal cultivars may increase vulnerability of the species to new pests with ability to induce greater or more widespread injury. Couch (1995) and Smiley, Dernoeden, and Clarke (1993) listed important diseases of bermudagrass used for turf. Martin et al. (2001) and Iriarte et al. (2004) noted that spring dead spot caused by *Ophiosphaerella herpotrica* (Fr.) Walker, *O. korrae* Walker and Smith, and *O. narmari* Walker and Smith has been an important root-rotting disease of turf bermudagrass in the transition zone in the United States. Bermudagrass decline caused by *Gaeumannomyces graminis* (Saac.) Arx & D. Olivier, dollar spot by *Sclerotina homoeocarpa* F. T. Bennett, and brown patch by *Rhizoctonia* spp. are common diseases of turf bermudagrass (Couch 1995; Smiley et al. 1993). Pratt (2000, 2005) reported that foliage diseases caused by *Bipolaris* spp., *Curvularia* spp., and *Exserohilum* spp. resulted in stand decline and substantial yield losses in forage bermudagrasses. Pratt and Brink (2007) reported substantial variation among forage bermudagrass cultivars for tolerance to the dematiaceous hyphomycetous pathogens. Root-feeding nematodes, including rootknot (*Meloidogyne* sp.), root lesion (*Pratylenchus* sp.), burrowing (*Radopholus similes* Thorne), spiral (*Helicotylenchus* sp.), sting (*Belonolaimus* sp.), stunt (*Tylenchorhynchus* sp.), ring (*Criconemella* sp.), pin (*Paratylenchus* sp.), stubby root (*Paratrichodorus* sp.), and lance (*Hoplolaimus* sp.), are capable of causing serious damage to the root systems of bermudagrass and consequently impairing the overall growth of susceptible bermudagrass plants (Couch 1995).

Fall armyworms (*Spodoptera frugiperda* J. E. Smith), grasshoppers (*Melanoplus* spp.), leafhoppers (*Carneocephala flaviceps* Riley, *Exitianus exitiosus* Uhler, *Graminella* spp., and *Draeculacephala* spp.), planthoppers (*Delphacodes propinqua* Fieber), black cutworms (*Agrotis ipsilon* Hufnagel), and some other chewing insects defoliate bermudagrass plants and reduce forage yield in bermudagrass pasture and hay crops (Leuck et al. 1968; Pitman, Croughan, and Stout 2002; Salehi et al. 2005; Lynch and Burton 1993). Mirid (*Trigonotylus doddi* Distant) is also a common insect pest on forage bermudagrasses. Two-lined spittlebugs (*Prosapia bicincta* Say.) injure susceptible bermudagrass by injecting a toxin into the plants as they feed (Byers and Wells 1966). Taliaferro, Leuck, and Stimmann (1969) and Stimmann and Taliaferro (1969) reported the resistance in *C. dactylon* and *C. transvaalensis* to two-lined spittlebugs. More insect pests were listed in damaging turf bermudagrasses by Taliaferro (1995). In Arizona and California, where commercial bermudagrass seed production is centered, Rethwisch et al. (1995) reported reduction in seed yield of common bermudagrass by infestations of mirid (*Trigonotylus tenuis* Rueter) and grass thrips (*Chirothrips* spp.). They observed that high numbers of *T. tenuis* were highly positively correlated with reduced numbers of inflorescence spikes and plant height. Mirid caused stunting and delayed flowering.

8.2.3 Use and Importance

Cynodon plants, especially common bermudagrass, have received great attention in scientific research and agricultural application because they are widely used in tropical and warmer temperate regions of the world for livestock herbage, turf and soil stabilization, and spoil site remediation (Taliaferro 1995; Taliaferro, Rouquette, et al. 2004).

Bermudagrass is an attractive long-lived perennial as a cultured crop because of enormous genetic diversity within the species. This trait allows plants to adapt to a wide range of climatic and edaphic conditions. Natural selection and scientific breeding have produced high-performance bermudagrass cultivars with excellent tolerance of heat, drought, and wear. Bermudagrass has few widely damaging insect and disease pests. Its sod-forming growth habits, aggressive spreading ability, and significant responsiveness to management practices for increasing productivity add substantially to its economic value. Breeders have developed specifically designed cultivars (Burton et al. 1954; Taliaferro, Rouquette, et al. 2004; Taliaferro, Martin, et al. 2004). The importance of bermudagrass has significantly increased over the last century.

8.2.3.1 Forage Grass

Bermudagrass and stargrass are the most important introduced warm-season perennial forage grasses in the southern United States. Bermudagrass is grown for grazing or stored forage on an estimated 10–12 million ha in the United States (Taliaferro, Rouquette, et al. 2004). A history of bermudagrass in the United States was recently reviewed by Taliaferro, Rouquette, et al. (2004). Bermudagrass was recognized in the early 1800s as one of the most important warm-season grasses in the southern United States, as described in *A Geological Account of the United States* by Mease (1807) (cited by Burton and Hanna 1995). Although its value for forage and soil stabilization was recognized by early agriculturists, farmers of cultivated row crops considered it a pernicious weed (Burton and Hanna 1995). Use of bermudagrass as a tame pasture grass in the south dramatically increased in the mid-twentieth century shortly after the end of World War II. Harlan (1970) cited several reasons for increased popularity of bermudagrass:

- the release of Coastal, which yielded roughly twice the amount of forage as unselected common bermudagrass;
- increased availability of fertilizer, especially nitrogen; and
- the decline of the cotton industry in the South and concurrent increase in the livestock industry.

Bermudagrass is very responsive to application of nitrogen and requires other fertilizer elements to maintain high productivity. Intensive research on improvement of bermudagrass management systems and development of specialized sprig harvesting and planting equipment provided scientific and technological support to the increased wide use of bermudagrass.

Bermudagrass development and use as an intensively managed pasture grass is largely an American phenomenon (Taliaferro, Rouquette, et al. 2004). There are two basic types of forage bermudagrass cultivars: "grazing type" and "hay type." Hay type bermudagrass is taller and less dense in stand density relative to grazing type. Grazing type bermudagrass normally establishes more quickly than hay type cultivars. Tifton 44 and Midland 99 are examples of hay type varieties. Greenfield and Quickstand are grazing type varieties. Both hay and grazing bermudagrass varieties can be used for grazing and haying. However, after bermudagrass stands are fully established, grazing varieties have lower forage yield but better resistance to weed encroachment (Taliaferro, Rouquette, et al. 2004).

Among many inherent characteristics of bermudagrass plants, its sod-forming habit and tolerance of close grazing greatly facilitate its use as a grazed pasture grass. Bermudagrass can be defoliated via grazing or haying to a greater extent than many other introduced and native perennial pasture grasses. Indeed, higher nutritive value of bermudagrass foliage can be maintained by close grazing combined with other management criteria that continuously stimulate new growth (Taliaferro, Rouquette, et al. 2004). The importance of *Cynodon* species for forage use is distinct.

Harlan (1970) discussed the relative importance of the *Cynodon* species for use as livestock herbage. Of the six taxonomic varieties of *C. dactylon,* only two, vars. *dactylon* and *aridus,* are of major importance. Of the remaining four taxonomic varieties, *C. dactylon* var. *afghanicus* is a robust form that provides some pasture for livestock near irrigation projects in the lowland steppes in Afghanistan, but it can hardly be considered a major fodder, even in the local areas (Harlan 1970). *Cynodon dactylon* var. *coursii* is a robust, nonrhizomatous form endemic to Madagascar and a moderately important grazing resource on the island, but of no significance elsewhere. The var. *polevansii* is known from only one site near Baberspan, South Africa. The var. *elegans* provides considerable grazing in South Africa. The var. *aridus* as a fodder grass has not been characterized widely. *Cynodon dactylon* var. *dactylon,* the most widely distributed, provides an important forage resource for grazing and the most important genetic resource for breeding improved hay and pasture grass cultivars suitable for intensive management (Harlan 1970; Burton and Hanna 1995).

Harlan (1970) indicated that the robust species *C. aethiopicus, C. plectostachyus, C. nlemfuensis* var. *nlemfuensis,* and var. *robustus* are rather widely used as forage grasses in East Africa because of their natural abundance and high productivity. These stargrasses are used in the United States chiefly in tropical Florida. Stargrasses are mainly used for grazing, but their use as a hay crop for horses has increased in recent years in Florida (Taliaferro, Rouquette, et al. 2004). Taliaferro, Rouquette, et al (2004) also indicated that stargrasses are extensively used in Central and South America, the Caribbean, and tropical Africa.

Among the *Cynodon* species, *C. arcuatus, C. barberi, C. transvaalensis, C. incompletus* var. *incompletus,* var. *hirsutus,* and *C. × magennisii* are of minor value as herbages for grazing and hay because of very small size, little yield, or endemic or limited distribution (Harlan 1970; Taliaferro, Rouquette, et al. 2004).

8.2.3.2 *Turfgrass*

Bermudagrass is used for turf throughout the world, where adapted forms are available. According to Beard (1973), bermudagrass turf is used on home and institutional lawns; parks; athletic fields; golf course fairways, tees, putting greens, and roughs; cemetery grounds; roadside right-of-ways; and other similar areas. Breeding of turf bermudagrass cultivars started in the 1940s in the United States (Burton 1991).

The taxa of predominant turf importance are *C. dactylon* (L.) Pers. var. *dactylon* and *C. transvaalensis* Burtt-Davy. Most of the turf bermudagrasses of economic importance emanate from these two taxa. Forms of the narrowly endemic *C. × magennisii* Hurcombe and *C. incompletus* Nees var. *hirsutus* (Stent) de Wet et Harlan have had fairly wide use as turfgrasses. Taxa of minor turf importance are *C. arcuatus* J. S. Presl. ex C. B. Presl., *C. barberi* Rang. et Tad., and *C. dactylon* var. *polevansii* (Stent) de Wet et Harlan. Other taxa have little or no value as turf but in some cases will hybridize with the turf species, thereby representing potentially important contributors of genes for the breeding improvement of turf bermudagrass (Taliaferro 2003).

8.2.3.3 Soil Conservation Grass

Bermudagrass provides excellent protection against soil erosion from wind and water. Characteristics making it so useful in stabilizing soils are its aggressive growth, sod-forming capability, and excellent drought and flood tolerance. The densely web-like sod of a healthy bermudagrass stand provides a continuous cover that protects the underlying soil from possible erosion caused by wind and water. Its extensive underground rhizomes and fibrous root system effectively bind the soil from washing away by moving water. The ability of bermudagrass to tolerate water logging and drought for an extended period increases its usefulness in the areas subject to periodic inundation and drought. In fact, bermudagrass always grows in highly disturbed areas, such as along water ditches, river banks, roadsides, and animal paths in many natural settings of tropical and warmer temperate regions in the world. Its ecological function in those areas of protecting the disturbed soils is not fully appreciated. However, the grass is widely and intentionally planted solely for the purpose of controlling soil erosion. Tracy (1917) noted that bermudagrass was the best plant found to control soil erosion and that it was extensively used along almost the entire course of the Mississippi River from Cairo, Illinois, southward to the Gulf of Mexico to protect levees.

8.2.3.4 Animal Effluent Receiver Grass and Remediation of Spoil Sites

In the southern United States, bermudagrass and stargrass are used in some areas adjacent to intensive animal-feeding facilities as a "receiver crop" or "sink crop" to remove excess nutrients from soils receiving large amounts of animal effluent. Bermudagrass has been demonstrated to protect against surface runoff or leaching of nutrients from applied animal wastes that could contaminate lakes, waterways, and groundwater with P and N. Burns et al. (1985) reported that application of effluent from animal manure lagoons to Coastal bermudagrass pasture is an effective method of utilizing the nutrients in the effluent and minimizing groundwater and surface water pollution.

Burns, King, and Westerman (1990) conducted an 11-year study of application of different rates of swine lagoon effluent on Coastal bermudagrass. Dry matter yields of 13.1, 19.0, and 19.6 Mg ha^{-1} were achieved at effluent application levels providing 335, 670, and 1,340 kg of N ha^{-1} year^{-1}, respectively. However, soil NO_3–N concentrations in the eleventh year of the study indicated the low rate did not pose a hazard for NO_3 pollution of the groundwater, but the high rate did; the middle rate could pose a hazard in some years (Burns et al. 1990; King, Burns, and Westerman 1990). Brinks, Sistani, and Rowe (2004) investigated differences in N and P uptakes among bermudagrass cultivars Alicia, Brazos, Coastal, Russell, Tifton 44, and Tifton 85 and common bermudagrass fertilized with broiler litter. Hybrid cultivars showed better N and P uptakes than common bermudagrass. They indicated that among forage crops used in the southeastern United States, bermudagrass has the greatest potential to recover nutrients from soil routinely fertilized with broiler litter.

Stargrass is also an excellent receiver crop for removing nutrients from soils overdisposed with animal manures and municipal waste. Effluent from dairies and other animal facilities is routinely applied to stargrass through irrigation systems with the dual benefits of satisfying regulatory disposal requirements and meeting many of the nutritional requirements of the stargrass (Taliaferro, Rouquette, et al. 2004).

Bermudagrass and stargrass have great potential for use in revegetation of drastically disturbed sites, like mining and landfill sites. In Florida, stargrass is commonly used in reclamation of spoil sites damaged by phosphate mining. Mislevy et al. (2000) indicated that stargrass provides the aggressive growth needed to maintain stand density sufficient to minimize soil erosion if appropriate fertility amendments are added to the spoil soils.

8.3 ORIGIN, DOMESTICATION, AND DISPERSION

8.3.1 Geographic Origin

Geographic origin of bermudagrass is not fully resolved, although this grass is so important in numerous ways and scientific curiosity concerning its origin has existed for a relatively long time. Tracy (1917) stated that bermudagrass is a native grass of the Old World, probably of India. Burton (1951a) speculated that the Indian origin of bermudagrass was based on the fact that it has existed ubiquitously in the nation for many centuries. However, bermudagrass germplasm introduced to the United States from Africa showed a greater amount of genetic diversity than the germplasm from India. Burton (1951a) indicated that Africa, rather than India, is the likely primary center of origin of bermudagrass. Similarly, Juska and Hanson (1964) assumed that bermudagrass originated in tropical Africa. They reached this conclusion from observations made by Stephens that bermudagrass is most prevalent in the Great Rift Valley in Africa. Beard (1973) indicated that the origin of bermudagrass species is centered on the Indian Ocean, ranging from eastern Africa to the East Indies, and that East Africa is the center of origin for most turf-type bermudagrass.

Harlan and colleagues described the geographic distribution of the species of the genus *Cynodon* L. C. Rich. and their ploidy levels (Harlan 1970; Harlan, de Wet, Huffine, et al. 1970; Harlan et al. 1970a; de Wet and Harlan 1970; Harlan and de Wet 1969). It has been known that diploid *Cynodon* are distributed over a large geographic area from India, Afghanistan, and west Asia and then turning southward to the African continent. All other higher ploidy ($>2n = 2x = 18$) *Cynodon,* including triploid, tetraploid, pentaploid, and hexaploid forms, probably have evolved from diploid forms. Therefore, the geographic area from southern and western Asia to Africa along the Indian Ocean probably is the center of bermudagrass origin.

8.3.2 Domestication

Undoubtedly, bermudagrass has provided benefits to man for many centuries, but its domestication as a cultured plant probably occurred largely over the past century. Compared to major food crops (wheat, rice, maize, etc.), breeding and cultivating bermudagrass are a recent activity. It is well known that the domestication of forage plants is basically different from the domestication of grain crops. Harlan (1983) indicated that nonshattering of cereal seeds and nondehiscence of legume pods are primary features separating domesticated forms from their wild relatives. Other features of domesticated seed crops include an increase in seed size, a reduction of seed dormancy, a decrease in seed protein content, and increased uniformity in ripening.

For forage or turf use, the significantly domesticated features of food crops are not the primary focus of bermudagrass genetic improvement, although the features are of some importance. The important features of improved bermudagrass forage cultivars encompass, but are not limited to, significantly improved forage yield, better nutritional quality, and good response to nitrogen application and other management practices. Normally, domesticated seed crops become dependent on environmental conditions because some of their surviving abilities in wild nature were lost during the domestication process. It is desirable and important that improved bermudagrass cultivars have good adaptation to different conditions from low-management grazing lands to medium-input hay fields to intensively managed golf putting greens.

8.3.3 Dispersion

As indicated by Taliaferro (2003), the global dispersion of bermudagrass, particularly the cosmopolitan taxon, *Cynodon dactylon* var. *dactylon,* resulted from the combined actions of nature and man over many centuries. Over the past several hundred years, human activities probably have played increasingly significant roles in dispersing bermudagrass far from areas of geographic origin. In 2006, Clayton, Harman, and Williamson reported one species in Europe, seven in Africa, three in temperate Asia, three in tropical Asia, six in Australasia, four in Pacific regions, three in North America, four in South America, and one in Antarctica.

Harlan (1970) and Harlan and de Wet (1969) suggested that cosmopolitan *C. dactylon* var. *dactylon* originated as a Eurasian grass probably from the diploid progenitors *C. dactylon* var. *aridus* and *C. dactylon* var. *afghanicus,* with an evolutionary center from West Pakistan to Turkey. Then, from that center, the taxon was distributed to other locations. The high polymorphism within the species likely resulted from internal fragmentation. The cosmopolitan and ubiquitous species probably spread widely in Asia, Africa, and Europe "only after agriculture had churned up the landscape and to North and South Americas and remote oceanic islands only in post-Columbia times" (Harlan and de Wet 1969; Harlan 1970; Harlan, de Wet, Huffine, et al. 1970). Langdon (1954) reported *C. dactylon* specimens were collected in Australia during the period 1802–1805. Early records indicated *C. dactylon* was widely distributed in Australia at the time of European settlement, and the grass was found in remote areas (Taliaferro, Rouquette, et al. 2004).

The introduction of bermudagrass into America can be traced with greater precision than its earlier geographic dispersion (Taliaferro 2003). Bermudagrass may have been introduced to the New World, first in West Indies, after its discovery by Columbus or possibly on ships commanded by Columbus (Kneebone 1966a; Taliaferro 2003). Bermudagrass with seed may have been carried on ships as bedding material or livestock feed. Bermudagrass seeds do not fully shatter easily. Burton and Hanna (1995) and Kneebone (1966a) noted that Henry Ellis first introduced bermudagrass to Savannah, Georgia, in 1751; at that time he was governor of Georgia, although where Governor Ellis obtained the bermudagrass is unknown. Kneebone (1966a) added that "very probably, Robert Miller, a botanist hired by the Lords Proprietor from 1733 to 1738, intentionally or unintentionally brought the grass into Savannah somewhat earlier with collections made in the Caribbean and Central America."

Early records indicated that bermudagrass spread in the southern United States rapidly, probably coinciding with the migrations of settlers and westward expansion of agriculture (Taliaferro 2003). By the early 1800s, bermudagrass was the most widely used, best pasture grass of the South (Kneebone 1966a). During this period, there is no evidence that bermudagrass seed was commercially available. Plantings for pasture and turf were predominantly or perhaps solely from vegetative parts. As an example of how bermudagrass was dispersed in the United States, Kneebone (1966a) cites an account of "a whole boat load of sprigs [that] was shipped from Baton Rouge to Ft. Smith, Arkansas, in the early 1830s to grass over the parade ground (and incidentally provide a later source of planting stock for many Arkansas pastures)." By this or many other ways, bermudagrass became the most important pasture grass in the South by the end of the 1800s (Burton and Hanna 1995).

In the early 1900s, bermudagrass was recognized as one of the most desirable pasture and lawn grasses in the southern states (Tracy 1917). Bermudagrass was of the same relative importance in the southern region that Kentucky bluegrass (*Poa pratensis* L.) was in more northern states (Tracy 1917). Kneebone (1966a) related that bermudagrass seed was imported from Australia and sold in the 1890s and early 1900s in the southern states.

Commercial production of bermudagrass seed started in the southwestern United States in the early twentieth century. The world supply of bermudagrass seed has largely been produced over the past century in Yuma County, Arizona, and the California Imperial Valley (Kneebone 1966a; Baltensperger et al. 1993; Taliaferro 2003). The early history of how this enterprise developed is interesting. In the latter half of the nineteenth century, bermudagrass was brought to the southwestern United States as a lawn grass, as a contaminant of Mexican sesbania seed from West Mexico, and from California via cattle carrying seed in their digestive tracts (Kneebone 1966a). It soon became a principal weed in alfalfa seed production fields. Bermudagrass seed was easily cleaned and separated from alfalfa seed by fan and sieve. The bermudagrass seed was initially sold as a by-product of the effort. As alfalfa seed production became more challenging due to decreased presence of natural insect pollinators, increasing soil salinity from irrigation from the Colorado River, and the encroachment of the bermudagrass in the alfalfa production fields, bermudagrass seed gradually changed from a by-product to a cash crop (Kneebone 1973).

The passing of the Hatch act in 1870 and the subsequent establishment of agricultural experiment stations in the respective states hastened the dispersion of bermudagrass in the United States. For example, in Oklahoma, bermudagrass seed was planted along with 100 other grasses in 1892 after the Oklahoma Agricultural Experiment Station was established (Elder 1953). Fields (1902) indicated that, of all the grasses tried on the station farm, bermudagrass alone showed the qualities that must be possessed by a pasture grass in Oklahoma. In 1904 and 1905, rootstocks (sprigs) of winter-hardy bermudagrasses were distributed to at least one farmer in every county in the Oklahoma Territory before statehood (Anonymous 1902; Elder 1953). Currently, bermudagrass is found on an estimated 1.6 million ha (4 million acres, about 10% of Oklahoma land area) in the state and provides benefits such as livestock herbage and turf or soil stabilization. In the United States, bermudagrass is ubiquitous in the southern states, where it is best adapted; it is sparsely distributed in most or all of the contiguous northern states. The introduction of the East African stargrass into the United States probably occurred at a much later time than the introduction of bermudagrass (Taliaferro, Rouquette, et al. 2004).

8.4 TAXONOMY AND SPECIES DISTRIBUTION

The genus *Cynodon* consists of a small biosystematic group in the tribe Cynodonteae Dumort., subfamily Chloridoideae Rouy, and family Gramineae Juss. (Poaceae Barnh.) (Clayton and Renvoize 1986). Harlan and de Wet (1969), de Wet and Harlan (1970), Clayton and Harlan (1970), Harlan, de Wet, Huffine, et al. (1970), and Harlan et al. (1970a) reported an extensive biosystematic analysis on morphological, anatomical, and ecological differences; ease of hybridization; cytological behavior of species; and geographic distribution of more than 700 *Cynodon* accessions collected in several trips to Asia, Africa, Australia, and Europe, as well as specimen studies in herbaria. Harlan et al. (1970a) investigated specimens of *Cynodon* plants in many major herbaria, including Kew, British Museum, Edinburgh, Paris, Brussels, Berlin, Geneva, Florence, Pretoria, Ibadan, Nairobi, Khartoum, Tokyo, Los Banos, Bangkok, Honolulu, and Washington. The prominent study resulted in the revised taxonomy of *Cynodon* genus (Harlan, de Wet, Huffine, et al. 1970).

The revised species classification of *Cynodon* L. C. Rich. by Harlan, de Wet, Huffine, et al. (1970) includes nine species and 10 varieties (Table 8.1). Clayton et al. (2006) of the Royal Botanic Gardens, Kew, also list nine species, without indication of any botanical varieties, omitting *C. × magennisii* Hurcombe but adding *C. parviglumis* Ohwi. Harlan, de Wet, Huffine, et al. (1970) indicated that *C. × magennisii* is a natural triploid hybrid between *C. dactylon* and *C. transvaalensis*. In the species list of Clayton et al. (2006), *C. radiatus* is used to replace *C. arcuatus,* which is in the list of Harlan, de Wet, Huffine, et al. (1970). The two names are synonymous (Clayton et al. 2006). Clayton et al. (2006) list *C. radiatus* as an annual species.

Although the species number in the genus is small, their distribution patterns and economic values used as forage and turf grasses are distinct from each other and significant as a group. Among those species, *C. dactylon* has been recognized as the most important species because of its enormous economic value as a turf grass and for hay crop and animal grazing, as well as its cosmopolitan distribution in the world. The three robust East African species, *C. aethiopicus, C. nlemfuensis, and C. plectostachyus,* are collectively important for forage production and grazing, but their value as turf grasses is low or nonexistent. In contrast to the robust stargrasses, *C. transvaalensis* is important as a parental grass in the development of turf hybrid bermudagrass cultivars by crossing with *C. dactylon* var. *dactylon;* however, its value is small for forage or grazing use. Juska and

Table 8.1   Revised Species List and Their Distribution of the Genus *Cynodon*

Species	Distribution
C. aethiopicus Clayton et Harlan	East Africa: Ethiopia to Transvaal
C. arcuatus J. S. Presl ex C. B. Presl	Northern Madagascar and adjacent islands, India, southeast Asia, south Pacific to Australia
C. barberi Rang. et Tad.	India
C. dactylon (L.) Pers.	
var. *dactylon*	Cosmopolitan
var. *afghanicus* Harlan et de Wet	Afghanistan
var. *aridus* Harlan et de Wet	South Africa northward to Palestine, east to south India
var. *coursii* Harlan et de Wet	Central plateau of Madagascar
var. *elegans* Rendle	Mozambique, Rhodesia, Zambia, Angola, and southward to the Cape
var. *polevansii* Harlan et de Wet	Near Barberspan, South Africa
C. incompletus Nees	
var. *incompletus*	S. Africa, as far north as 23° S latitude
var. *hirsutus* de Wet et Harlan	S. Africa below 23° S latitude
C. nlemfuensis Vanderyst	
var. *nlemfuensis*	Ethiopia to Rhodesia and Angola
var. *robustus* Clayton et Harlan	Ethiopia to Rhodesia and westward to Angola
C. plectostachyus (K. Schum.) Pilg.	East tropical Africa
C. transvaalensis Burtt-Davy	South Africa
C. × magennisii Hurcombe	South Africa

Source: Harlan, J. R., de Wet, Huffine, et al. 1970. Bulletin B-673. Oklahoma Agricultural Experiment Station.

Hanson (1964) indicated that *C. transvaalensis* cultivars have been used for intensively managed sporting turfs such as golf course putting greens, bowling greens, and tennis courts, but the use is limited by the high water and fertility requirements, summer decline in turf quality, and intolerance to sustained, very low mowing heights (≤3.2 mm). The narrowly endemic *C. × magennisii* and *C. incompletus* have fairly wide use as turfgrasses (Taliaferro 1995). *C. arcuatus* and *C. barberi* have limited value for forage and grazing; they are also of minor importance for turf purposes (Harlan 1970). More details in geographic dispersion, economic value, and morphological characteristics of each *Cynodon* species follow.

Cynodon aethiopicus Clayton et Harlan has two ploidy levels: diploid ($2n = 2x = 18$) and tetraploid ($2n = 4x = 36$) (Clayton and Harlan 1970). Morphologically, the two ploidy forms are difficult to distinguish. The species is exceptionally robust and can grow up to 2 m tall (Taliaferro, Rouquette, et al. 2004). When mature, the culms of the species are shiny and so woody and hard that they rattle in the wind (Clayton and Harlan 1970). Plants of the species do not produce underground rhizomes, but their stolons are always coarse, with long internodes. Glabrous or sparsely hairy leaf blades of the species are from linear to lanceolate in shape and are coarse, stiff, and harsh in texture (Harlan, de Wet, Huffine, et al. 1970). The species is easily recognized by its multiple (usually two to three whorls) stiff, stout, spreading, dark red or purple racemes (Harlan, de Wet, Huffine, et al. 1970; Clayton and Harlan 1970). Harlan (1970) observed that when plants are heavily grazed or frequently mowed, the number of raceme whorls in the inflorescence is usually reduced to one and the plant could easily be confused with *C. dactylon* whenever the rhizomatous character is ignored. Spikelets are medium dense to lax on a raceme. Subequal glumes are about three fourths of the spikelet length (Harlan, de Wet, Huffine, et al. 1970). Plants of the species have distribution from

Ethiopia to Transvaal, primarily along the Great Rift Valley and East African highlands. Harlan et al. (1970a) indicated that the species occurs in greatest abundance and in the purest stands along the Western Rift but is rather widespread in East Africa. Because of its abundance and high biomass potential, the species has been used considerably as a forage plant (Harlan 1970). Like the other two stargrasses, *C. aethiopicus* has no winter hardiness (Harlan, de Wet, Huffine, et al. 1970).

Cynodon arcuatus J. S. Presl ex C. B. Presl is a tetraploid ($2n = 4x = 36$) species (Harlan, de Wet, Huffine, et al. 1970; Harlan, de Wet, Rawal, Felder, et al. 1970). *Cynodon arcuatus* has fairly wide geographic distribution present on the Malagasy Islands and in India, South Asian countries, and Australia. The species is also present in some African countries and China (Clayton et al. 2006). *Cynodon arcuatus* is indigenous to the islands of southern China, where plants reach 50 cm in height (Anonymous 1990). Internodes are 1–1.5 mm in diameter and up to 8 cm long. Harlan, de Wet, Huffine, et al. (1970) indicated that *C. arcuatus* plants form low, open mats and are nonrhizomatous perennials without winter hardiness. The most distinctive morphological feature of the species is its large ovate-lanceolate leaf blades compared to leaf blades of other *Cynodon* species (de Wet and Harlan 1970). Its leaf blade surface is normally scaberulous, glabrous, and hairless (Clayton et al. 2006). Each inflorescence has one whorl including four to eight drooping, flexuous, slender, spreading, unilateral racemes. Clayton et al. (2006) described spikelets as forming two overlapping rows on one side of the raceme rachis. Each spikelet contains one fertile floret and no extension of rachilla. Glumes are persistent and shorter than the spikelet. Caryopsis is laterally compressed and close to trigonous in shape. Harlan (1970) stated that plants of the species are rather weedy in wet and low habitats. The species might be used as forage or for protecting soil from water erosion, but economic value is minor.

Cynodon barberi Rang. et Tad. is a diploid species with $2n = 2x = 18$ chromosomes. The species is endemic to South India (Harlan et al. 1970a). Harlan (1970) described the plants as small, rare, and too unproductive in forage yield to be of interest as a forage grass. However, in some local areas, it can be sufficiently abundant to be a turfgrass. The plant grows in low, wet areas and along edges of permanent waterways (Harlan et al. 1970a). Its slender stolons can form loose mats less than 10 cm in height. In general morphology and growth habit, *C. barberi* resembles *C. arcuatus*. However, the inflorescence of the former has very short, straight racemes in one whorl, and its long glumes may exceed the length of the lemma. Plants of the species are perennials, with no rhizomes and no winter hardiness (Harlan, de Wet, Huffine, et al. 1970).

Cynodon dactylon (L.) Pers. is the most important species in the genus because of its prevalence, wide geographic distribution, and uses as livestock herbage, as turf, and for soil stabilization and soil remediation. The species contains plants that are stoloniferous, sod forming, and perennial. Rhizomes, when present, vary from slender to fleshy thick and may grow to about 20 cm in depth (Zheng et al. 2003). Leaf blades are flat and linear, with or without sparse hairs, and range in length from a few centimeters to 13.5 cm and in width from 2 to 4 mm. Leaves have membranous ligules. Racemes are normally four to six in a single whorl, rarely more than one. Subequal glumes are about half or three fourths of the length of the spikelet. Caryopses are laterally compressed.

Harlan, de Wet, Huffine, et al. (1970) subdivided the species into six taxonomic varieties on the basis of cytogenetic behavior, morphology, ecology, and geographic distribution. The six varieties are var. *afghanicus* Harlan et de Wet, var. *aridus* Harlan et de Wet, var. *coursii* (A. Camus) Harlan et de Wet, var. *dactylon*, var. *elegans* Rendle, and var. *polevansii* (Stent) Harlan et de Wet. The species has diploid ($2n = 2x = 18$), tetraploid ($2n = 4x = 36$), and hexaploid ($2n = 6x = 54$) cytological forms, as well as triploid ($2n = 3x = 27$) and pentaploid ($2n = 5x = 45$) derivatives of the three even ploidy forms (Harlan, de Wet, Huffine, et al. 1970; Harlan, de Wet, Rawal, Felder, et al. 1970; Forbes and Burton 1963; Hanna, Burton, and Johnson 1990; Hoff 1967; Wu, Taliaferro, Bai, et al. 2006).

Cynodon dactylon var. *afghanicus* comprises two ploidy forms: diploid ($2n = 2x = 18$) and tetraploid ($2n = 4x = 36$). Harlan, de Wet, Huffine, et al. (1970) indicated that rhizomes are lacking in diploids and a very short rhizome-like structure is present in tetraploids. The inflorescences are stout and stiff and the spikelets are closely overlapped. Leaf color is bluish green, and leaf blades are rather long and linear lanceolate. The variety is endemic only to Afghanistan. The variety has robust forms providing some pasture in lowland steppes and along water channels in Afghanistan, but it is not considered of major importance as a forage grass. Though lacking rhizomes, plants of the variety have relatively good winter hardiness (Harlan and de Wet 1969). It can be used as a genetic resource for the development of forage type cultivars.

Cynodon dactylon var. *aridus* Harlan et de Wet is a diploid ($2n = 2x = 18$) with fairly wide distribution in arid regions. Harlan, de Wet, Huffine, et al. (1970) and Harlan et al. (1970a) indicated that specimens of the variety have been collected from South Africa, Zambia, Tanzania, Israel, and southern and western India. Plant size varies with geographic origin (Harlan et al. 1970a). Plant size is smallest in India and gradually increases; it is somewhat larger in southwest Asia and northeast Africa and largest in South Africa. Plants in South Africa are typically large and robust, with remarkably long, fast-growing stolons (Harlan and de Wet 1969). This grass was introduced to the Yuma region of Arizona, where it is called giant bermudagrass (Harlan, de Wet, Huffine, et al. 1970). Seed of giant bermudagrass has for many years been commercially produced in Arizona and California and is widely distributed. Cooper and Burton (1965) indicated that giant bermudagrass yielded no more forage than common bermudagrass in Georgia and was inferior to common bermudagrass as a turf grass. The variety has little winter hardiness but often survives winters with freezing temperatures by virtue of having deep rhizomes (Cooper and Burton 1965).

Cynodon dactylon var. *dactylon* is the cosmopolitan, ubiquitous taxon so widely distributed in the world (Harlan 1970; Kneebone 1966a). Harlan and de Wet (1969) described the variety as extremely valuable and enormously variable. Plants of var. *dactylon* vary from very small, fine turf-grasses suitable for putting greens to large, robust, leafy types harvested as a hay crop or grazed as a pasture grass (Harlan and de Wet 1969). However, plants of var. *dactylon* are never as fine textured as African bermudagrass (*C. transvaalensis*), and large plants of the variety are never as large as the robust East African species, *C. aethiopicus, C. nlemfuensis,* and *C. plectostachyus.*

Variety *dactylon* has a remarkable distribution; it is one of the most widely geographically distributed of all plants (Harlan et al. 1970a). de Wet and Harlan (1971) observed that the variety is distributed continuously across all continents between about 45° S and 45° N latitudes. In Europe, the variety extends to regions at approximately 53° N latitude. In south and southeast Asian regions, like north Nepal and southwest China, plants of var. *dactylon* have been found growing at about 3,000 m elevation above sea level; in North Africa, the Jordan rift, and California (United States), it occurs below sea level (Harlan and de Wet 1969). It is found on many islands, even some remote isolated rocks and atolls in the Atlantic, Pacific, and Indian Oceans. It also occurs on sub-Antarctic islands (Clayton et al. 2006).

The inflorescence of var. *dactylon* has one whorl (on rare occasions two), with four to six racemes. Glumes are three fourths of the length or more of a spikelet. The plant has rhizomes ranging from slender to stout and fleshy; these typically are at shallow soil depth in tropical regions and often quite deep in temperate regions. Leaf blades are flat and linear lanceolate. Variety *dactylon* prefers disturbed habitats. As described by Harlan and de Wet (1969), var. *dactylon* plants rarely form pure stands over a large area in natural settings and do not invade natural grasslands or forest vegetation. The plants occurring in temperate areas in Europe, Asia, and America have cold hardiness, and plants originating in tropical regions may have little or no winter hardiness. Although the variation in morphology is fairly continuous from plants in pantropical areas to the plants occurring in temperate regions, Harlan and de Wet (1969) described three major races in var. *dactylon*—tropical, temperate, and seleucidus—on the basis of appearance, adaptation, and geographic distribution. The three races are reasonably different but blend and outcross easily with each other (Harlan, de Wet, Rawal, Felder, et al. 1970).

Tropical race is distributed in pantropics. Plants are short in stature and adapted to acid soils of low fertility. They are normally found in damp and disturbed habitats, such as along roadsides, drainage ditches, field sides, and walking paths in villages and towns where disturbed soils exist. Harlan and de Wet (1969) indicated that the race is widely distributed on the Pacific islands. Plants in this race have no or little freeze tolerance but can tolerate waterlogged conditions for a considerable period as well as prolonged drought. Plants of the tropical race can form a relatively fine-textured turf. The temperate race is similar to the tropical race in appearance and plant size but has winter hardiness. Plants of the temperate race can form a denser turf. The temperate race is common in highly disturbed habitats throughout the warm temperate parts of the world.

The seleucidus race is so named because its distribution roughly coincides with the boundaries of the Seleucid Empire (now Pakistan to Turkey) (Harlan et al. 1970b). Typical plants of the seleucidus race found in Afghanistan, Pakistan, Iran, Iraq, and Turkey are strikingly coarse, robust, glaucous in color, usually hairy, and strongly winter hardy. Specimens from Greece, Bulgaria, Romania, Yugoslavia, Italy, Southern France, Spain, and Portugal appear reduced in size and vigor, and seleucidus plants in Austria, Hungary, Czechoslovakia, and Germany are still smaller. The temperate race and seleucidus race converge in central Europe (Harlan et al. 1970b).

The plants of the seleucidus race spread aggressively by thick and heavy stolons and rhizomes having short internodes. The blades of stolon leaves are described as small flaps by Harlan et al. (1970b). The stolon tip is sharp and strong and can penetrate into soils. Usually, stolons and rhizomes are interchangeable. Stolons can grow into soils and convert to rhizomes, and rhizomes can emerge from the soil and change to stolons as well. Harlan and de Wet (1969) reported that the Seleucid bermudagrass is tall and productive in biomass, especially on fertile soils. The characteristics of the race, including vigor, cold hardiness, and biomass productivity, are very valuable for forage production as a hay crop or grazing grass. Harlan et al. (1970b) and Harlan and de Wet (1969) indicated that the strong vigor of the Seleucid race derives from its hybrid origins between var. *dactylon* temperate race and tetraploid var. *afghanicus* and from genetic interactions between var. *afghanicus* and the Seleucid race. They observed a series of graduations and intermediates between the tetraploid var. *afghanicus* and the Seleucid race. Leaf length and spikelet density of artificial hybrids between plants of the temperate race var. *dactylon* and var. *afghanicus* resemble those of plants of the Seleucid race found in Afghanistan and Iran (Harlan et al. 1970b). Harlan and de Wet (1969) suggested that the seleucidus race came from the interactions of tetraploid var. *afghanicus* and the temperate race of var. *dactylon*.

Cynodon dactylon var. *coursii* (*A. Camus*) Harlan et de Wet is tetraploid ($2n = 4x = 36$) and endemic to the central plateau of Madagascar (Harlan and de Wet 1969). Harlan, de Wet, Huffine, et al. (1970) described morphological and taxonomic characteristics. Plants of the variety have no rhizomes and exhibited no winter hardiness in Stillwater, Oklahoma. Inflorescence normally has one (rarely two) whorls. Racemes are stout and spreading. Subequal glumes are three fourths of the length of a spikelet. There are some hairs on margins of lemma. Leaves are linear lanceolate. The grass can grow up to 50 cm tall. The robust plant is a fairly important grazing resource on the island but of no significance elsewhere (Harlan 1970).

Cynodon dactylon var. *elegans* Rendle is a tetraploid ($2n = 4x = 36$) variety (Harlan and de Wet 1969). The variety is indigenous to and the common type in southern Africa, south of 12° S latitude. Var. *elegans* is abundant within its range from Mozambique, Zambia, Zimbabwe, Angola, and southward to the Cape (Harlan, de Wet, Huffine, et al. 1970; Harlan et al. 1970a). Its habitat is from coastal plains to high lands. The grass is rhizomatous but had no winter hardiness in Stillwater, Oklahoma (Harlan, de Wet, Huffine, et al. 1970). The morphology of the variety has been described as distinctive to other *Cynodon* taxa (Harlan, de Wet, Huffine, et al. 1970). Culms of the plants are distinctly decumbent, ascending in an arc with the narrow leaves pressed close to it. Plants form a lax, loose sod rather than a dense turf. In southern Africa, the variety is part of natural vegetation. Harlan (1970) noted that the variety is of considerable importance in grazing in southern African countries.

Cynodon dactylon var. *polevansii* (Stent) Harlan et de Wet is a tetraploid ($2n = 4x = 36$) found only near Barberspan in South Africa (Harlan, de Wet, Huffine, et al. 1970; Harlan et al. 1970a). The variety has been described as having small rhizomes and good winter hardiness (Harlan, de Wet, Huffine, et al. 1970). Its inflorescence is in one whorl with small and spreading racemes. Stolons are rather fine and branched with short internodes. Leaves are linear lanceolate and fine in texture. Plants form a dense turf with very harsh, rigid, and erect leaves. Harlan and de Wet (1969) believed that the variety may have value for turf use. Though grazed by animals, var. *polevansii* is relatively low in biomass production and consequently of little value as a forage (Taliaferro, Rouquette, et al. 2004).

Cynodon incompletus Nees is a small, nonrhizomatous species strictly endemic to South Africa (de Wet and Harlan 1971). The taxon often is a diploid ($2n = 2x = 18$) and rarely tetraploid ($2n = 4x = 36$) (Harlan, de Wet, Huffine, et al. 1970). Plants can form a low and dense sod. The species is far more winter hardy than its habitats currently demand. Therefore, along with sympatric taxa *C. transvaalensis* and *C. dactylon* var. *polevansii, C. incompletus* is assumed to have survived the Pleistocene climates of South Africa (Harlan et al. 1970a). Inflorescence of *C. incompletus* has one whorl of racemes, which are slender and stiff and vary from two to several. Subequal glumes are less than half to three fourths of the length of a spikelet. Lemmas are blunt with hairy keel. Plants have fine and reddish stolons with short internodes. Leaves are linear lanceolate, with variable blade length. The species is characterized by hairy leaves and has been subdivided into two varieties: var. *incompletus* and var. *hirsutus* (Stent) de Wet et Harlan (de Wet and Harlan 1970). Plants having leaf-blades with coarse stiff hairs on the lower surface and some pilose on the upper surface are classified as *C. incompletus* var. *incompletus* (de Wet and Harlan 1970). In contrast, plants with coarse stiff hairs on both sides of a leaf are assigned to *C. incompletus* var. *hirsutus.*

Var. *incompletus* is widely distributed in the more arid regions across the Karroo, Orange Free State, and West Transvaal; var. *hirsutus* is widely distributed in the Bushveld but also extends into the Highveld and the Karroo (de Wet and Harlan 1970). Var. *incompletus* is quite drought tolerant; it is often found in dry sites forming a dense turf less than 20 cm tall and provides some grazing forage for sheep. Compared to other *Cynodon* species, *C. incompletus* is of minor importance for grazing and of no value as a hay grass (Harlan 1970). However, the species is rather widely used as a turf grass (Beard 1973). The taxon has been introduced into some European countries for turf purposes (Harlan 1970). Harlan (1970) noted that one hybrid form (used to be *C. bradleyi*) of the species has been introduced to many areas of the world as a turfgrass.

Cynodon nlemfuensis Vanderyst is one of the three stargrass species. There are two ploidy forms: diploid ($2n = 2x = 18$) and tetraploid ($2n = 4x = 36$). Clayton and Harlan (1970) and Harlan, de Wet, Huffine, et al. (1970) provided a description of morphological characteristics of the taxon. The species is a stoloniferous perennial grass without rhizomes. Plants form loose mats. Stolons are stout and quite woody; culms grow up to 60 cm tall and are fairly slender and soft. Leaf blades are soft, flat, linear lanceolate, mostly 5–16 cm in length and 2–6 cm in width, with a scabrous blade surface. Racemes (4–13) of an inflorescence are slender, even flexuous in one or sometimes two whorls. Spikelets constituting one fertile floret are 2–3 mm long and appressed to rachis in two rows. Glumes are persistent and subequal, about half to three fourths of the length of a spikelet. Lemma is silky pubescent to softly ciliate on the keel, and palea is glabrous or ciliate on the keels.

Two botanical varieties of *C. nlemfuensis* are var. *nlemfuensis* and var. *robustus* Clayton et Harlan, distinguished on the basis of their morphology and distribution. Var. *nlemfuensis* is finer and less robust as compared to var. *robustus*. The general morphology of var. *nlemfuensis* is similar to that of large *C. dactylon* except that the former has no rhizomes and the latter develops rhizomes. Var. *nlemfuensis* has culms with width of 1.0–1.5 mm in diameter and frequently has glaucous foliage. There are four to nine racemes in its inflorescence, with each raceme 4–7 cm long. Plants of var. *nlemfuensis* are able to tolerate somewhat more heat and drought than the plants of the variety *robustus*. Var. *robustus* is characterized by large and robust plants; each inflorescence comprises 6–13 long (6–10 cm) and slender racemes in a single whorl, and foliage is rather soft and tender. However, the two varieties are similar in growth habits and found without winter hardiness.

The distributions of the two varieties are sympatric in many places of East Africa but are somewhat different in other locations (Harlan, de Wet, Huffine, et al. 1970; Harlan et al. 1970a). Var. *robustus* is abundant in high rain areas in many parts from Ethiopia to Zambia and East Congo in East Africa. It has been introduced to some parts of West Africa, such as West Congo and Nigeria

(Harlan 1970). Var. *nlemfuensis* is widely distributed in East African countries. It appears more abundant in Tanzania and spreads to savanna zones in western Angola. Var. *nlemfuensis* often occurs in damp, shady areas and along the banks of streams from coastal plains to the highlands (Harlan, de Wet, Huffine, et al. 1970). Harlan (1970) noted that the two varieties were included in a breeding program by H. R. Chheda at the University of Ibadan, Nigeria. As indicated by Harlan (1970) and Harlan et al. (1970a), *C. nlemfuensis* is the most promising forage grass of all stargrasses, probably because it has favorable features as a forage grass and is genetically related to *C. dactylon*. *Cynodon nlemfuensis* has been introduced to Australia, many countries in America, and the Pacific region (Clayton et al. 2006).

    Cynodon plectostachyus (K. Schum.) Pilger is a large and robust diploid ($2n = 2x = 18$) East African stargrass species. Clayton and Harlan (1970) indicated that the most diagnostic feature of the species in morphology is tiny to nearly absent glumes. The minute glumes are at most one fourth of the length of a spikelet (de Wet and Harlan 1971). Harlan, de Wet, Huffine, et al. (1970); Clayton and Harlan (1970), and de Wet and Harlan (1970) described morphological characteristics and taxonomy of *C. plectostachyus* in detail. The grass is a stoloniferous perennial without any rhizomes. The stout and woody stolons have arching internodes. Culms are quite erect up to 90 cm tall. The plants have large leaf blades that are linear lanceolate, 10–30 cm long, and 4–7 mm wide. The other characteristics of its leaves are conspicuously soft hairs on both sides of leaf blades, which are soft and tender. Ligules are hyaline. Racemes are stout and 3–7 cm long; their number varies from 7 to 20 per inflorescence. At maturity, racemes tend to curl inward. Racemes form two to seven whorls, rarely only a single whorl on each inflorescence. Lemma is pointed and hairy on keel and margins. Plants often are found in relatively high rain areas, particularly along the highlands of East Africa. The large and robust plants have no winter hardiness. The natural distribution of the species is along the Eastern Great Rift in Ethiopia, Kenya, Tanzania, Zambia, Malawi, Uganda, and East Congo. Harlan (1970) noted that *C. plectostachyus* is most abundant on the floor of the Rift. The species has been introduced as a forage grass to Zimbabwe, South Africa, Ghana, Nigeria, Madagascar, Sri Lanka, India, and the United States.

    Plants of *Cynodon transvaalensis* Burtt-Davy are small in size, turf forming, diploid ($2n = 2x = 18$), and endemic to South Africa (de Wet and Harlan 1970; Harlan, de Wet, Huffine, et al. 1970). Although the grass is genetically allied to *C. dactylon,* it is recognized as a species because of its distinct and unique morphology (de Wet and Harlan 1970). The morphological characteristics that best distinguish plants of *C. transvaalensis* from those of other *Cynodon* species are their small size, erect and fine linear leaf blades (less than 2 mm wide), and yellowish-green foliage color (de Wet and Harlan 1971). Spikelets are loosely arranged on a short, stiff, fine raceme. Inflorescences of the species normally consist of two to four racemes in a single whorl. Stolons are slender and often red in color, with short and fine internodes. The species is endemic to southwestern Transvaal, Orange Free State, and the northern part of the central Cape Province of South Africa (Harlan, de Wet, Huffine, et al. 1970; Harlan et al. 1970a). Plants in its indigenous area are often found in damp sites around water sources and along stream banks. The plants are so small and nonproductive as to be of minor value for hay production and grazing. However, African bermudagrass is a valuable turfgrass because of its fine texture, dense sod, and tolerance of relatively low mowing heights. More importantly, African bermudagrass has been used in interspecific hybridizations with *C. dactylon* var. *dactylon* to produce hybrid turf grass cultivars.

8.5 CYTOGENETICS AND REPRODUCTIVE BEHAVIOR

8.5.1 Chromosome Number and Ploidy Level

    Several ploidy levels and two base chromosome numbers have been reported in *Cynodon.* Hurcombe (1947) reported the basic chromosome number of the genus as 10 and that *C.*

transvaalensis is the diploid form, *C. magennisii* the triploid form, and *C. dactylon* the tetraploid form. She correctly reported $2n = 18$ chromosomes for *C. bradleyi* (now *C. incompletus*) but incorrectly assumed it to be an aneuploid number. Moffett and Hurcombe (1949) reported base chromosome numbers of $x = 9$ and $x = 10$, *C. dactylon* with 36 somatic chromosomes, and *C. plectostachyus* plants with 18 or 54 chromosomes. More recent investigation by Forbes and Burton (1963) and Harlan, de Wet, Rawal, Felder, et al. (1970) firmly established that nine is the correct base chromosome number in *Cynodon;* the existence of the second base number $x = 10$ is questionable (Harlan, de Wet, Huffine, et al. 1970; Harlan, de Wet, Rawal, Felder, et al. 1970; Forbes and Burton 1963). Forbes and Burton (1963) noted that a part or all of the chromosome numbers in multiples of 10 reported in earlier literature were based on plants with $2n = 18$, 27, or 36, in which satellites of the nucleolus organizer chromosomes were counted as whole chromosomes. de Silva and Snaydon (1995) added that technical difficulties may cause problems in accurate counts of bermudagrass chromosomes because the chromosomes of *Cynodon* plants are small and fragments commonly occur in root-tip preparations.

The diploid ($2n = 18$) species are *C. barberi, C. dactylon* var. *aridus, C. incompletus* var. *incompletus, C. plectostachyus,* and *C. transvaalensis.* The tetraploid ($2n = 36$) species are *C. arcuatus, C. dactylon* var. *dactylon, C. dactylon* var. *coursii, C. dactylon* var. *elegans,* and *C. dactylon* var. *polevansii. C. aethiopicus, C. dactylon* var. *afghanicus, C. incompletus* var. *hirsutus, C. nlemfuensis* var. *nlemfuensis,* and *C. nlemfuensis* var. *robustus* have both tetraploid and diploid chromosome numbers (Harlan, de Wet, Rawal, Felder, et al. 1970).

Hexaploid ($2n = 6x = 54$) and pentaploid ($2n = 5x = 45$) *Cynodon* plants occur rarely. Powell, Burton, and Taliaferro (1968) reported a hexaploid plant among putative progeny of a cross between tetraploid *C. dactylon* var. *dactylon* by diploid *C. transvaalensis.* Felder (1967) reported a hexaploid plant among putative progeny of a cross of *C. dactylon* ($4x$) by *C. plectostachyus* ($2x$). Malik and Tripathi (1968) and Johnston (1975) also reported hexaploid *Cynodon* plants. Goldman et al. (2004) reported large numbers of hexaploid plants derived from tissue culture and bombardment transformation of triploid TifEagle (*C. transvaalensis* × *C. dactylon*). It is believed that, as the auxin source during tissue culture, 2,4-dichlorophenoxyacetic acid (2,4-D) may increase the chance of recovering hexaploid plants either by inducing polyploidy or by selecting for polyploidy cell growth (Goldman et al. 2004). Moffett and Hurcombe (1949) observed hexaploid *C. plectostachyus* (K. Schum) Pilg. from Rhodesia (now Zimbabwe). *C. dactylon* cv. Tifton 10 is a natural hexaploid plant collected in Shanghai, China, in 1974 by Burton (Burton 1991; Hanna et al. 1990).

More recently, six hexaploid bermudagrasses were reported in germplasm collections from Shanghai, Jiangsu, and Zhejiang provinces in eastern China by Wu, Taliaferro, Bai, et al. (2006); they also reported three pentaploid bermudagrasses collected in the southern China provinces of Fujian and Hainan. However, most (88%) of the 110 accessions from 11 China provinces studied by Wu, Taliaferro, Bai, et al. (2006) were tetraploids. Thomas and Murray (1978) reported a pentaploid *C. dactylon* plant from Malta. Tifton 85, a common bermudagrass by stargrass (*C. nlemfuensis*) interspecific F_1 hybrid developed by Burton, Gates, and Hill (1993), is a pentaploid. Harlan and de Wet (1969) observed a high frequency of unreduced male and female gametes in *C. dactylon.* Taliaferro, Rouquette, et al. (2004) noted that hexaploids may originate through the functioning of unreduced gametes, through chromosome doubling in early embryonic stages of triploid zygotes, or through both.

De Silva and Snaydon (1995) reported that *Cynodon* populations from Sri Lanka differed in ploidy based on ecological origin. Populations from the arid and intermediate regions, roadsides, lawns, and grasslands in the hill country (pH > 6.5) were entirely tetraploid; however, populations from roadsides and lawns in the wet region and from forests in the hill country (pH < 5.0) were entirely diploid. Mixtures of diploid and tetraploid occurred only in the areas with a soil pH between 6.0 and 6.5. Their results suggested that soil acidity seems to be the main factor determining the distribution of the two cytotypes.

Chromosome morphology in the genus *Cynodon* has been reported by Ourecky (1963) and Brilman, Kneebone, and Endrizzi (1982). Both reports described pachytene chromosome morphology of diploid *C. dactylon,* probably var. *aridus,* var. *afghanicus,* and their hybrids. Chromosomes were numbered from one through nine on the basis of morphological length; chromosome 1 was the shortest and 9 the longest. Chromosome 4 contains the nucleolus organizing region. Arm ratios ranged from 1.3 to 2.1 in the study of Brilman et al. (1982) and from 1.1 to 1.9 in the Ourecky (1963) study. However, numbering identities of the pachytene chromosomes specified in the report of Brilman et al. (1982) are not necessarily the same as those reported by Ourecky (1963).

Nuclear DNA content is a characteristic of a species (Brown 1999). Knowledge of genome size of a species is very useful in estimating ploidy levels and is essential for assessing the coverage of a genomic library, estimating the copy number of a gene in the genome, and developing strategies for gene cloning based on genome mapping. Flow cytometry has proven to be a fast and reliable method of estimating ploidy in bermudagrass. Taliaferro et al. (1997) developed a flow cytometry protocol for estimating nuclear DNA content in *Cynodon* and reported mean nuclear DNA contents of 1.11 ± 0.04 pg, 1.60 ± 0.04 pg, 2.25 ± 0.13 pg, and 2.80 ± 0.14 pg for $2x, 3x, 4x,$ and $6x$ cytotypes, respectively. Arumuganathan et al. (1999) reported 2C nuclear DNA contents of 1.03 ± 0.01 pg, 1.61 ± 0.00 pg, 1.37 ± 0.01 pg, and 1.95 ± 0.01 pg for diploid African bermudagrass DTC 95, triploid Tifgreen, triploid Tifway, and tetraploid Savannah, respectively. The flow cytometry results of Chinese *Cynodon* accessions by Wu, Taliaferro, Bai, et al. (2006) indicated a larger genome size range (2.90–3.13 pg/2C) for hexaploid plants and a genome size of 2.37–2.49 pg/2C for pentaploids; the results also confirmed the genome size data for tetraploid and triploid bermudagrasses by Taliaferro et al. (1997) and Arumuganathan et al. (1999).

8.5.2 Chromosome Behavior in Meiosis

Meiosis is fundamental to sexual reproduction of higher organisms. Chromosomal behavior in meiosis can reflect the phylogenic relationships among parental genomes, and it affects pollen viability and seed set (i.e., fertility). Forbes and Burton (1963); Malik and Tripathi (1968); Harlan and de Wet (1969); Harlan, de Wet, Rawal, Felder, et al. (1970); Gupta and Srivastava (1970, 1971); Chheda and Rawal (1971); Rawal and Chheda (1971); and Hanna and Burton (1977) investigated chromosomal associations and behaviors in meiosis of various *Cynodon* species.

Forbes and Burton (1963) reported that chromosomes in the diploid taxa *C. incompletus, C. plectostachyus, C. transvaalensis,* and *C. dactylon* mainly formed bivalents (IIs) in prophase I and that two accessions of *C. transvaalensis* and one of *C. dactylon* had slightly irregular chromosome pairing, with univalents (Is) averaging 1.6 or less. They also observed irregular meiosis in *C. bradleyi.* According to de Wet and Harlan (1970), *C. bradleyi* is classified as a hybrid of *C. incompletus* var. *incompletus* × var. *hirsutus.* Forbes and Burton (1963) reported unpaired chromosomes at diakinesis and metaphase I, chromosome laggards in anaphase I, and micronucei in tetrad spores in the irregular meiosis of the *Cynodon* diploids used in their investigation. Malik and Tripathi (1968) observed regular meiosis in diploid *C. dactylon,* a native to India. Harlan, de Wet, Rawal, Felder, et al. (1970) conducted an extensive cytogenetic study in hybrids of *Cynodon* taxa including all diploid forms. Collectively, chromosome pairing is regular in all possible diploid cross combinations, with average bivalent formation ranging from 8.4 to 9.0. They believed that *C. nlemfuensis* var. *nlemfuensis* is genetically closely related to *C. dactylon* vars. *aridus* and *afghanicus.*

Harlan, de Wet, Rawal, Felder, et al. (1970) reported the poorest pairing in *aridus* × *hirsutus, aridus* × *incompletus, afghanicus* × *hirsutus,* and *C. transvaalensis* × *C. nlemfuensis.* Chheda and Rawal (1971) reported regular chromosomal behavior in meiosis of diploid *C. aethiopicus, C. nlemfuensis,* and *C. transvaalensis.* Meiotic behavior of two hybrids between *C. aethiopicus* and *C. nlemfuensis* was slightly irregular, averaging up to four univalents, which had a tendency to lag behind at anaphase I. Chromosome pairing in hybrids between *C. nlemfuensis* and *C. transvaalensis* was also irregular, with mean univalents ranging from two to eight; hybrids were completely sterile if fewer than eight bivalents formed. In a similar study on *C. nlemfuensis, C. dactylon* vars. *Afghanicus* and *aridus,* diploid taxa had regular chromosomal pairing in meiosis with occasional univalents, while diploid hybrids averaged 8.6 or more bivalents with up to two univalents, and occasionally up to six univalents. Gupta and Srivastava (1970) recorded normal bivalent association in a diploid *C. dactylon* accession.

Chromosomal behavior in meiosis of various triploid *Cynodon* hybrids has been described. Forbes and Burton (1963) observed a maximum of eight trivalents (IIIs) per cell at diakinesis in a triploid hybrid of tetraploid *C. dactylon* cv. Coastal × diploid *C. dactylon* PI 142280. Their results suggested a high homology between chromosomes of diploid and tetraploid *C. dactylon* parents. Similarly, Tiffine, a hybrid of *C. transvaalensis* and *C. dactylon,* had a maximum of nine IIIs per cell at diakinesis, suggesting homology between the genomes of the two species. Some cells of three of the triploid hybrids, including Tiffine, tetraploid *C. dactylon* cv. Coastal × diploid *C. dactylon* PI 142280, contained one or two multivalents and had lagging chromosomes at anaphase I. They observed frequent stickiness of meiotic chromosomes in triploid hybrids and in some diploid plants from Africa. Malik and Tripathi (1968) reported mainly nine IIs and nine Is in an intraspecific *C. dactylon* triploid hybrid derived from crossing 2x by 4x races native to Udaipur, India. This indicates that the triploid hybrid consists of two genomes. They also observed one to three quadrivalents.

Harlan, de Wet, Rawal, Felder, et al. (1970) reported the most common chromosomal configuration of nine II + nine I in triploid hybrids. The triploid hybrids were derived from reciprocal combinations of diploid *C. dactylon* var. *afghanicus,* var. *aridus, C. incompletus* var. *incompletus,* var. *hirsutus, C. nlemfuensis* var. *nlemfuensis,* and var. *robustus,* and *C. transvaalensis,* and tetraploid *C. dactylon* var. *dactylon,* var. *afghanicus,* var. *coursii,* var. *elegans,* var. *polevansii,* and *C. nlemfuensis* var. *robustus.* They found a significant excess of univalents in hybrids between *C. dactylon* var. *afghanicus* and *C. incompletus* var. *hirsutus* and between *C. transvaalensis* and *C. dactylon* var. *elegans* and var. *coursii.* A low frequency of trivalents and quadrivalents existed in most hybrids except those combinations involving *C. dactylon* var. *dactylon* and *C. transvaalensis, C. dactylon* var. *afghanicus,* and *C. incompletus* var. *hirsutus, C. dactylon* var. *coursii,* and *C. nlemfuensis* var. *nlemfuensis, C. dactylon* var. *elegans,* and *C. transvaalensis,* and *C. dactylon* var. *coursii,* and *C. transvaalensis. Cynodon dactylon* triploid hybrids native to Gorakhpur, India, had chromosomal associations of univalents + bivalents + trivalents to various extents and laggard chromosomes at anaphases I and II (Gupta and Srivastava, 1970, 1971). Rawal and Chheda (1971) observed high frequencies of univalents and bivalents and a very low frequency of trivalents in hybrids of 4x race var. *afghanicus* by 2x var. *aridus,* and in 4x race of *C. nlemfuensis* var. *robustus* by 2x var. *aridus.*

Information on the chromosome pairing and meiotic behavior of tetraploid *Cynodon* species and their hybrids is limited. Forbes and Burton (1963) reported that Coastal bermudagrass, a hybrid from a cross of Tift bermudagrass by a tall growing bermudagrass strain from Africa, had mean chromosome association at diakinesis of 0.15 I + 16.00 II + 0.96 IVs. At metaphase I, more than 36% of cells had nonoriented univalents. At anaphase I, about 49% of the cells had laggards, and 50% of tetrads contained micronuclei. In the same study, a tetraploid from Rhodesia (PI 226011) showed average chromosomal associations of 1.36 I + 15.69 II + 0.18 III + 0.68 IV at diakinesis. They also observed that 39% of the cells contained laggard and occasional micronuclei in tetrads. They noted that the formation of quadrivalents in tetraploids could result either from chromosome homology or from reciprocal translocation heterozygosity.

The investigation of chromosome pairing of a tetraploid bermudagrass from India by Malik and Tripathi (1968) indicated that chromosomes paired as 18 IIs and meiosis was regular; however, one to two quadrivalents were observed in some cells. Colchicine-induced autotetraploids had a higher frequency (three to nine IVs) of quadrivalents than those observed for the natural tetraploids. Results of nine reciprocal combinations and three directional crosses of six tetraploid *Cynodon* taxa by Harlan, de Wet, Rawal, Felder, et al. (1970) revealed complex chromosome association patterns depending on parents. Hybrids of *C. nlemfuensis* var. *robustus* by *C. dactylon* vars. *dactylon, coursii,* and *elegans* had 4 to 10 Is. High Is also were observed in hybrids of vars. *coursii* and *elegans.* Hybrids of *afghanicus* × *robustus, dactylon* × *afghanicus,* and *dactylon* by *robustus* contained high frequencies of multivalents. Although meiosis was relatively regular in the hybrids of *polevansii* × *coursii, dactylon* × *elegans,* and *dactylon* × *polevansii,* there were some occurrences of Is, IIIs, and IVs at low frequencies.

Hybrids of var. *afghanicus* × var. *robustus* had chromosome pairings of 12–18 IIs and zero to five IVs. Harlan and de Wet (1969) studied meiotic behavior of 50 hybrids of three *C. dactylon* var. *dactylon* races involving 23 tropical, five temperate, and 11 seleucidus parents. Meiosis of the var. *dactylon* was regular in 11, slightly irregular in 13, irregular in 19, and very irregular in 7 hybrids. The most regular meioses occurred in hybrids between parents of the most similar origins, and the most irregular meioses were found in hybrids from parents of widely divergent geographic sources, with few exceptions. In irregular meioses, cells contained univalents, quadrivalents, and, more rarely, trivalents. Rawal and Chheda (1971) examined two hybrids of var. *afghanicus* by *C. nlemfuensis* var. *robustus* and reported completely regular meiosis; eight other hybrids from the same parental taxa had mainly bivalent pairings and occasionally a few quadrivalents at metaphase I.

Hanna and Burton (1977) reported cytology and fertility of four tetraploid forage type bermudagrass cultivars: Coastal, Coastcross-1, Midland, and Suwannee. Coastcross-1 showed very irregular meiosis with mean associations of 5.07 I + 13.94 II + 0.41III + 0.51 IV per cell. A high frequency (94%) of cells of Coastcross-1 had laggards and micronuclei in microspores and very low (2%) stainable pollen grains. In Coastal, 43, 20, and 39% of microsporocytes had 18 bivalents, one or two tetravalents, and univalents, respectively. In Midland, mean chromosome association was 1.71 I + 16.13 II + 0.15 III + 0.39 IV. Suwannee had the most regular meiosis among the four cultivars, and mean chromosome association per cell was 0.10 I + 17.45 II + 0.0 III + 0.34 IV.

8.5.3 Reproductive Traits

Cynodon plants reproduce both asexually and sexually. Individual plants are easily asexually propagated by planting vegetative parts such as rhizomes, stolons, shoots, and crown buds—collectively termed "sprigs" (Taliaferro, Rouquette, et al. 2004). The planting of sprigs has been widely used to establish bermudagrass for pasture, turf, and other uses (Burton 1954). Intact cut sod of many clonal turf bermudagrass cultivars is used widely to establish new plantings, particularly on small areas such as lawns and athletic fields. Not all bermudagrass plants have no sufficient sod tensile strength to allow mechanical harvesting of intact sod. Accordingly, sufficient sod tensile strength is important to the market success of clonal turf bermudagrass cultivars.

However, the enormous genetic variability within *Cynodon* has been derived from sexual reproduction. The sexual reproduction capability of individual bermudagrass plants varies from none to very high. In general, seed production capability is low, but most plants have potential to produce some seed, thereby allowing for genetic recombination and segregation. Results by Burton (1947), Burton and Hart (1967), and Taliaferro and Lamle (1997) indicated that cross-pollination and self-incompatibility result in strong outcrossing in the species. Burson and Tischler (1980) demonstrated that bermudagrass mature embryo sacs are of the normal polygonium type containing an egg and two polar, two synergid, and three antipodal nuclei. Mature pollen grains contain a vegetative nucleus and a generative nucleus (Taliaferro, Rouquette, et al. 2004). The generative nucleus divides into two nuclei, which combine with the egg and two fused polar nuclei, resulting in production of the $2n$ embryo and $3n$ endosperm. Results of Taliaferro and Lamle (1997) indicated that self-pollination results in pollen tubes that grow more slowly and seldom reach the micropylar region of ovaries in comparison to tubes from cross-pollination. Self-pollinated seed set in bermudagrass varies, typically from 0.5 to 3.0% (Taliaferro, Rouquette, et al. 2004). Cross-pollinated seed set varies, with much larger variation among bermudagrass plants, from extremely low to those with high seed-producing capability (Burton and Hart 1967; Richardson, Taliaferro, and Ahring 1978; Kenna, Taliaferro, and Richardson 1983; Wu, Taliaferro, Martin, et al. 2006).

Bermudagrass seed yield is a quantitative trait regulated by complex genetic factors and heavily affected by environmental conditions and genotype by environment interaction. Kneebone (1966b) reported that bermudagrass seed yield was strongly positively correlated with seedheads per plot and seed set. Kenna et al. (1983) reported a positive and significant correlation (r = 0.52) between open-pollinated seed set percentage and seed yield. Ahring, Taliaferro, and Morrison (1974) reported that high seed set contributed significantly to seed yield. Ahring et al. (1974) and Wu, Taliaferro, Martin, et al. (2006) reported that correlations of seed yield with raceme number inflorescence^{-1} and raceme length inflorescence^{-1} were negligible, indicating that the two seed yield components had no substantial effects on seed yield. Using path coefficient analysis to partition correlation coefficients among seed yield and its components into direct and indirect effects, Wu, Taliaferro, Martin, et al. (2006) reported that inflorescence prolificacy and percent seed set had the largest and second largest direct positive effects on seed yield, respectively. The two seed yield components had small but positive correlation (r = 0.14), suggesting that concurrent selection of the two traits would increase seed yield.

8.6 GERMPLASM ENHANCEMENT—CONVENTIONAL BREEDING

The breeding of superior bermudagrass cultivars for pasture, hay, and turf use over the past seven decades has revolutionized the livestock and turf industries in the southern United States and many other climatically similar regions in the world. Harlan (1970) described the effort as one of the success stories of modern plant breeding. The potential for future development of new cultivars is enormous.

8.6.1 Breeding Objectives

There are many important decisions to be made by a plant breeder during the long process of developing commercially successful cultivars. Among them, breeding objectives should be carefully and clearly defined for a specific breeding program at an early stage, although some specific objectives can be added or changed later as programs are in progress.

Improved adaptation in targeted environments is always a major breeding objective in a bermudagrass breeding program. Bermudagrass is a perennial requiring an ability of cultivars to maintain stands and performance while enduring multiple cycles of year-long climatic variations. Many environmental factors may cause serious damage or even kill bermudagrass stands; consequently, dependability of a cultivar and longevity of its stands will decrease. In the transition zone of the United States, added cold hardiness will significantly increase commercial value of bermudagrass cultivars by ensuring their dependable use. Midland bermudagrass developed by G. W. Burton in Tifton, Georgia, and released cooperatively by Oklahoma Agricultural Experiment Station and Georgia Coastal Plain Experiment Station and Plant Science Research Division, USDA-ARS, in 1953 in Oklahoma is one of the early successful attempts to expand the use of bermudagrass further north.

Increasing winter hardiness to enhance bermudagrass adaptability in the transition zone was one of the main objectives at the onset of the bermudagrass breeding program directed by C. M. Taliaferro at Oklahoma State University (OSU) (Taliaferro, Richardson, and Ahring 1977). The OSU program focused on the winter hardiness improvement and developed several improved forage and turf cultivars (Taliaferro and Richardson 1980a, 1980b; Taliaferro, Ahring, and Richardson 1983; Taliaferro et al. 2002, 2003; Alderson and Sharp 1994). These and other winter-hardy cultivars have helped move bermudagrass for commercial use further north in the United States.

Increasing drought tolerance is an important breeding goal, particularly as water for irrigation becomes less available in arid and semiarid regions. Compared to other perennial grasses, the relative drought resistance of bermudagrass is excellent (Beard 1973). Some physiological and rooting characteristics of bermudagrass contribute to its drought tolerance (Taliaferro 2003). Deep rooting is one of the characteristics that enhance drought tolerance in bermudagrass. Substantial genetic variability exists in bermudagrass germplasm for the drought tolerance traits. Substantial genetic diversity for evapotranspiration rate was observed in bermudagrass cultivars by Beard, Green, and Sifers (1992), indicating a potential of saving irrigation water by selecting cultivars with lower rates.

Coastal is a good example of a bermudagrass cultivar with superior drought tolerance, having produced six times as much dry matter as common bermudagrass in the 1954 drought summer (Burton 1973). Coastal has a root system capable of extracting available water from the soil more efficiently, especially in deep, sandy soil. Adams et al. (1966) indicated that the quantity of Coastal bermudagrass roots was at least twofold that of common bermudagrass in the soil profile below 45.7 cm, and it appeared that the root system of Coastal penetrated more deeply and was more evenly distributed throughout the soil than the root system of common bermudagrass at a high fertilization level. Adams et al. (1966) revealed that water use efficiency of bermudagrass increased as fertilizer levels increased. Improved drought tolerance of Coastal and Suwannee was demonstrated by less forage yield fluctuation from dry years to wet years than unimproved common bermudagrass exhibited (Burton et al. 1957).

Increasing forage yield potential is a major breeding objective in every forage bermudagrass breeding program. In commercial use, criteria for selecting a crop depend on its economic value. The economic value of a forage bermudagrass cultivar is mainly associated with its production level. This driving force makes high-yield potential one of the major focuses for the forage breeder. Increased forage yield potential can make bermudagrass cultivars more economically competitive with other potential or alternative crops. The forage bermudagrass breeding program initiated by Burton in 1936 released Coastal bermudagrass in 1943 (Alderson and Sharp 1994). Coastal yields two to four times as much as naturalized common bermudagrass strains. About 4 million ha across the southern states have been planted with Coastal, which produced annual forage yield gains valued at $300 million compared to common bermudagrass on the same acres (Burton 1973). This and other high-yielding cultivars have provided a solid foundation for the animal industry in the southern United States.

However, animal performance trials in bermudagrass repeatedly revealed that, as forage, it has a high production potential per unit area, but its forage quality is relatively low, as demonstrated by low production potential per grazing animal (Harlan 1970; Taliaferro et al. 1977). Accordingly, increasing the nutritive value of bermudagrass has been a focus of breeding programs over the past 50 years. Burton, Hart, and Lowrey (1967) observed substantial variability in Coastal, Midland, Kenya 56, and their progeny for forage quality traits, including dry matter digestibility and crude protein content. Their results further indicated that one hybrid of Coastal by Kenya 56 produced 12.3% more digestible dry matter than Coastal over a 4-year period. Steers had 30% higher average daily gains grazing on the highly digestible hybrid grass over Coastal with the same amount of forage intake (Burton et al. 1967; Chapman et al. 1971). Among many forage quality traits, forage crude protein content, lignin content, palatability or intake, dry matter digestibility, and animal performance should be given more attention.

Breeding cultivars to biotic resistance such as diseases, insects, and weeds will protect bermudagrass yield potential and/or quality, increase longevity of grass stands, and consequently enhance economic value of new cultivars. Many diseases and insect pests can cause damage to bermudagrass. *Bipolaris cynodontis* (Marig.) Shoemaker, a fungal pathogen, needs special emphasis in the development of cultivars for humid regions because it can cause serious leaf, crown, and root diseases of susceptible plants; reduce forage quality; and even kill the stands. Tifton 85, a high-performance cultivar, was derived from a cross of Tifton 68 by Tifton 292 (PI 290884), a clone introduced from South Africa that has strong resistance to fall armyworm feeding (Leuck et al., 1968; Lynch et al. 1983; Burton 2001). The resistance of Coastal bermudagrass to foliage diseases caused by leaf spot fungi provides protection against biomass loss and contributes better-quality feed than it would as a susceptible bermudagrass.

In addition to the development of clonal or vegetatively propagated cultivars, breeding of seeded forage bermudagrass has achieved significant progress since 1982, when Guymon, the first cold hardy and seed-propagated cultivar, was developed by Taliaferro et al. (1983). In the past 25 years, the interest in the development of seeded bermudagrass cultivars has increased dramatically (Taliaferro 2003; Taliaferro, Rouquette, et al. 2004). Baltensperger et al. (1993) noted that this interest derives from favorable market opportunities for seeded varieties. The increased need of seeded cultivars also relates to the ease and efficiency of shipping seed and the relative ease and economy of seed versus vegetative propagation. High seed set percentage, seed head prolificacy, good seed quality, low seed dormancy, and other components affecting seed yield and quality become important objectives for the development of seeded bermudagrass cultivars. Currently, more seed-producing cultivars are developed for turf use compared to clonal cultivars (Taliaferro 2003; Baltensperger et al. 1993).

8.6.2 Germplasm Collection and Maintenance

Scientists in South Africa and the United States began collecting *Cynodon* germplasm for scientific research and breeding use around the start of the twentieth century (Taliaferro 2003). These germplasm pools were enlarged during the twentieth century and provided genetic variation for selection and programmed breeding efforts. In the bermudagrass breeding program initiated by G. W. Burton in 1936 in Tifton, Georgia, bermudagrass and stargrass germplasm from Africa were accumulated via collecting trips, gifts, and exchange (Burton 1951b, 1965). The African germplasm sources included the collections by A. V. Bogdan at the Kitale Research Station, Kitale, Kenya; at the Frankenwald Botanical Research Station, University of Witwatersrand, Johannesburg, South Africa; and at the Henderson Research Station, near Salisbury, Rhodesia (now Harare, Zimbabwe) (Harlan 1970; Taliaferro, Rouquette, et al. 2004). The efforts by Burton, Hanna, and colleagues have accumulated a worldwide collection of more than 600 forage bermudagrass accessions maintained at USDA-ARS, Coastal Plain Experiment Station, Tifton, Georgia (Anderson 2005).

For efficient and feasible assessments of variability for the chemical composition or stress tolerance of the worldwide collection, Anderson (2005) developed a core collection using plant phenotype evaluations to capture the full phenotypic diversity in the collection. At OSU, a world collection was amassed in the late 1950s and early 1960s by Jack R. Harlan, Wayne W. Huffine, J. M. J. de Wet, and others. The large collection contained about 700 accessions and was used in a comprehensive biosystematic study of *Cynodon*. Charles M. Taliaferro initiated and directed the bermudagrass breeding program to breed forage and turfgrass cultivars at OSU from 1968 to 2006. He and colleagues added more *Cynodon* germplasm accessions from Africa, Australia, Europe, and Asia to the OSU collection (Figure 8.1). The increased germplasm collection is the key to the successful releases of several cultivars at OSU (Taliaferro, Rouquette, et al. 2004). Although bermudagrass germplasm in current collections is diverse, the potential to increase the diversity by more comprehensive germplasm collection is enormous. Harlan (1970) noted that only a small portion of the total bermudagrass germplasm had been collected at that time at which he was writing and even less had been used in breeding programs.

Long-term maintenance of a large number of bermudagrass germplasm accessions is challenging but necessary to breeding programs and related scientific endeavors. In addition to the collections maintained by individual breeding programs, the USDA-ARS Plant Germplasm Resource Conservation Unit in Griffin, Georgia, now maintains 431 *Cynodon* accessions from 40 countries current to May 2007 (NPGS 2007). As a common practice, bermudagrass germplasm accessions

Figure 8.1   (See color insert following page 274.) Part of Oklahoma State University forage bermudagrass germplasm nursery in Stillwater, Oklahoma.

are clonally maintained as potted plants in a greenhouse and/or as individual plants in field plots. Regular treatments such as applying herbicides to control alleys of field plots and trimming excessive growth from the tops of potted plants are necessary to minimize contamination. Seed samples can be stored in a cold room (5°C or lower) for many years.

However, preservation of clonally propagated bermudagrass germplasm in the greenhouse or field is difficult due to contamination from adjacent vegetative propagules of other accessions and from volunteer seedlings of its seeds derived from cross-pollination. Alternative forms of preservation methods have been developed. Bermudagrass tissues cultured on medium can be stored healthy at 4°C from 4 months to more than 1 year. Reed et al. (2005) developed a cryopreservation protocol for bermudagrass germplasm maintenance in liquid nitrogen (−196°C), which was used to store 25 *Cynodon* accessions at the National Clonal Germplasm Repository in Corvallis, Oregon, and at the National Center for Germplasm Resources Preservation in Fort Collins, Colorado, as a long-term backup.

8.6.3 Breeding Methods

Bermudagrass, especially the common type, follows the human footsteps to be cosmopolitan as it prefers disturbed habitats. Harlan (1970) speculated that common bermudagrass spread widely in the Old World following agricultural activities that caused agitation to soil. However, the development of bermudagrass cultivars as a hay or pasture grass and as a turfgrass for intensive management is a relatively recent event that primarily occurred in the United States. Scientists in China, Australia, and other countries have assembled bermudagrass germplasm and are conducting breeding and research programs.

Similar to many other crops or forage species used in agriculture, initial cultivars of bermudagrass were selections from naturally occurring variation. In 1936, the first forage bermudagrass breeding program was initiated at the University of Georgia, Coastal Plain Experiment Station, Tifton, Georgia, by USDA-ARS geneticist G. W. Burton. Subsequently, breeding methods were developed in the Tifton and other programs to produce clonal cultivars. Breeding approaches were also created to develop seeded bermudagrass cultivars.

8.6.3.1 Ecotype Selection

Ecotypes are plants that have evolved by the genetic mechanisms of selection, recombination, mutation, migration, and random drift in specific environments. Bermudagrass ecotypes are diverse. Early agriculturists selected naturally occurring bermudagrass plants that they deemed superior for agricultural purposes before the 1940s (Taliaferro, Rouquette, et al. 2004). The plants superior in forage yield were selected from local adapted populations and tested in small plots in specific regions. If some of the selected plants continued to show desirable performance and adaptation, then they were distributed to public users on a larger scale. In most cases, the cultivars were released without any additional breeding work. Even at present some early selections or cultivars are in use in permanent pastures in the southern United States, although many new and improved cultivars have been released since the 1950s.

Tracy (1917) described the early bermudagrass selections in the United States. Russell, released by Alabama and Louisiana Agricultural Experiment Stations in 1994, is an example of ecotype selection (Ball et al. 1996). Liu, He, and Liu (1996) released a cultivar from a clonal selection of bermudagrass collected in a roadside population in the east suburbs of Nanjing, Jiangsu Province, in China (Liu et al. 2006). One typical example of bermudagrass introduction and subsequent ecotype selection is recorded in Oklahoma Agricultural Experiment Station (OAES) Bulletin No. 55 (1902) and by Moorhouse, Burlison, and Ratcliff (1909). Plantings of many grasses, including bermudagrass, were made soon after the OAES was established in 1892 (Moorhouse et al. 1909). Natural selection quickly reduced the seeded bermudagrass populations to plants that had sufficient freeze tolerance to survive Oklahoma winters and were otherwise adapted.

In July 1904, a large planting of selected plants was made to further test for their winter hardiness. Some of the planting rootstocks were taken from the original plants grown in 1892, and other rootstocks were from the plants recently grown. In the severe winter of 1904 and 1905, a marked difference was observed in the selected plants. On March 29, 1905, all of the bermudagrasses grown from hardy selections were green and grew vigorously, but the plants established from seed in the previous year suffered severe winter injury. Sprigs of the hardy plants were distributed to a large number of districts in the Oklahoma Territory. Although no specific selections were made, the hardy bermudagrass plants were diverse in morphology and over time further subjected to forces of natural selection. Elder (1955) selected Greenfield bermudagrass among a large number of common strains growing at the Agronomy Research Station, Oklahoma State University, where early plantings had been made. The cultivar has been widely used in Oklahoma and adjacent regions as a pasture grass due to its winter hardiness, establishment capability, and ability to grow on less fertile soils (Alderson and Sharp 1994).

Seven of eight released stargrass cultivars were derived from ecotype selection (Taliaferro, Rouquette, et al. 2004). Ecotype selection of stargrass has been made by evaluation for adaptation to target environments; performance traits relating to forage yield, quality, and stand persistence under grazing; and hay production. Stargrass cv. no. 2 (*C. aethiopicus*) and Muguga (*C. nlemfuensis* var. *nlemfuensis*) were ecotype selections in Zimbabwe (Clatworthy 1985), IB 8 (*C. nlemfuensis* var. *nlemfuensis*) in Nigeria (Harlan 1970), and McCaleb, Ona, Florico, and Florona (*C. nlemfuensis*) in Florida. The Florida releases followed introduction and extensive evaluation of germplasm accessions by the University of Florida Agricultural Research and Education Center, Ona, Florida (Alderson and Sharp 1994; Mislevy, Brown, Caro-Costas, et al. 1989; Mislevy, Brown, Dunavin, et al. 1989).

8.6.3.2 Development of Clonal Hybrid Cultivars

Clonal hybrid cultivars are directionally created by artificial hybridizations, which lead to gene recombination; ecotype selections are naturally occurring genotypes. Breeders can take advantage of asexual and sexual reproduction mechanisms of bermudagrass to develop clonal hybrids. Because bermudagrass plants are highly heterozygous due to their high cross-pollination behavior, progeny plants of any crosses of two different genotypes typically are morphologically and genetically heterogeneous. Each progeny plant is a potential new cultivar. Therefore, development of clonal bermudagrass hybrid cultivars has been achieved by hybridization of selected parental plants and selection in resultant progeny populations. Initially, selected plants must be evaluated in controlled experiments through time (i.e., multiple years) and usually through space (i.e., multiple locations) to characterize performance accurately and identify the most elite and to compare them to standard cultivars. Intraspecific hybrids are derived from crosses of two parents from the same species; interspecific hybrids are from two parents belonging to two different species.

A number of forage type bermudagrass clonal hybrid cultivars have been developed since 1943 when Coastal was cooperatively released by the Georgia Coastal Plain Experiment Station and Plant Science Research Division, USDA-ARS (Burton 1947; Alderson and Sharp 1994). The basic breeding procedure of Burton (1947) is exemplified in the development of Coastal bermudagrass. The procedural steps follow.

In 1937, two common bermudagrass strains and two tall growing strains from South Africa were interplanted to facilitate crossing. Of the two common strains, one was Tift bermudagrass, discovered by J. L. Stevens in an old cotton patch near Tifton, Georgia, in 1929. The two African introductions were subsequently lost, but Harlan (1970) believed they were accessions of *C. dactylon* var. *elegans*. Enough open pollinated seed was collected from these clones in the same year to produce more than 5,000 seedlings. The seedling plants were space transplanted to a field nursery on 1.52 m centers in 1938. During the 1938 growing season, plants were visually rated for vigor, morphological traits, resistance to leaf spot, inflorescence abundance, and seed set. In the fall of 1938 and early spring of 1939, 128 and 19 plants were, respectively, selected from the 5,000 plants on the basis of ratings and measurements of performance characteristics. The 147 selected plants were advanced to a replicated small plot performance trial in 1939. From 1939 to 1946, these plots were uniformly managed and a large number of performance traits were measured. This experiment and other experiments established during the period evaluated response of the 147 plants to different nitrogen application treatments, comparative palatability to grazing cattle, and forage yield under simulated grazing conditions. Nine best plants were selected from the 147 initial selections. Planting stock of these elite selections was sent to a number of the experiment stations in the Southeast in 1941. By 1943, selection no. 35 was identified as the very best of the 5,000 progeny plants according to accumulated information and was released as Coastal bermudagrass.

Although Coastal was outstanding in forage yield and widely adapted to the lower southern states, its forage quality is relatively low and winter hardiness is inadequate for use in the transition zone of the United States (Burton et al. 1967). More cultivars with improved traits and performance features were developed for different bermudagrass production environments after the release of Coastal, but the breeding procedures used for the development of the additional hybrid cultivars are basically the same as that of Coastal.

A controlled cross between a cold-hardy common bermudagrass from Indiana and Coastal was made at the Georgia Coastal Plain Experiment Station by Burton in 1942. Among the 66 F_1 hybrids of the crossing, selection no. 13 was winter hardy in Indiana and Oklahoma. Later, the selection was released as Midland in Oklahoma because of good performance and adaptation in the state (Harlan, Burton, and Elder 1954). Tifton 44 was one winter-hardy F_1 hybrid with better disease resistance and higher forage yield than Midland. The cultivar was derived from crossing Coastal and a bermudagrass accession that had survived winters in Berlin, Germany, for 15 years before it was collected in 1966 (Burton and Monson 1978). Coastcross-1 was an interspecific F_1 hybrid of Coastal and Kenya PI 255455 (Kenya #14) from Kitale, Kenya (Burton 1972). The Kenyan parent plant was an accession indicated by Harlan (1970) to be *C. nlemfuensis* var. *robustus*. Coastcross-1 grew taller and had wider leaf blades than Coastal, and its forage was more digestible (11–12%) than Coastal. Average daily gains of steers consuming Coastcross-1 were 30% greater than steers consuming Coastal bermudagrass (Alderson and Sharp 1994).

The basic breeding procedure has been used to develop additional bermudagrass cultivars:

Suwannee, released in 1953 by the Georgia Coastal Plain Experiment Station and Plant Science Research Division, USDA-ARS (Burton 1962);

Hardie, released by Oklahoma Agricultural Experiment Station in 1974 (Taliaferro and Richardson 1980a);

Brazos, released by the Texas Agricultural Experiment Station in cooperation with Soil Conservation Service, ARS, and the Louisiana Agricultural Experiment Station in 1982 (Alderson and Sharp 1994);

Tifton 78, released by USDA-ARS and Georgia Coastal Plain Experiment Station in 1984 (Burton and Monson 1988);

Grazer, released by Louisiana Agricultural Experiment Station and USDA-ARS in 1985 (Eichhorn et al. 1986);

Tifton 85, released by USDA-ARS and Georgia Coastal Plain Experiment Station in 1992 (Burton et al. 1993);

Florakirk, released by Florida Agricultural Experiment Station in 1994 (Mislevy et al. 1999);

Midland 99, released by the Agricultural Experiment Stations of Oklahoma, Kansas, Missouri, Arkansas, the Samuel Noble Foundation, and the USDA-ARS in 1999 (Taliaferro et al. 2002); and

Goodwell bermudagrass, released by Oklahoma Agricultural Experiment Station in 2007.

8.6.3.3 *Development of Seed-Propagated Cultivars*

Kneebone (1966a) indicated that more than 2,724,000 kg of bermudagrass seed was produced annually as a special crop in Arizona in the 1960s. A major portion of the bermudagrass seed produced was of cv. Arizona Common, which was naturalized in the southwestern states. NK 37 giant bermudagrass (*C. dactylon* var. *aridus*) was developed by Northrup, King & Co. (Hanson 1972). Kneebone initiated a bermudagrass research program at the University of Arizona in 1959 to develop seeded cultivars (Kneebone 1973). However, it was not until the last 25 years that new seeded cultivars began to be released (Taliaferro et al. 1983; Baltensperger 1989).

Guymon, released in 1982 by the Oklahoma Agricultural Experiment Station and the USDA-ARS, was the first seeded cultivar developed for the U.S. transition region. Subsequently, Wrangler was released in 1999 by the Johnston Seed Co., Enid, Oklahoma (Taliaferro, Rouquette, et al. 2004). Wrangler has similar adaptation and yield performance as Guymon but has approximately twice the seed yield. Use of self-incompatibility in bermudagrass to produce commercial seed-propagated F_1 hybrids was proposed by Burton and Hart (1967). They proposed growing two clonal bermudagrass parent plants with high specific combining ability in alternate rows to facilitate cross-pollination. The protocol was used in developing Guymon (Taliaferro et al. 1983). The seeded cultivar was derived from crossing two winter-hardy and self-incompatible accessions from Yugoslavia and Guymon, Oklahoma, respectively. The outstanding feature of Guymon relative to seeded common bermudagrass produced in the Yuma and Imperial Valleys of Arizona and California, respectively, is substantially greater cold tolerance. Annual forage yields of Guymon were 25–40% less than yields of clonal type cultivars such as Midland and Hardie.

Breeding efforts to develop seed-propagated bermudagrass cultivars have dramatically increased over the past 25 years, with most of that effort focused on cultivars for turf use. Phenotypic, genotypic, and phenotypic–genotypic recurrent selection procedures have been used in breeding seeded turf bermudagrass (Baltensperger et al. 1993).

8.7 MOLECULAR VARIATION

DNA molecular marker technology provides unprecedented, precise tools to quantify genetic variation in bermudagrass at the genus, species, population, and within-population levels. The molecular variation revealed by DNA markers provides essential information for the efficient selection of superior plant material for breeding, an adequate management of genetic resources, and the effective preservation of biodiversity (Caetano-Anolles 1998a). Among a wide range of molecular marker systems available, restriction fragment length polymorphism (RFLP) of Botstein et al. (1980); random amplified polymorphic DNA (RAPD) of Williams et al. (1990); DNA amplification fingerprinting (DAF) of Caetano-Anolles, Bassam, and Gresshoff (1991); and amplified fragment length polymorphism (AFLP) of Vos et al. (1995) have been used in bermudagrass to differentiate genotypes or identify off-types, to quantify genetic variation or diversity, to measure genetic relatedness, and to establish linkage maps (Yaneshita et al. 1993; Caetano-Anolles et al. 1995; Caetano-Anolles, Callahan, and Gresshoff 1997; Caetano-Anolles 1998b; 1999; Assefa et al. 1998; Zhang et al. 1999; Anderson et al. 2001; Karaca et al. 2002; Roodt, Spies, and Burger 2002; Wu et al. 2004, 2005; Yerramsetty et al. 2005; Bethel et al. 2006).

Comparing four DNA profiling techniques—AFLP, chloroplast-specific simple sequence repeat length polymorphism (CpSSRLP), RAPD, and directed amplification of minisatellite-region DNA (DAMD)—in forage bermudagrass fingerprinting, Karaca et al. (2002) reported that the AFLP technique produced the highest number of polymorphic bands per primer combination, and CpSSRLP produced the lowest number of polymorphic bands. Zhang et al. (1999) and Wu et al. (2004) reported that the AFLP technique is highly reliable in respective bermudagrass genetic analysis (Figure 8.2). Reproducibility and resolution of the AFLP and CpSSRLP techniques were higher than in the other two (Karaca et al. 2002).

Phylogenetic dendrograms of Yaneshita et al. (1993), based on chloroplast DNA RFLP patterns of 10 cool-season and warm-season turfgrasses, clearly indicated that *C. dactylon* was grouped with *Zoysia japonica;* both species belonged to the same subfamily Eragrostoideae but were separated from the turfgrass species of Panicoideae and Pooideae, with agreement with conventional taxonomy. Using the DAF profiling technique to assess genetic relatedness among 62 accessions of eight *Cynodon* species, Assefa et al. (1999) observed 92% of scored bands were polymorphic, indicating extremely high genetic diversity in the genus (Figure 8.3). High species similarities were between *C. aethiopicus* and *C. arcuatus, C. transvaalensis* and *C. plectostachyus,* and *C. incompletus* and *C. nlemfuensis.* They also reported that within-species molecular variation was the least for *C. aethiopicus, C. arcuatus,* and *C. transvaalensis* and the largest for the cosmopolitan *C. dactylon.*

Wu et al. (2004) reported genetic similarity coefficients ranging from 0.53 to 0.98 among 28 *C. dactylon* var. *dactylon* accessions originating from 11 countries of Africa, Asia, Australia, and Europe. Groupings of the accessions in cluster analysis were consistent with their geographic origins (Figure 8.4). Clearly, geographic origin was a significant factor in their genetic differentiation. Genetic diversity of *C. dactylon* var. *dactylon* evolved under different climatic and edaphic conditions (Wu et al. 2004). Consequently, comprehensive germplasm collection in different continents is required to sample the full extent of the available variation. AFLP investigation with 114 Chinese *C. dactylon* accessions by Wu, Taliaferro, Bai, et al. (2006) indicated that genetic similarity ranged from 0.65 to 0.99, and genetic differentiation among the Chinese indigenous accessions was evidently associated with ploidy levels (tetraploid, pentaploid, and hexaploid). It was not surprising that tetraploid accessions were predominant in the collection and had the greatest genetic variation, while pentaploid accessions were fewest in number and had the least genetic variation.

Cynodon transvaalensis plants were described as being very uniform in appearance (de Wet and Harlan 1971). Using AFLP markers, Wu et al. (2005) quantified genetic variation of 14 African bermudagrass accessions and reported genetic similarity coefficients ranging from 0.66 to 0.99, indicating molecular variation within the species. Substantial variability for morphological and adaptation traits has been observed in segregating populations of *C. transvaalensis* (Taliaferro 1992). Recently, Kenworthy et al. (2006) reported significant genetic variation ($P < 0.05$) for 17 of 21 agronomic traits in *C. transvaalensis.* Comparative molecular variation assessments for *C. transvaalensis, C. dactylon,* and their triploid and tetraploid hybrids by Caetano-Anolles et al.

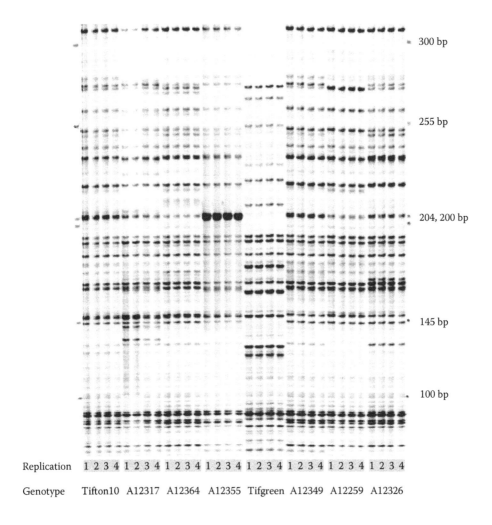

Replication 1 2 3 4 1 2 3 4 1 2 3 4 1 2 3 4 1 2 3 4 1 2 3 4 1 2 3 4 1 2 3 4

Genotype   Tifton10 A12317 A12364 A12355 Tifgreen A12349 A12259 A12326

Figure 8.2 AFLP fingerprints of eight bermudagrass genotypes with four replications showing high reliability of the DNA marker technology. The AFLP patterns of replications 1 and 2 are from different DNA isolations of the same bermudagrass genotypes. The AFLP patterns of replication 3 are from different gel loadings of the same PCR products of replication 2, and those of replication 4 are from different PCR reactions with the same DNA samples as replication 2. (From Wu, Y. Q. et al. 2004. *Genome* 47:689–696. With permission.)

(1995), Zhang et al. (1999), and Wu et al. (2005) indicated that the two species consistently grouped separately, and their hybrids either grouped with the latter species or formed groupings between the two species.

Karaca et al. (2002) characterized genetic variation and relationships within 31 forage bermudagrass genotypes including released cultivars such as Tifton 85, Hardie, Tifton 78, Tifton 44, Coastal, Alicia, Callie, Grazer, Russell, P. Rico, and ecotype selections identified from existing bermudagrass fields by DNA markers. Genetic similarity coefficients of the forage bermudagrasses ranged from 0.61 to 0.98. They observed that many ecotypes and cultivars had close genetic relatedness and concluded that the forage bermudagrass genotypes had a narrow genetic base. They suggested a genetic similarity value of 0.85 as a sufficient cutoff line to identify a unique genotype as genetic justification for a cultivar release. Although numerous investigations have indicated enormous genetic diversity exists in bermudagrass, only a small fraction of bermudagrass germplasm has been used in breeding forage bermudagrass cultivars.

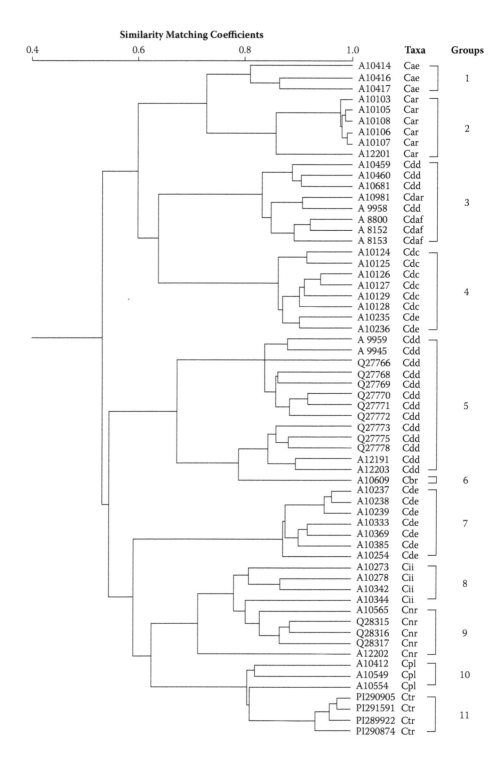

Figure 8.3 Dendrogram of *Cynodon* species showing their relatedness. Cae = *Cynodon aethiopicus*; Car = *C. arcuatus*; Cdd = *C. dactylon* var. *dactylon*; Cdaf = *C. dactylon* var. *afghanicus*; Cdar = *C. dactylon* var. *aridus*; Cdc = *C. dactylon* var. *coursii*; Cde = *C. dactylon* var. *elegans*; Cii = *C. incompletus* var. *incompletus*; Cnr = *C. nlemfuensis* var. *robustus*; Cpl = *C. plectostachyus*; Ctr = *C. transvaalensis*; Cba = *C. barberi*. (Assefa, S. et al. 1998. *Genome* 42:465–474. With permission.)

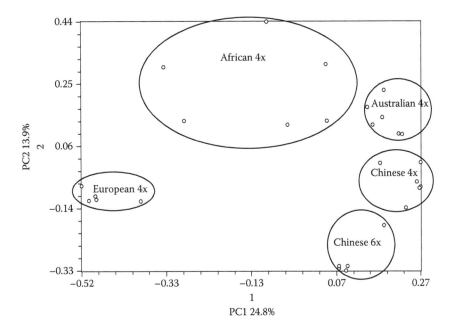

Figure 8.4 Principal coordinate (PC) map for the first and second coordinates estimated for 443 AFLP markers using the genetic similarity matrix for 28 *Cynodon dactylon* var. *dactylon* accessions. (From Wu, Y. Q. et al. 2004. *Genome* 47:689–696. With permission.)

The genetic linkage map is a valuable tool for studying genetic control mechanisms for important agronomic traits, gaining new knowledge of genome structure in a species, and comparison with genome structures of other species. Linkage maps are extremely valuable in mapping quantitative trait loci (QTLs) useful in marker-assisted selection and map-based cloning. Bethel et al. (2006) described the first linkage maps of *C. dactylon* and *C. transvaalensis* based on single dose restriction fragments. The *C. dactylon* map consisted of 35 linkage groups with 172 markers mapped; the *C. transvaalensis* map included 77 markers on 18 linkage groups. It was estimated that the recombinational lengths of the *C. dactylon* and *C. transvaalensis* genomes are 3,012 and 1,569 cM, respectively (Bethel et al. 2006).

8.8 TISSUE CULTURE AND GENETIC TRANSFORMATION

Genetic transformation constitutes the integration of genes from one species to the genome of another species overcoming biologically interspecific gene flow barriers. Along with major crops and other important forage species, genetic transformation research in bermudagrass is increasing. In the 1980s, Ahn, Huang, and King (1985, 1987) and Artunduaga, Taliaferro, and Johnson (1988, 1989) conducted bermudagrass tissue culture investigations for the development of somatic embryogenic callus and plant regeneration. In the past few years, significant progress in tissue culture and genetic transformation of bermudagrass has been made in the following areas: (1) improvement in tissue culture protocols, (2) generation of transgenic plants by projectile bombardment and *Agrobacterium*-mediated transformation, and (3) development of direct transformation of stolons bypassing the callus formation (Goldman et al. 2004; Zhang et al. 2003; Li and Qu 2004; Hu et al. 2005; Jain et al. 2005; Li et al. 2005; Wang and Ge 2005).

Ahn et al. (1985, 1987) reported induction of embryogenic callus from immature inflorescence explants on N_6 medium supplemented with 1 mg L^{-1} 2,4-D and germination of the somatic embryos on hormone-free N_6 medium. Artunduaga et al. (1988, 1989) reported regeneration from embryogenic callus cultures of immature floral tissue explants of several bermudagrass genotypes on Murashige and Skoog (MS) medium containing 1–3 mg L^{-1} 2,4-D, with adding casein hydrolysate (200 mg L^{-1}) in addition to 3 mg L^{-1} 2,4-D stimulating induction of embryogenic callus for cv. Zebra bermudagrass. Using scanning electron microscopy, Chaudhury and Qu (2000) found that somatic embryogenesis was the major route for bermudagrass regeneration. Li and Qu (2002) reported stimulatory effects of abscisic acid on repetitive somatic embryogenesis and improved germination of embryos on medium supplemented with gibberellic acid using triploid Tifgreen bermudagrass. Jain et al. (2005) reported an optimized protocol for tissue culture of young inflorescence explants of Tifton 85 and regeneration. Tifton 85 embryogenic callus was obtained on MS basal medium containing 30 g L^{-1} sucrose, 4 mg L^{-1} 2,4-D, 0.01 mg L^{-1} 6-benzylaminopurine (BAP), and 200 mg L^{-1} casein hydrolysate.

Zhang et al. (2003) established a transformation system for TifEagle triploid bermudagrass using a biolistic bombardment delivery system. Embryogenic callus was induced from stolons on MS medium supplemented with 30 μM dicamba, 20 μM BAP, and 100 mg myo-inositol L^{-1}. Transgenic plants expressing the *hpt* marker gene were recovered. Goldman et al. (2004) also reported biolistic transformation of TifEagle regenerate plants expressing the *bar* gene; sliced nodes were explants to initiate embryogenic tissues. Li and Qu (2004) reported that yellowish, compact calli were achieved from immature inflorescence tissue culture of line J1224 when the concentration of BAP was elevated to 0.9 μM in the culture medium, and highly regenerable callus lines were established. They obtained four transgenic plants via biolistic transformation expressing the *gusA* and/or *bar* genes.

Agrobaterium-mediated transformation as a means to produce transgenic plants has achieved success in major cereal crops, spurring its applications in forage species including bermudagrass (Wang and Ge 2006). Major advantages of *Agrobacterium*-mediated transformation over other gene delivery systems include stable integration of transgenes into the plant genome and a lower copy number of transgenes, fewer rearrangements, and an improved stability of transgenic expression over generations. In the transformation system, embryogenic calli were co-cultivated with *Agrobacterium* harboring an exotic vector containing target and marker genes, and resistant calli were obtained after antibiotic selection. Hu et al. (2005) reported integration of the *bar* gene into TifEagle by *Agrobacterium tumefaciens*-mediated transformation. Li et al. (2005) reported achievement of two green independent transgenic plants of common bermudagrass (*C. dactylon*) using *Agrobacterium*-mediated transformation. Concurrently, Salehi et al. (2005) reported *Agrobacterium*-mediated transformation of Arizona Common bermudagrass with the *Bacillus thuringiensis* Berliner *cryIAc* gene encoding an endotoxin active against black cutworm (*Agrotis ipsilon* Hufnagel). They used mature seed for producing embryogenic callus. An insect feeding assay indicated that the larvae of black cutworms fed transgenic plant leaves experienced more than 80% mortality.

Callus induction and plant regeneration are costly and laborious and may lead to somaclonal variation (Goldman et al. 2004; Wang and Ge 2006). A new breakthrough is the "direct transformation of stolons bypassing the callus formation phase" developed by Wang and Ge (2005) for bermudagrass. They reported 4.8–6.1% transgenic plants developed from bermudagrass stolon nodes infected and co-cultivated with *A. tumefaciens* harboring binary vectors. The direct transformation procedure does not need callus production from explants. In the method, green shoots directly grew out from infected stolon nodes 4–5 weeks after antibiotic selection, and transgenic plants developed in 7–9 weeks. Biolistic and traditional *Agrobacterium*-mediated transformation systems need 22 or more weeks. Southern, northern, and GUS staining analyses provided positive response to the direct transformation protocol. Wang and Ge (2005) have successfully transformed forage type bermudagrass cultivars Midland 99 and Tifton 85 by the same protocol. To date, no cultivar derived from genetic transformation has been released for commercial use.

8.9 INTERSPECIFIC AND INTERGENERIC HYBRIDIZATION

The extensive, systematic research by Harlan and colleagues in the 1960s provided detailed insight on interspecific hybridization potentials between *Cynodon* taxon forms (de Wet and Harlan, 1970, 1971; Harlan and de Wet 1969; Harlan, de Wet, and Richardson 1969; Harlan, de Wet, Rawal, Felder, et al. 1970). *Cynodon arcuatus, C. barberi,* and *C. plectostachyus* are genetically well isolated from each other and from reminder species in the genus. Despite a large number of spikelets of *C. arcuatus* being emasculated and pollinated with pollen from other species, no interspecific hybrids were produced (Harlan et al. 1969). Similarly, a massive, unsuccessful effort was made to cross *C. plectostachyus* with other species in *Cynodon*. Although the crossing produced 1,529 plump and well-formed seeds, only two plants from these seeds were considered hybrids; the authors concluded that species isolated well genetically with the other taxon in *Cynodon* (Harlan et al. 1969). *Cynodon aethiopicus* is rather strongly isolated from the other *Cynodon* species. However, Harlan et al. (1969) produced five hybrids in a cross between one diploid *C. aethiopicus* and one tetraploid *C. nlemfuensis* plant. Diploid *C. aethiopicus* crossed sparingly with tetraploid *C. dactylon* var. *dactylon* and produced six putative triploids. No hybrid plants were obtained from the seed of crosses between *C. aethiopicus* and other species, including *C. arcuatus, C. plectostachyus,* diploid *C. dactylon, C. transvaalensis,* and *C. × magennisii*.

Diploid *C. nlemfuensis* var. *nlemfuensis* crosses readily with *C. dactylon* var. *aridus* and var. *afghanicus* and *C. transvaalensis*. Harlan et al. (1969) indicated that hybrids between *C. nlemfuensis* var. *nlemfuensis* and the diploid variety of *C. dactylon* are among the most vigorous and productive ones. However, interspecific crosses between *C. nlemfuensis* var. *nlemfuensis* and *C. arcuatus, C. plectostachyus, C. aethiopicus,* and *C. incompletus* did not produce any hybrid plants although the crosses did produce some seeds (Harlan et al. 1969; Harlan, de Wet, Rawal, Felder, et al. 1970). The crossability of *C. nlemfuensis* var. *robustus* with other *Cynodon* species is different from that of the former variety (*C. nlemfuensis* var. *nlemfuensis*). *Cynodon nlemfuensis* var. *robustus* showed good crossability with *C. dactylon*. Harlan et al. (1969) reported that 126 hybrid plants resulted from 386 seeds of 124 crosses between *C. nlemfuensis* var. *robustus* and tetraploid *C. dactylon,* and only one hybrid plant was produced from many crosses with diploid *C. dactylon*. Seed from crosses between *C. nlemfuensis* var. *robustus* and *C. incompletus* rarely produced hybrid plants, and crosses of var. *robustus* with the remaining species produced rare seeds and no hybrid plants. The two varieties *incompletus* and *hirsutus* of *C. incompletus* cross easily, but only a few artificial hybrids were made in crosses of the two varieties with diploid and tetraploid *C. dactylon* forms. *C. transvaalensis* crosses easily with tetraploid *C. dactylon* var. *dactylon,* but with varying degrees with other *C. dactylon* varieties.

Breeding programs in the United States have developed numerous superior forage and turf cultivars by interspecific hybridization. Coastcross-1 and Tifton 85 are examples of commercially released cultivars that are hybrid plants involving tetraploid *C. dactylon* and *C. nlemfuensis*. Similarly, numerous interspecific hybrids derived from crosses between *C. dactylon* var. *dactylon* and *C. transvaalensis* have been released as turf cultivars (Alderson and Sharp 1994).

No significant investigations have been conducted on the hybridization potential of *Cynodon* with other genera; consequently, little information is available. Clayton and Renvoize (1986) indicated that *Cynodon* is "related to Chloris, with which it will hybridize" but provided no references.

8.10 FUTURE DIRECTIONS

Bermudagrass is among the most important and widely used warm-season forage grasses for livestock grazing and hay production in warmer parts of the world. In the United States alone, forage bermudagrass is grown on more than 10 million ha in the southern states (Burton and Hanna 1995).

Permanent bermudagrass pastures have provided and continue to provide a major part of low-cost feed to animal husbandry, especially the beef industry in the southern United States. Bouton (2007) estimated that the current economic value of the southeastern forage systems for beef and calves (the main livestock group in the region) alone is approximately US $11.6 billion annually. Forage bermudagrass plays a major role in the forage-livestock systems in the southern United States. Further improvements in bermudagrass will significantly contribute to the agricultural economy of the southern United States and many other warmer regions in the world. Future directions of forage bermudagrass breeding programs should focus on the development of new cultivars enhanced for performance traits, including yield and quality of biomass and tolerance of stresses caused by biotic and abiotic agents. Efforts should include enhancement of germplasm collections, characterization and preservation of accessions, employment of conventional breeding technology, and application of biotechnology.

Among the nine *Cynodon* species and 10 varieties, *C. dactylon* and its six varieties (*aridus, afghanicus, dactylon, coursii, elegans,* and *polevansii*), *C. nlemfuensis* and two varieties (*nlemfuensis* and *robustus*), *C. aethiopicus,* and *C. plectostachyus* have been evaluated as major forage species. Over the past 75 years, a series of cultivars developed from *C. dactylon, C. nlemfuensis, C. aethiopicus,* or interspecific hybrids of *C. dactylon* by *C. nlemfuensis* was developed using introduced germplasm accessions and their derivatives. Currently, in addition to some germplasm maintained in individual bermudagrass breeding programs, there are 315, 1, 3, and 32 accessions of *C. dactylon, C. aethiopicus, C. nlemfuensis,* and *C. plectostachyus* collected from 38, 1, 2, and 6 countries, respectively, in the *Cynodon* collection of the USDA National Plant Germplasm System (NPGS 2007). It is likely that this collection does not fully represent the genetic diversity in the genus, even for *C. dactylon,* which constitutes 90% of the total collection.

Harlan (1970) indicated that only a tiny fraction of the germplasm pool of *Cynodon* was collected and characterized and even less used in the breeding programs. After more than three decades, Taliaferro, Rouquette, et al. (2004) indicated that Harlan's statement of *Cynodon* germplasm remains true. They further noted that the scope of future efforts in procuring, characterizing, and maintaining *Cynodon* germplasm, particularly in publicly accessible collections like that of the USDA National Plant Germplasm System, will strongly impact future improvement success. To exploit the maximum genetic diversity of the major forage bermudagrass species for breeding improvement, efforts to procure, evaluate, and maintain more comprehensive germplasm accessions from original geographic regions, including Africa and Asia, are necessary to add the collections like that of the NPGS.

In the predictable future, conventional breeding procedures, which have been proven to be highly effective, will continue to be used as principal techniques to develop new forage bermudagrass cultivars. New techniques developed by biotechnology will not replace conventional breeding procedures but, rather, complement them.

As a main avenue to develop new cultivars, the breeders may need to focus on creating new genetic diversity. Clonal F_1 hybrid cultivars can be derived from intra- and interspecific hybridization by crossing parental plants from natural selections, enhanced germplasm, or combinations of both. Natural selections encompass original collections and discovered plants from cultivated settings and naturalized wild populations. Enhanced germplasm includes commercial cultivars and selections derived from breeding efforts of population improvements or crosses. The hybrid breeding procedure allows heterosis to be exploited. Therefore, specific combining ability among parental plants is important. Chances of success increase as number of crosses and size of progeny population increase. In the past, only a few seeded varieties were developed. Phenotypic and genotypic recurrent selections have been powerful breeding tools in the development of numerous turf-type cultivars in recent years. The techniques should be valuable to developing seeded forage type cultivars.

Molecular biology is revolutionizing biological science, first in model organisms, such as *Arabidopsis thaliana* of plants, then in major food crops, such as rice (*Oryza sativa* L.), and finally in more applied species including bermudagrass. Molecular biology creates new channels for bermudagrass scientific research. In recent years, flow cytometry has been applied to measure nuclei DNA in bermudagrass to estimate ploidy levels (Taliaferro et al. 1997). Flow cytometry is a rapid and accurate method for large numbers of accessions. Biotechnology provides new challenges and opportunities for the development of next-generation, forage-type bermudagrass cultivars. Biotechnological techniques include genetic engineering (i.e., gene transformation), molecular marker applications (e.g., genetic diversity and relatedness analysis and marker-assisted selection), and emerging tools from postgenomic and future scientific investigations.

In addition to random, dominant DNA markers based on PCR reaction (i.e., AFLP, RAPD, and DAF) and co-dominant RFLP markers based on probe hybridization, co-dominant, PCR-based markers like simple sequence repeat (SSR), genomic region-specific markers like sequence tagged sites (STSs), and single nucleotide polymorphisms (SNPs) need to be developed for bermudagrass. The more reliable markers (SSR, AFLP, and RFLP) are very useful in molecular map (linkage groups) construction. Bethel et al. (2006) published a preliminary map for each of tetraploid *C. dactylon* and diploid *C. transvaalensis*. More markers to be added to the linkage map can set a dense framework for tracking genomic regions responsible for quantitative traits or termed as quantitative trait locus (QTL) by molecular markers (QTL mapping). Knowing QTL locations and the molecular markers tightly linked to the target QTLs is helpful for breeding applications such as marker-assisted selection.

Many important agronomic traits, such as forage yield, winter hardiness, or organic matter digestibility, require extensive testing to search for and confirm individuals that have desired traits in conventional breeding. If molecular markers are associated with a quantitative trait, breeders can likely identify desired individual progeny plants by PCR, amplifying the linked marker fragments in the laboratory. Although it is difficult to tag all the genes conditioning a quantitative trait, QTL mapping and marker-assisted selection are so powerful that major genetic variation can be mapped and monitored by markers. Molecular markers and marker-assisted selection potentially enable breeders to select quantitative traits more efficiently. To date, no QTL mapping of complex traits for bermudagrass has been published, so marker-assisted selection will be a hot topic for research.

Although molecular markers can be used to tag desired genes (i.e., gene markers) or QTL (i.e., linked markers) that are available in existing germplasm and can be incorporated or transferred by conventional hybridization, genetic transformation can integrate new genes that are not available in existing germplasm but, rather, from other organisms into adapted bermudagrass plants, generating value-added traits. Recently, significant progress has been accomplished in bermudagrass *Agrobacterium*-mediated transformation. The developed protocols will be valuable to transform desired and value-added traits into adapted germplasm for bermudagrass improvement.

One of the key traits for improving forage bermudagrass is to reduce lignin content to improve organic matter digestibility. It is well known that bermudagrass is potentially high yielding, but animal performance (i.e., daily body gain) is low because of its low digestibility. Using conventional breeding methods, several digestibility-improved bermudagrass cultivars have been released. Small improvement in forage digestibility can significantly increase animal gains and is more important than a proportional increase in forage yield (Redfearn and Nelson 2003). Chen et al. (2003, 2004) demonstrated improved digestibility of tall fescue (*Festuca arundinacea*) by transgenic down-regulation of cinnamyl alcohol dehydrogenase and caffeic acid 0-methyltransferase (COMT), respectively. Similar protocols can be used to improve forage digestibility in bermudagrass. Increasing bermudagrass abiotic and biotic resistance, such as winter hardiness and resistance to leaf spot caused by *Bipolaris cynodontis* (Shoemaker), is most likely achieved by transgenic expression of regulatory gene or genes involved in signal transduction and genetic engineering (Wang and Ge 2006).

In conclusion, comprehensive collection of naturally occurring genetic diversity of bermudagrass germplasm will provide larger genetic variation for future forage improvement. Integration of conventional breeding procedures and modern biotechnology will result in the development of new-generation cultivars in the twenty-first century.

REFERENCES

Adams, W. E., C. B. Elkins, and E. R. Beaty. 1966. Rooting habits and moisture use of Coastal and common bermudagrasses. *Journal of Soil and Water Conservation* 21:133–135.

Ahn, B. J., F. H. Huang, and J. W. King. 1985. Plant regeneration through somatic embryogenesis in common bermudagrass tissue culture. *Crop Science* 25:1107–1109.

———. 1987. Regeneration of bermudagrass cultivars and evidence of somatic embryogenesis. *Crop Science* 27:594–597.

Ahring, R. M., C. M. Taliaferro, and R. D. Morrison. 1974. Seed production of several strains and hybrids of bermudagrass, *Cynodon dactylon* (L.) Pers. *Crop Science* 14:93–95.

Ahring, R. M., and G. W. Todd. 1978. Seed size and germination of hulled and unhulled bermudagrass seeds. *Agronomy Journal* 70:667–670.

Alcordo, I. S., P. Misley, and J. E. Rechcigl. 1991. Effect of defoliation on root development of stargrass under greenhouse conditions. *Communications in Soil Science and Plant Analysis* 22:493–504.

Alderson, J., and W. C. Sharp. 1994. *Grass varieties in the United States*. USDA, SCS, Agriculture handbook no. 170.

Anderson, J. A., M. P. Kenna, and C. M. Taliaferro. 1988. Cold hardiness of 'Midiron' and 'Tifgreen' bermudagrass. *HortScience* 23:748–750.

Anderson, M. P., C. M. Taliaferro, D. L. Martin, and C. S. Anderson. 2001. Comparative DNA profiling of U-3 turf bermudagrass strains. *Crop Science* 41:1184–1189.

Anderson, W. 2005. Development of a forage bermudagrass (*Cynodon* sp.) core collection. *Japanese Society Grassland Science* 51:305–308.

Anonymous. 1902. Bermuda grass. Bulletin No. 55, Oklahoma Agricultural Experiment Station, Stillwater.

———. 1990. *China flora* 10:82–85. Beijing, China: China Scientific Press.

Artunduaga, I. R., C. M. Taliaferro, and B. L. Johnson. 1988. Effects of auxin concentration on induction and growth of embryogenic callus from young inflorescence explants of Old World bluestem (*Bothriochloa* spp.) and bermuda (*Cynodon* spp.) grasses. *Plant Cell, Tissue and Organ Culture* 12:13–19.

———. 1989. Induction and growth of callus from immature inflorescence of "Zebra" bermudagrass as affected by casein hydrolysate and 2,4-D concentration. *In Vitro Cellular and Developmental Biology* 25:753–756.

Arumuganathan, K., S. P. Tallury, M. L. Fraser, A. H. Bruneau, and R. Qu. 1998. Nuclear DNA content of thirteen turfgrass species by flow cytometry. *Crop Science* 39:1518–1521.

Assefa, S., C. M. Taliaferro, M. P. Anderson, B. G. de los Reyes, and R. M. Edwards. 1999. Diversity among *Cynodon* accessions and taxa based on DNA amplification fingerprinting. *Genome* 42:465–474.

Ball, D. M., M. M. Eichhorn, Jr., R. A. Burdett, Jr., and D. M. Bice. 1996. Registration of 'Russell' bermudagrass. *Crop Science* 36:467–467.

Baltensperger, A. A. 1989. Registration of 'NUMEX SAHARA' bermudagrass. *Crop Science* 29:1326–1326.

Baltensperger, A. A., B. Dossey, L. Taylor, and J. Klingenberg. 1993. Bermudagrass, *Cynodon dactylon* (L.) Pers., seed production and variety development. *International Turfgrass Society Research Journal* 7:829–838.

Beard, J. B. 1973. *Turfgrass: science and culture*, chap. 4. Englewood Cliffs, NJ: Prentice Hall.

Beard, J. B., R. L. Green, and S. I. Sifers. 1992. Evapotranspiration and leaf extension rates of 24 well-watered, turf-type *Cynodon* genotypes. *HortScience* 27:986–988.

Beard, J. B., and J. R. Watson. 1982. Recent turfgrass plant explorations in Africa. *USGA Green Section Record* 20 (4): 6–8.

Bethel, C. M., E. B. Sciara, J. C. Estill, J. E. Bowers, W. W. Hanna, and A. H. Paterson. 2006. A framework linkage map of bermudagrass (*Cynodon dactylon × transvaalensis*) based on single-dose restriction fragments. *Theoretical and Applied Genetics* 112:727–737.

Black, C. C., T. M. Chen, and R. H. Brown. 1969. Biochemical basis for plant competition. *Weed Science* 17:338–344.

Botstein, D., R. L. White, M. Skolnick, and R. W. Davis. 1980. Construction of a genetic linkage map in man using restriction fragment length polymorphisms. *American Journal of Human Genetics* 32:314–331.

Bouton, J. 2007. The economic benefits of forage improvement in the United States. *Euphytica* 154:263–270.

Brilman, L. A., W. R. Kneebone, and J. E. Endrizzi. 1982. Pachytene chromosome morphology of diploid *Cynodon dactylon* (L.) Pers. *Cytologia* 47:171–181.

Brink, G. E., K. R. Sistani, and D. E. Rowe. 2004. Nutrient uptake of hybrid and common bermudagrass fertilized with broiler litter. *Agronomy Journal* 96:1509–1515.

Brown, T. A. 1999. *Genomes, 1–13.* Oxford, U.K.: Bio Scientific Publishers Ltd.

Burns, J. C., L. D. King, and P. W. Westerman. 1990. Long-term swine lagoon effluent applications on 'Coastal' bermudagrass: I. Yield, quality, and element removal. *Journal of Environmental Quality* 19:749–756.

Burns, J. C., P. W. Westerman, L. D. King, G. A. Cummings, M. R. Overcash, and L. Goode. 1985. Swine lagoon effluent applied to Coastal bermudagrass: I. Forage yield, quality, and element removal. *Journal of Environmental Quality* 14:9–14.

Burson, B. L., and C. R. Tischler. 1980. Cytological and electrophoretic investigations of the origin of 'Callie' bermudagrass. *Crop Science* 20:409–410.

Burton, G. W. 1947. Breeding bermudagrass for the Southeastern United States. *Journal of American Society of Agronomy* 39: 551–569.

———. 1951a. Bermudagrass. In *Forages: The science of grassland and agriculture*, ed. H. D. Hughes, M. E. Heath, and D. S. Metcalfe. Ames: Iowa State Univ. Press.

———. 1951b. Intra- and interspecific hybrids in bermudagrass. *Journal of Heredity* 42:153–156.

———. 1954. Coastal bermuda grass for pasture, hay and silage. Bulletin N. S. 2. Georgia Coastal Plain Experiment Station, Tifton, GA.

———. 1962. Registration of varieties of bermudagrass. *Crop Science* 2:352–354.

———. 1965. Breeding better bermudagrasses. In *Proceedings of IX International Grassland Congress*, Brazil, 1:93–96.

———. 1972. Registration of Coastcross-1 bermudagrass. *Crop Science* 12:125–125.

———. 1973. Breeding better forages to help feed man and preserve and enhance the environment. *Bio Science* 23: 705–710.

———. 1991. A history of turf research at Tifton. *Green Section Record*, 29 (3): 12–14, U.S. Golf Association, Far Hills, NJ.

———. 2001. Tifton 85 bermudagrass—Early history of its creation, selection, and evaluation. *Crop Science* 41:5–6.

Burton, G. W., E. H. DeVane, and R. L. Carter. 1954. Root penetration, distribution and activity in southern grasses measured by yields, drought symptoms and P^{32} uptake. *Agronomy Journal* 46:229–233.

Burton, G. W., R. N. Gates, and G. M. Hill. 1993. Registration of 'Tifton 85' bermudagrass. *Crop Science* 33:644–645.

Burton, G. W., and W. W. Hanna. 1995. Bermudagrass. In *Forages: The science of grassland and agriculture*, ed. H. D. Hughes, M. E. Heath, and D. S. Metcalfe. Ames: Iowa State Univ. Press.

Burton, G. W., and R. H. Hart. 1967. Use of self-incompatibility to produce commercial seed-propagated F_1 bermudagrass hybrids. *Crop Science* 7:524–527.

Burton, G. W., R. H. Hart, and R. S. Lowrey. 1967. Improving forage quality in bermudagrass by breeding. *Crop Science* 7:329–332.

Burton, G. W., J. E. Jackson, and F. E. Knox. 1959. The influence of light reduction upon the production, persistence and chemical composition of Coastal bermudagrass, *Cynodon dactylon. Agronomy Journal* 51:537–542.

Burton, G. W., and W. G. Monson. 1978. Registration of Tifton 44 bermudagrass. *Crop Science* 18:911.

———. 1988. Registration of 'Tifton 78' bermudagrass. *Crop Science* 28:187–188.

Burton, G. W., G. M. Prine, and J. E. Jackson. 1957. Studies of drought tolerance and water use of several southern grasses. *Agronomy Journal* 49:498–503.

Byers, R. A., and H. D. Wells. 1966. Phytotoxemia of Coastal bermudagrass caused by the two lined spittlebug, *Prosapia bicincta* (Homoptera: Cercopidae). *Annals of Entomology Society of America* 59:1067–1071.

Caetano-Anolles, G. 1998a. DNA analysis of turfgrass genetic diversity. *Crop Science* 38:1415–1424.

———. 1998b. Genetic instability of bermudagrass (*Cynodon*) cultivars Tifgreen and Tifdwarf detected by DAF and ASAP analysis of accessions and off-types. *Euphytica* 101:165–173.

———. 1999. High genome-wide mutation rates in vegetatively propagated bermudagrass. *Molecular Ecology* 8:1211–1221.

Caetano-Anolles, G., B. J. Bassam, and P. M. Gresshoff. 1991. DNA amplification fingerprinting using very short arbitrary oligonucleotide primers. *Bio Technology* 9:553–557.

Caetano-Anolles, G., L. M. Callahan, and P. M. Gresshoff. 1997. The origin of bermudagrass (*Cynodon*) off-types inferred by DNA amplification fingerprinting. *Crop Science* 37:81–87.

Caetano-Anolles, G., L. M. Callahan, P. E. Williams, K. R. Weaver, and P. M. Gresshoff. 1995. DNA amplification fingerprinting analysis of bermudagrass (*Cynodon*): Genetic relationships between species and interspecific crosses. *Theoretical and Applied Genetics* 91:228–235.

Chapman, H. D., W. H. Marchant, G. W. Burton, W. G. Monson, and P. R. Utley. 1971. Performance of steers grazing Pensacola bahia, Coastal, and Coastcross-1 bermudagrasses. *Journal of Animal Science* 32:374–374.

Chaudhury, A., and R. Qu. 2000. Somatic embryogenesis and plant regeneration of turf-type bermudagrass: Effects of 6-benzyladenine in callus induction medium. *Plant Cell, Tissue and Organ Culture* 60:113–120.

Chen, L., C. Auh, P. Dowling, J. Bell, F. Chen, A. Hopkins, R. A. Dixon, and Z. Y. Wang. 2003. Improved forage digestibility of tall fescue (*Festuca arundinacea*) by transgenic down-regulation of cinnamyl alcohol dehydrogenase. *Plant Biotechnology Journal* 1:437–449.

Chen, L., C. Auh, P. Dowling, J. Bell, D. Lehmann, and Z. Y. Wang. 2004. Transgenic down-regulation of caffeic acid *O*-methyltransferase (COMT) led to improved digestibility in tall fescue (*Festuca arundinacea*). *Functional Plant Biology* 31:235–245.

Chen, T. M., R. H. Brown, and C. C. Black. 1969. CO2 compensation concentration, rate of photosynthesis, and carbonic anhydrase activity of plants. *Weed Science* 18:399–403.

Chheda, H. R., and K. M. Rawal. 1971. Phylogenetic relationships in *Cynodon*. 1. *C. aethiopicus, C. nlemfuensis* and *C. transvaalensis. Phyton* 28:15–21.

Chippindall, L. K. A. 1955. A guide to the identification of grasses in South Africa. In *The grasses and pastures of South Africa*, part 1, ed. D. Meredith. New York: Hafner Publishing Company.

Clatworthy, J. N. 1985. Pasture research in Zimbabwe: 1964–84. In *Pasture improvement research in eastern and southern Africa*, ed. J. A. Kategile, 25. Proceedings of 1st PANESA Workshop, September 17–21, 1984, Harare, Zimbabwe.

Clayton, W. D., and J. R. Harlan. 1970. The genus *Cynodon* L.C. Rich. in tropical Africa. *Kew Bulletin* 24:185–189.

Clayton, W. D., K. T. Harman, and H. Williamson. 2006. *GrassBase—The online world grass flora*, http://www.kew.org/data/grasses-db.html [accessed 08 November 2006; 15:30 GMT].

Clayton, W. D., and S. A. Renvoize. 1986. *Genera Graminum: Grasses of the world.* London: Her Majesty's Stationery Office.

Cooper, R. B., and G. W. Burton. 1965. Forage and turf potential of giant bermudagrass in the southeastern United States. *Agronomy Journal* 57:239–240.

Couch, H. B. 1995. *Diseases of turfgrasses,* 3rd ed. Malabar, FL: Krieger Publishing Co.

De Gruchy, J. H. B. 1952. Water fluctuation as a factor in the life of the higher plants of a 3300 acre lake in the Permian red beds of central Oklahoma. PhD thesis, Oklahoma State University, Stillwater.

de Silva, P. H. A. U., and R. W. Snaydon. 1995. Chromosome number in *Cynodon dactylon* in relation to ecological conditions. *Annals of Botany* 76:535–537.

de Wet, J. M. J., and J. R. Harlan. 1970. Biosystematics of *Cynodon* L.C. Rich. (Gramineae). *Taxon* 19:565–569.

———. 1971. South African species of *Cynodon* (Gramineae). *Journal of South African Botany* 37 (1): 53–56.

Doss, B. D., D. A. Ashley, and O. L. Bennett. 1960. Effect of soil moisture regimes on root distribution of warm-season forage species. *Agronomy Journal* 52:569–576.

Downton, W. J. S., and E. B. Tregunna. 1968. Carbon dioxide compensation—Its relation to photosynthetic carboxylation reactions, systematics of the Gramineae, and leaf anatomy. *Canadian Journal of Botany* 46:207–215.

Dudeck, A. E., and C. H. Peacock. 1992. Shade and turfgrass culture. In *Turfgrass,* ed. D. V. Waddington, R. N. Carrow, and R. C. Shearman, 269–284. Agronomy monograph no. 32. Madison, WI: ASA-CSSA-SSSA.

————. 1993. Salinity effects on growth and nutrient uptake of selected warm-season turf. *International Turfgrass Society Research Journal* 7:680–686.

Dudeck, A. E., S. Singh, C. E. Giordano., T. A. Nell, and D. B. McConnell. 1983. Effects of sodium chloride on *Cynodon* turfgrasses. *Agronomy Journal* 75:927–930.

Duell, R. W. 1961. Bermudagrass has multiple-leaved nodes. *Crop Science* 1:230–231.

Eichhorn, Jr., M. M., W. M. Oliver, W. B. Hallmark, W. A. Young, A. V. Davis, and B. D. Nelson. 1986. Registration of Grazer bermudagrass. *Crop Science* 26:835–835.

Elder, W. C. 1953. Bermudagrass (*Cynodon dactylon*). Forage Crops Leaflet No. 14, Oklahoma Agricultural Experiment Station, Stillwater.

————. 1955. Greenfield bermudagrass. Bulletin B-455. Oklahoma Agricultural Experiment Station, Stillwater.

Felder, M. R. 1967. Chromosome associations in triploid *Cynodon* hybrids. MS thesis, Oklahoma State University, Stillwater.

Fields, J. 1902. Bermudagrass. Bulletin No. 55, Oklahoma Agricultural Experiment Station, Stillwater.

Forbes, I., Forbes, I. Jr., and G. W. Burton. 1963. Chromosome numbers and meiosis in some *Cynodon* species and hybrids. *Crop Science* 3:75–79.

Francois, L. E. 1988. Salinity effects on three turf bermudagrasses. *HortScience* 23:706–708.

Gamble, M. D. 1958. Inundation tolerance of bermudagrass. In *Agronomy Abstracts,* 37. Lafayette, IN: ASA-SSSA-CSSA.

Gaussoin, R. E., A. A. Baltensperger, and B. N. Coffey. 1988. Response of 32 bermudagrass clones to reduced light intensity. *HortScience* 23:178–179.

Goldman, J. J., W. W. Hanna, G. H. Fleming, and P. Ozias-Akins. 2004. Ploidy variation among herbicide-resistant bermudagrass plants of cv. TifEagle transformed with the *bar* gene. *Plant Cell Reports* 22:553–560.

Gupta, P. K., and A. K. Srivastava. 1970. Natural triploidy in *Cynodon dactylon* (L.) Pers. *Caryologia* 23:29–35.

————. 1971. Formation of restitution nucleus and chromatin migration in a triploid clone of *Cynodon dactylon* (L.) Pers. *Journal of the Indian Botany Society* 50:132–137.

Hanna, W. W., and G. W. Burton. 1977. Cytological and fertility characteristics of some hybrid bermudagrass cultivars. *Crop Science* 17:243–245.

Hanna, W. W., G. W. Burton, and A. W. Johnson. 1990. Registration of 'Tifton 10' turf bermudagrass. *Crop Science* 30:1355–1356.

Hanna, W., and B. Maw. 2007. Shade-resistant bermudagrass: Research has produced an improved cultivar. *Green Section Record* March–April: 9–11.

Hanson, A. A. 1972. *Grass varieties in the United States*. USDA-ARS Agricultural Handbook 170. Washington, D.C.: U.S. Gov. Printing House.

Harlan, J. R. 1970. *Cynodon* species and their value for grazing and hay. *Herbage Abstract* 40:233–238.

————. 1983. The scope for collection and improvement of forage plants. In *Genetic resources of forage plants*, ed. J. G. McIvor and R. A. Bray, 3–14. Melbourne, Australia: Commonwealth Scientific and Industrial Research Organization.

Harlan, J. R., G. W. Burton, and W. C. Elder. 1954. Midland bermudagrass, a new variety for Oklahoma pastures. Bull. B-416, Oklahoma Agricultural Experiment Station.

Harlan, J. R., and J. M. J. de Wet. 1969. Sources of variation in *Cynodon dactylon* (L.) Pers. *Crop Science* 9:774–778.

Harlan, J. R., J. M. J. de Wet, W. W. Huffine, and J. R. Deakin. 1970. A guide to the species of *Cynodon* (Gramineae). Bulletin B-673, Oklahoma Agricultural Experiment Station.

Harlan, J. R., J. M. J. de Wet, and K. M. Rawal. 1970a. Geographic distribution of the species of *Cynodon* L.C. Rich. *East African Agricultural and Forestry Journal* 36:220–226,

————. 1970b. Origin and distribution of the seleucidus race of *Cynodon dactylon* (L.) Pers. var. *dactylon* (Gramineae). *Euphytica* 19:465–469.

Harlan, J. R., J. M. J. de Wet, K. M. Rawal, M. R. Felder, and W. L. Richardson. 1970. Cytogenetic studies in *Cynodon* L. C. Rich (Gramineae). *Crop Science* 10:288–291.

Harlan, J. R., J. M. J. de Wet, and W. L. Richardson. 1969. Hybridization studies with species of *Cynodon* from East Africa and Malagasy. *American Journal of Botany* 56:944–950.

Ho, C. Y., S. J. McMaugh, A. N. Wilton, I. J. McFarlane, and A. G. Mackinlay. 1997. DNA amplification variation within cultivars of turf-type couch grasses (*Cynodon* spp.). *Plant Cell Reports* 16:797–801.

Hoff, B. J. 1967. The cytology and fertility of bermudagrass, *Cynodon dactylon* (L.) Pers., in Arizona. *Dissertation Abstracts* 2227–B.

Holt, E. C., and F. L. Fisher. 1960. Root development of Coastal bermudagrass with high nitrogen fertilization. *Agronomy Journal* 52:593–596.

Hu, F., L. Zhang, X. Wang, J. Ding, and D. Wu. 2005. Agrobacterium-mediated transformed transgenic triploid bermudagrass (*Cynodon dactylon* X *C. transvaalensis*) plants are highly resistant to the glufosinate herbicide Liberty. *Plant Cell, Tissue and Organ Culture* 83:13–19.

Hurcombe, R. 1947. A cytological and morphological study of cultivated *Cynodon* species. *Journal of South African Botany* 13:107–116.

Iriarte, F. B., H. C. Wetzel, J. D. Fry, D. L. Martin, and N. A. Tisserat. 2004. Genetic diversity and aggressiveness of *Ophiosphaerella korrae*, a cause of spring dead spot of bermudagrass. *Plant Disease* 88:1341–1346.

Jain, M., K. Chengalrayan, M. Gallo-Meagher, and P. Mislevy. 2005. Embryogenic callus induction and regeneration in a pentaploid hybrid bermudagrass cv. Tifton 85. *Crop Science* 45:1069–1072.

Jiang, Y. W., R. R. Duncan, and R. N. Carrow. 2004. Assessment of low light tolerance of seashore paspalum and bermudagrass. *Crop Science* 44:587–594.

Johnston, R. A. 1975. Cytogenetics of some hexaploid × tetraploid hybrids in *Cynodon*. MS thesis, Oklahoma State University, Stillwater.

Juska, F. V., and A. A. Hanson. 1964. Evaluation of bermudagrass varieties for general purpose turf. USDA-ARS, Agricultural handbook no. 270, Washington, D.C.

Karaca, M., S. Saha, A. Zipf, J. N. Jenkins, and D. J. Lang. 2002. Genetic diversity among forage bermudagrass (*Cynodon* spp.): Evidence from chloroplast and nuclear DNA fingerprinting. *Crop Science* 42:2118–2127.

Kenna, M. P., C. M. Taliaferro, and W. L. Richardson. 1983. Comparative fertility and seed yields of parental bermudagrass clones and their single cross F_1 and F_2 populations. *Crop Science* 23:1133–1135.

Kenworthy, K. E., C. M. Taliaferro, B. F. Carver, D. L. Martin, J. A. Anderson, and G. E. Bell. 2006. Genetic variation in *Cynodon transvaalensis* Burtt-Davy. *Crop Science* 46:2376–2381.

King, L. D., J. C. Burns, and P. W. Westerman. 1990. Long-term swine lagoon effluent applications on Coastal bermudagrass: II. Effect on nutrient accumulation in soil. *Journal of Environmental Quality* 19:756–760.

Kneebone, W. R. 1966a. Bermuda grass—Worldly, wily, wonderful weed. *Economic Botany* 20:94–97.

———. 1966b. Genotypic and environmental variances of seed yield components in bermudagrass *Cynodon dactylon*. In *Agronomy Abstracts,* ASA, CSSA, SSSA Annual Meetings, American Society of Agronomy, Madison, WI, 9.

———.1973. Breeding seeded varieties of bermudagrass for turfgrass use. *Proceedings of Scotts Turfgrass Research Conference* 4:149–153.

Krans, J. V., J. B. Beard, and J. F. Wilkinson. 1979. Classification of C3 and C4 turfgrass species based on CO_2 compensation concentration and leaf anatomy. *HortScience* 14:183–185.

Langdon, R. F. N. 1954. The origin and distribution of *Cynodon dactylon* (L.) Pers. In *University of Queensland papers*, III, 42–44. Brisbane, Australia: University of Queensland Press.

Leuck, D. B., C. M. Taliaferro, G. W. Burton, R. L. Burton, and M. C. Bowman. 1968. Resistance in bermudagrass to the fall armyworm. *Journal of Economic Entomology* 61:1321–1322.

Li, L., R. Li, S. Fei, and R. Qu. 2005. Agrobacterium-mediated transformation of common bermudagrass (*Cynodon dactylon*). *Plant Cell, Tissue and Organ Culture* 83:223–229.

Li, L., and R. Qu. 2002. In vitro somatic embryogenesis in turf-type bermudagrass: Roles of abscisic acid and gibberellic acid and occurrence of secondary somatic embryogenesis. *Plant Breeding* 121:155–158.

———. 2004. Development of highly regenerable callus lines and biolistic transformation of turf-type common bermudagrass [*Cynodon dactylon* (L.) Pers.]. *Plant Cell Reports* 22:403–407.

Liu, J. X., S. H. He, and Y. D. Liu. 2006. Registration of 96-C-106 germplasm clone of bermudagrass. *Crop Science* 46:2341–2341.

Lundberg, P. E., O. L. Bennett, and E. L. Mathias. 1977. Tolerance of bermudagrass selections to acidity. *Agronomy Journal* 69:913–916.

Lynch, R. E., and G. W. Burton. 1993. Relative abundance of insects on bermudagrasses and bahiagrasses. *Journal of Entomological Science* 29:120–129.

Lynch, R. E., W. G. Monson, B. R. Wiseman, and G. W. Burton. 1983. Bermudagrass resistance to the fall armyworm (Lepidoptera: Noctuidae). *Environmental Entomology* 12:1837–1840.

Malik, C. P., and R. C. Tripathi. 1968. Cytological evolution within the *Cynodon dactylon* complex. *Biologisches Zentral blatt B* 87:625–627.

Marcum, K. B. 2000. Growth and physiological adaptations of grasses to salinity stress. In *Handbook of plant and crop physiology,* 2nd ed., ed. M. Pessaraki. New York: Marcel Dekker, Inc.

Marcum, K. B., and M. Pessarakli. 2006. Salinity tolerance and salt gland excretion efficiency of bermudagrass turf cultivars. *Crop Science* 46:2571–2574.

Martin, D. L., G. E. Bell, J. H. Baird, C. M. Taliaferro, N. A. Tisserat, R. M. Kuzmic, D. D. Dobson, and J. A. Anderson. 2001. Spring dead spot resistance and quality of seeded bermudagrasses under different mowing heights. *Crop Science* 41:451–456.

McBee, G. G., and E. C. Holt. 1966. Shade tolerance studies on bermudagrass and other turfgrasses. *Agronomy Journal* 58:523–525.

Mislevy, P., W. G. Blue, J. A. Stricker, B. C. Cook, and M. J. Vice. 2000. Phosphate mining and reclamation. In *Reclamation of drastically disturbed lands,* ed. R. I. Barnhisel, 961–1005. Agronomy Monograph 41. Madison, WI: ASA, CSSA, and SSSA.

Mislevy, P., W. F. Brown, R. Caro-Costas, J. Vicente-Chandler, L. S. Dunavin, D. W. Hall, R. S. Kalmbacher, et al. 1989. Florico stargrass. Circ. S-361, Florida Agricultural Experiment Station, Gainesville.

Mislevy, P., W. F. Brown, L. S. Dunavin, D. W. Hall, R. S. Kalmbacher, A. J. Overman, O. C. Ruelke, et al. 1989. Florona stargrass, Circ. S-362, Florida Agricultural Experiment Station, Gainesville.

Mislevy, P., W. F. Brown, R. S. Kalmbacher, L.S. Dunavin, W. S. Judd, T. A. Kucharek, O. C. Ruelke, et al. 1999. Registration of 'Florakirk' bermudagrass. *Crop Science* 39:587–587.

Moffett, A. A., and R. Hurcombe. 1949. Chromosome number of South African grasses. *Heredity* 3:369–373.

Moorhouse, L. A., W. L. Burlison, and J. A. Ratcliff. 1909. Bermuda grass. Bulletin no. 85, Oklahoma Agricultural Experiment Station, Stillwater.

NPGS (National Plant Germplasm System). 2007. http://www.ars-grin.gov/cgi-bin/npgs/html/stats/genus.pl?C*

Ourecky, D. K. 1963. Pachytene chromosome morphology in *Cynodon dactylon* (L.) Pers. *Nucleus* 6:63–82.

Pitman, W. D., S. S. Croughan, and M. J. Stout. 2002. Field performance of bermudagrass germplasm expressing somaclonal variation selected for divergent responses to fall armyworm. *Euphytica* 125:103–111.

Powell, J. B., G. W. Burton, and C. M. Taliaferro. 1968. A hexaploid clone from a tetraploid × diploid cross in *Cynodon. Crop Science* 8:184–185.

Pratt, R. G. 2000. Diseases caused by dematiaceous fungal pathogens as potential limiting factors for production of bermudagrass on swine effluent application sites. *Agronomy Journal* 92:512–517.

———. 2005. Variation in occurrence of dematiaceous hyphomycetes on forage bermudagrass over years, sampling times, and locations. *Phytopathology* 95:1183–1190.

Pratt, R. G., and G. E. Brink. 2007. Forage bermudagrass cultivar responses to inoculations with *Exserhilum rostratum* and *Bipolaris spicifera* and relationships to field persistence. *Crop Science* 47:239–244.

Rawal, K. M., and H. R. Chheda. 1971. Phylogenetic relationships in *Cynodon*. II. *C. nlemfuensis, C. dactylon* vars. *Afghanicus* and *aridus. Phyton* 28:121–130.

Redfearn, D. D., and C. J. Nelson. 2003. Grasses for southern areas. In *Forages: An introduction to grassland agriculture,* 6th ed., ed. R. F. Barnes, C. J. Nelson, M. Collins, and K. J. Moore. Ames: Iowa State Univ. Press.

Reed, B. M., L. Schumacher, N. Wang, J. D'Achino, and R. E. Barker. 2005. Cryopreservation of bermudagrass germplasm by encapsulation dehydration. *Crop Science* 46:6–11.

Rethwisch, M. D., E. T. Natwick, B. R. Tickes, M. Meadows, and D. Wright. 1995. Impact of insect feeding and economics of selected insecticides on early summer bermudagrass seed production in the desert southwest. *Southwestern Entomologist* 20:187–201.

Richardson, W. M., C. M. Taliaferro, and R. M. Ahring. 1978. Fertility of eight bermudagrass clones and open-pollinated progeny from them. *Crop Science* 18:332–334.

Rochecouste, E. 1962. Studies on the biotypes of *Cynodon dactylon* (L.) Pers.: I. botanical investigations. *Weed Research* 2:1–23.

Roodt, R., J. J. Spies, and T. H. Burger. 2002. Preliminary DNA fingerprinting of the turf grass *Cynodon dactylon* (Poaceae: Chloridoideae). *Bothalia* 32:117–122.

Salehi, H., Z. Seddighi, A. N. Kravchenko, and M. B. Sticklen. 2005. Expression of the cry1Ac in 'Arizona Common' common bermudagrass via Agrobacterium-mediated transformation and control of black cutworm. *Journal of American Society of Horticultural Science* 130:619–623.

Smiley, R. W., P. H. Dernoeden, and B. B. Clarke. 1993. *Compendium of turfgrass diseases,* 2nd ed. St. Paul, MN: American Phytopathological Society.

Stimmann, M. W., and C. M. Taliaferro. 1969. Resistance of selected accessions of bermudagrass to phytotox-emia caused by adult two-lined spittlebugs. *Journal of Economic Entomology* 62:1189–1190.

Taliaferro, C. M. 1992. Out of Africa—A new look at "African" bermudagrass. *USGA Green Section Record* 30:10–12.

———. 1995. Diversity and vulnerability of Bermuda turfgrass species. *Crop Science* 35:327–332.

———. 2003. Bermudagrass. In *Turfgrass biology, genetics and breeding*, ed. M. D. Casler and R. Duncan, 235–256. New York: John Wiley & Sons.

Taliaferro, C. M., R. M. Ahring, and W. L. Richardson. 1983. Registration of Guyman bermudagrass. *Crop Science* 23:1219–1219.

Taliaferro, C. M., J. A. Anderson, W. L. Richardson, J. L. Baker, S. W. Coleman, W. A. Phillips, L. J. Sandage, et al. 2002. Registration of Midland 99 forage bermudagrass. *Crop Science* 42:2212–2213.

Taliaferro, C. M., A. A. Hopkins, J. C. Henthorn, C. D. Murphy, and R. M. Edwards. 1997. Use of flow cytom-etry to estimate ploidy level in *Cynodon* species. *International Turfgrass Society Research Journal* 8:385–392.

Taliaferro, C. M., and J. T. Lamle. 1997. Cytological analysis of self-incompatibility in *Cynodon dactylon* (L.) Pers. *International Turfgrass Society Research Journal* 8:393.

Taliaferro, C. M., D. B. Leuck, and M. W. Stimmann. 1969. Tolerance of *Cynodon* clones to phytotoxemia caused by the two-lined spittlebug. *Crop Science* 9:765–767.

Taliaferro, C. M., D. L. Martin, J. A. Anderson, M. P. Anderson, G. E. Bell, and A. C. Guenzi. 2003. Registration of Yukon bermudagrass. *Crop Science* 43:1131–1132.

Taliaferro, C. M., D. L. Martin, J. A. Anderson, M. P. Anderson, and A. C. Guenzi. 2004. Broadening the hori-zons of turf bermudagrass. *USGA Turfgrass and Environmental Research Online* 3 (2): 1–9.

Taliaferro, C. M., and W. L. Richardson. 1980a. Registration of Hardie bermudagrass. *Crop Science* 20:413.

———. 1980b. Registration of Oklan bermudagrass. *Crop Science* 20:414.

Taliaferro, C. M., W. L. Richardson, and R. M. Ahring. 1977. The Oklahoma bermudagrass breeding program: Objectives and approaches. In *Proceedings of the 34th Southern Pasture and Forage Crop Improvement Conference*, April 12–14, Auburn University, Auburn, Alabama, 113–119.

Taliaferro, C. M., F. M. Rouquette, Jr., and P. Mislevy. 2004. Bermudagrass and stargrass. In *Warm-season (C4) grasses*, ed. L. E. Moser, chap. 12. Agronomy Monograph 45. Madison, WI: ASA, CSSA, and SSSA.

Thomas, S. M., and B. G. Murray. 1978. Herbicide tolerance and polyploidy in *Cynodon dactylon* (L.) Pers. (Gramineae). *Annals of Botany* 42:137–143.

Tracy, S. M. 1917. Bermuda grass. *Farmer's Bulletin 814,* United States Department of Agriculture.

Vos, P., R. Hogers, M. Bleeker, M. Reijans, T. van de Lee, M. Hornes, A. Frijters, et al. 1995. AFLP: A new technique for DNA fingerprinting. *Nucleic Acids Research* 23:4407–4414.

Wang, Z. Y., and Y. Ge. 2005. Rapid and efficient production of transgenic bermudagrass and creeping bent-grass bypassing the callus formation phase. *Functional Plant Biology* 32:769–776.

———. 2006. Recent advances in genetic transformation of forage and turf grasses. *In Vitro Cellular Development and Biology—Plant* 42:1–18.

Watson, L., and M. J. Dallwitz. 1992. *The grass genera of the world.* Cambridge, Great Britain: C. A. B. International, Univ. Press.

Williams, J. G. K., A. R. Kubelik, K. J. Livak, J. A. Rafalski, and S. V. Tingey. 1990. DNA polymorphisms amplified by arbitrary primers are useful as genetic markers. *Nucleic Acids Research* 18:6531–6535.

Wu, Y. Q., C. M. Taliaferro, G. H. Bai, and M. P. Anderson. 2004. AFLP analysis of *Cynodon dactylon* (L.) Pers. var. *dactylon* genetic variation. *Genome* 47:689–696.

———. 2005. Genetic diversity of *Cynodon transvaalensis* Burtt-Davy and its relatedness to hexaploid *C. dactylon* (L.) Pers. as indicated by AFLP markers. *Crop Science* 45:848–853.

Wu, Y. Q., C. M. Taliaferro, G. H. Bai, D. L. Martin, J. A. Anderson, M. P. Anderson, and R. M. Edwards. 2006. Genetic analyses of Chinese *Cynodon* accessions by flow cytometry and AFLP markers. *Crop Science* 46:917–926.

Wu, Y. Q., C. M. Taliaferro, D. L. Martin, C. L. Goad, and J. A. Anderson. 2006. Genetic variability and relationships for seed yield and its components in Chinese *Cynodon* accessions. *Field Crops Research* 98:245–252.

Yaneshita, M., T. Ohmura, T. Sasakuma, and Y. Ogihara. 1993. Phylogenetic relationships of turfgrasses as revealed by restriction fragment analysis of chloroplast DNA. *Theoretical and Applied Genetics* 87:129–135.

Yerramsetty, P. N., M. P. Anderson, C. M. Taliaferro, and D. L. Martin. 2005. DNA fingerprinting of seeded bermudagrass cultivars. *Crop Science* 45:772–777.

Zhang, G., S. Lu, T. A. Chen, C. R. Funk, and W. A. Meyer. 2003. Transformation of triploid bermudagrass (*Cynodon dactylon* × *C. transvaalensis* cv. TifEagle) by means of biolistic bombardment. *Plant Cell Reports* 21:860–864.

Zhang, L.-H., P. Ozias-Akins, G. Kochert, S. Kresovich, R. Dean, and W. Hanna. 1999. Differentiation of bermudagrass (*Cynodon* spp.) genotypes by AFLP analyses. *Theoretical and Applied Genetics* 98:895–902.

Zheng, Y. H., J. X. Liu, and S. Y. Chen. 2003. Studies on rhizome characteristics of germplasm resources of *Cynodon dactylon* in China. *Acta Prataculturae Sinica* 12:76–81 (in Chinese with English abstract).

CHAPTER 9

Ryegrass

Scott Warnke, Reed E. Barker, and Geunhwa Jung

CONTENTS

9.1 INTRODUCTION

The ryegrasses are a very diverse group of species used throughout the world for both turf and forage purposes. In the United States, the two primary cultivated types are annual or Italian ryegrass and perennial ryegrass. Both species are used by the turf and forage industries; however, the vast majority of seed is used for overseeding of warm-season turfgrasses. This chapter covers aspects of *Lolium* taxonomy and genetic mapping. Readers are encouraged also to examine a recent chapter by Thorogood (2002) that contains an excellent discussion of conventional breeding aspects of this important cool-season grass genus.

9.2 TAXONOMY

Linnaeus classified the ryegrasses to the genus *Lolium* of the Poeae tribe in the Gramineae family. Essad (1954) recognized five *Lolium* species grouped as allogamous (*L. perenne* L., *L. multiflorum* Lam., and *L. rigidum* Gaud.) or autogamous (*L. temulentum* L. and *L. remotum* Schrank.).

Based on an extensive examination of plant morphologies from widely collected herbaria specimens and living plants, Terrell (1968) revised the genus by adding *L. persicum* Boiss. & Hohen. ex Boiss. as the possible prototype of the autogamous group and *L. subulantum* Vis. and *L. canariense* Steud. to the allogamous group.

Probably the most studied of the grasses used worldwide for forage and turf, the taxonomic classification of the *Lolium* genus and its species seems still to be evolving (the genus arising some 6,000 years ago). The ryegrasses are closely related to the more diverse fescues. In fact, in recent years the subgenus *Schedonorus* of the fescues that include tall fescue and meadow fescue has been placed in the ryegrass taxonomic grouping (Darbyshire 1993). In some taxonomic databases, the currently accepted alignment is (USDA, NRCS. 2004. The PLANTS Database, Version 3.5 <http://plants.usda.gov>. National Plant Data Center, Baton Rouge, LA 70874-4490 USA):

> *Lolium arundinaceum* (Schreb.) S. J. Darbyshire—tall fescue
> *Lolium × festucaceum* Link—*Festuca pratensis × Lolium perenne*
> *Lolium giganteum* (L.) S. J. Darbyshire—giant fescue
> *Lolium perenne* L.—perennial ryegrass
>> *Lolium perenne* L. ssp. *multiflorum* (Lam.) Husnot—Italian ryegrass
>>> *Lolium perenne* L. ssp. *multiflorum* var. *italicum*
>>> *Lolium perenne* L. ssp. *multiflorum* var. *westerwoldicum*—Westerwolds
>>> *Lolium perenne* L. ssp. *perenne*—perennial ryegrass
> *Lolium persicum* Boiss. & Hohen. ex Boiss.—Persian ryegrass
> *Lolium pratense* (Huds.) S. J. Darbyshire—meadow ryegrass
> *Lolium rigidum* Gaudin—Wimmera ryegrass
> *Lolium temulentum* L.—Darnel ryegrass

Although we prefer the preceding alignment with *Lolium* for practical breeding and genomic relatedness reasons, *Schedonorus* was raised to its own genus level (Holub, 1998a). However, the species designations remain somewhat disputed (Soreng et al. 2001) and the hybrids with *Lolium* designated × *Schedololium* (Soreng and Terrell 1997; Holub 1998b).

Of the traditional species classified as ryegrass, four are of agricultural economic importance. Perennial ryegrass, Italian ryegrass (including the Westerwolds), and Wimmera ryegrass are outcrossing grasses that are self-incompatible and readily interbreed. Because of this intermixing, there is continuous variation among these three ryegrasses, and their growth types overlap (Figure 9.1). The fourth species of concern is Darnel ryegrass, a serious weed.

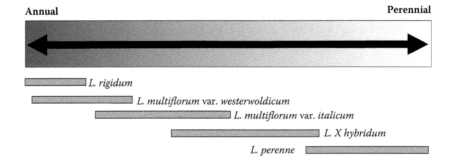

Figure 9.1 Diagrammatic representation of growth type among the main outbreeding ryegrasses. Types range from truly annual to fully perennial. This relationship is based on reported intercrossing, plant morphology, and molecular data.

9.3 DESCRIPTION OF THE FOUR MAIN RYEGRASSES

9.3.1 Perennial Ryegrass (*L. perenne* ssp. *perenne*)

This grass is widely used as a turf and as a high-quality forage grass. The inflorescence is a spike with alternately arranged spikelets attached edgewise directly to the central axis. Leaves of perennial ryegrass are usually folded in the bud with blades that are bright green, prominently ridged on the upper surface, and sharply taper pointed.

9.3.2 Italian Ryegrass (*L. perenne* ssp. *multiflorum*)

This grass is also known as annual ryegrass in the United States. It should not be confused with the true annual, Wimmera ryegrass (*L. rigidum*). Two types of Italian ryegrass are botanically the same: the widespread Italian type—a short-lived perennial that can survive 2–3 years in some environments—and the Westerwolds, with a growth type more closely that of an annual. The Westerwolds came from the Westerwolde area in the province of Groningen, the Netherlands, as an ecotype selected for earliness. The Italian type was reported in France in 1818 and is the more common of the two types grown in the United States. It is a fast-growing, competitive winter annual cool-season grass (and thus commonly recognized as annual). It is used as a cover crop, as a forage, and as a winter pasture and golf course overseeder in southern U.S. climates. The inflorescence is a solitary spike with alternately arranged spikelets attached edgewise directly to the central axis. Leaf blades are bright green, rolled in the bud, and sharply taper pointed.

9.3.3 Wimmera, Annual, or Rigid Ryegrass (*L. rigidum*)

Although it is used as a forage crop in Australia, wimmera, annual, or rigid ryegrass (*L. rigidum*) is considered a weed in many places in the world. In the United States, it is primarily found in waste places in states on the West Coast. It is a highly mutable grass, rapidly developing resistance and cross-resistance to many herbicides. Wimmera ryegrass is a true annual that has a dominant upright growth habit with rigid stems, sprawling after maturity. Leaves are shiny below and evenly ribbed above; they are broader than those of perennial ryegrass and rolled in bud on young plants. Plants have a general purplish tinge with a reddish coloration common at the base of the stem. The inflorescence is a spike similar to perennial ryegrass but usually longer and with a lower number of spikelets more widely spaced. Six- to eight-flowered spikelets are generally longer and narrower than those of perennial ryegrass, with only one outer glume, which is much shorter than the spikelet but longer than the lemma above it. The lemma is long, rigid, and awnless, although it is occasionally awned.

9.3.4 Darnel or Poison Ryegrass (*L. temulentum*)

Darnel or poison ryegrass (*L. temulentum*) is a mimic weed (thought to be the "tare" of the Bible) and is reputed to be poisonous. The species grows in grain fields and waste places in most areas of the United States and is listed as a noxious weed in 43 states. It usually has long awns, is self-pollinated, and does not naturally cross with other ryegrasses.

9.4 SPECIES SEPARATION

Because of its use as a high-quality turfgrass, there has been considerable interest in distinguishing perennial ryegrass from Italian ryegrass. Differences between the two types are trivial at

Table 9.1   Selected Taxonomic Characteristics of Ryegrass That Are
             Different for Perennial and Italian Ryegrass

	Species	
Taxonomic Descriptors	Perennial Ryegrass	Italian Ryegrass
Spikelet flower numbers	Low	High
Awn	Typically absent	Typically present
Plant height	Short	Tall
Longevity	Perennial	Annual or perennial
Leaf blade width	Narrow	Wide
Leaf blade length	Short	Long
Leaf venation	Folded	Rolled
Auricles	Present (short) or absent	Usually present (often long)
Ligules	Truncate or erose	Rounded, truncate, or erose
Root fluorescence	Typically nonfluorescent	Typically fluorescent

Source: Jung, G. A. et al. 1996. In Cool-season forage grasses, ed. L. E. Moser, D. R.
Buxton, and M. D. Casler, 605–641. Madison, WI: ASA, CSSA, and SSSA.

best as seed, and the two can be distinguished only by a number of traits as the plants grow. Jung et al. (1996) provided a list of selected taxonomic characteristics that are different between the two species (Table 9.1). Each of the traits listed has continuous variation and overlaps the other species in distribution. No one trait, not even the annual characteristic itself, provides distinctiveness. The close genetic similarity of these two species is of concern to seed certification agencies because genetic or physical contamination of turf-type perennial ryegrass by forage-type Italian ryegrass results in an objectionable turf.

U.S. seed regulatory agencies utilize the seedling root fluorescence test to identify Italian ryegrass contamination of perennial ryegrass seed lots. The seedling root fluorescence test is based on the finding that seedling roots of Italian ryegrass secrete an alkaloid compound called annuloline that produces a blue fluorescence under ultraviolet light; however, perennial ryegrass roots do not normally fluoresce (Gentner 1929). Since Genter's discovery, the seedling root fluorescence test has been used to determine the purity of perennial ryegrass seed samples. Nyquist (1963) was able to develop fluorescent perennial ryegrass, indicating that the fluorescent trait does not influence growth habit and is not tightly linked to any trait that conditions annuality.

The search for alternative tests for seedling root fluorescence has resulted in the identification of several traits associated with the Italian or perennial growth type, but none has been developed into a suitable alternative to seedling root fluorescence. A seed esterase isoenzyme (Nakamura 1979; Payne, Scott, and Koszykowski 1980; Griffith and Banowetz 1992), like the electrophoresis of other general seed proteins and isoenzymes, was able to detect species differences; however, these associations were based on bulked samples and lacked the necessary sensitivity to be an adequate replacement test for seedling root fluorescence (Larsen 1966; Ferguson and Grabe 1984). A phosphoglucose isomerase locus (Pgi-2) and a superoxide dismutase locus (Sod-1) have shown allele frequency differences between perennial and Italian ryegrass; the Sod-1 locus has been studied as a possible replacement of or supplement to the seedling root fluorescence test (Warnke et al. 2002). Approximately 99% of cultivated perennial ryegrass is homozygous for the Sod-1b allele, and this allele is homozygous in only about 20% of annuals. This percentage is lower if the cultivar Gulf is not considered. Unfortunately, Gulf is commonly grown in the Willamette Valley seed production area of Oregon. A sod test like the seedling root fluorescence test can be used as an indicator of potential seed lot contamination.

9.5 CYTOGENETICS

All of the ryegrasses are diploids with $2n = 2x = 14$ with disomic inheritance (Evans 1926). The allogamous species readily interbreed and, as a group, were described once as having been a "hybrid swarm" with a continuum from the annual types to the perennial types at the two extremes (Tyler et al. 1987). Even now that the members have emerged as species, they readily intermate with each other (Jenkin 1933, 1935, 1953, 1954a, 1954b, 1955; Jenkin and Thomas 1939; Charmet, Balfourier, and Chatard 1996). Induced autotetraploids have been developed in perennial and Italian ryegrasses as commercial cultivars for forage use.

Intergeneric hybrids have been produced with a number of species from the Bovinae section of the *Festuca* genus. More information on intergeneric hybrids as well as a discussion of *L. perenne-F. pratensis* substitution lines can be found in Thorogood (2002).

Recent advancement of DNA marker technology has led to the development of genetic maps in many cereal crops and also in ryegrass during the past decade. Genetic maps have unveiled inheritance of many important complex traits within the same taxa through quantitative trait loci (QTLs) mapping analysis. They have further enabled the dissection of important traits common among the diverse species within the same families through comparative QTLs mapping because those species share similar gene order over large chromosomal segments.

9.5.1 Genetic Mapping

Genetic maps have been used for crop improvement on the basis of tagging major genes of interest, genetic dissection of agronomically important complex traits, comparative traits mapping within and between genomes of grass species, and gene isolation by map-based cloning. Genetic maps have also been used for an understanding of genome organization. However, the genetic research on ryegrass has been relatively slow compared with other major cereals due to complex genetic systems such as self-incompatibility, outcrossing, and a relatively large genome size. The individual offspring of a ryegrass mapping population are genetically unique due to self-incompatibility and must be clonally propagated because inbred lines and doubled haploids are difficult to produce. Therefore, maintaining mapping populations clonally and exchanging genetic materials among researchers have been major obstacles.

Most commonly used mapping populations for linkage map construction in self-incompatible ryegrass are "pseudotestcross F_1" populations. These populations are derived from a cross between two highly heterozygous individual genotypes. This genetic system has been used in forest trees for many years (Ritter, Gebhardt, and Salamini 1990; Grattapaglia and Sederoff 1994). The disadvantage of this method is the complexity of the genetic analysis due to a mixture of different segregating genotypes in the offspring that contain both BC_1 and F_2 types of segregation. A one-way pseudotest-cross population has also been developed by crossing a multiple heterozygous individual and a doubled haploid genotype (Bert et al. 1999; Jones, Mahoney, et al. 2002). Two-way pseudotestcross populations have been developed through crosses between two heterozygous individuals (Hayward et al. 1998; Warnke et al. 2004; Inoue et al. 2004). In the case of the two-way cross, both parents of a mapping population are highly heterozygous so that up to four different combinations of four alleles can segregate at a given locus.

For example, co-dominant restriction fragment length polymorphism (RFLP) markers can be segregated in one of five segregation types (shown in Figure 9.2A): a heterozygous locus in both parents with four different alleles (Figure 9.2B), a heterozygous locus in both parents with one common allele (Figure 9.2C), a heterozygous locus in both parents with two common alleles (Figure 9.2D), a heterozygous locus in one parent (Figure 9.2E), and a heterozygous locus in the other parent. Linkage phase relationships among alleles are not easily resolved in these population structures,

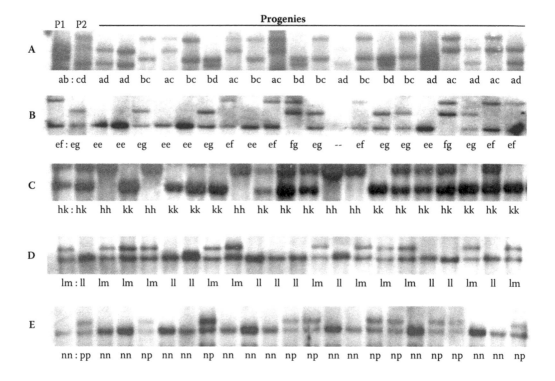

Figure 9.2 Autoradiograms showing the five segregation types of EST-RFLP loci segregating in ryegrass mapping population.

but they can be determined in a three-generation population structure. Warnke et al. (2004) demonstrated the development of a three-generation mapping population (MFA × MFB) by crossing two parent plants randomly selected from each of two F_1 populations, which were derived from two separate crosses between two Italian ryegrass plants from the cultivar Floregon and two perennial ryegrass plants from the cultivar Manhattan. This same population was successfully used for QTL mapping of resistances to gray leaf spot and crown rust (Curley and Jung 2004; Sim et al. 2007) caused by *Magnaporthe grisea* and *Puccinia coronata,* respectively, and provided strong evidence for the resistance source in Italian ryegrass, grandparent of the original mapping population.

During the past decade, various DNA marker systems have been developed for constructing genetic maps of ryegrass. For dominant markers such as amplified fragment length polymorphisms (AFLPs) and random amplified polymorphic DNA (RAPD), two separate data sets are prepared that contain the segregating marker data from each parent, allowing independent linkage maps to be constructed for each parent (Figure 9.3). Co-dominant markers such as EST (expressed sequence tags)-simple sequence repeat (SSR), sequence tagged site (STS), and EST-restriction fragment length polymorphism (RFLP) segregating from both parents can then be used to align linkage groups of the two parent maps and to construct an integrated map representing both parent maps. Because co-dominant marker systems can allow the detection of all alleles at any given locus and are more informative in QTL mapping, they are essentially required for the construction of detailed linkage maps in self-incompatible outcrossing species. EST-RFLP marker-based maps have an additional advantage, in that markers mapped as common between ryegrass and well-studied grasses such as rice, wheat, and oat can be used to align ryegrass linkage groups to those of other published grass maps and to understand syntenic genome relationships between ryegrass and other grasses.

The first low-density ryegrass genetic map was constructed by Hayward et al. (1994, 1998) based on isozyme, RAPD, and RFLP markers. High-density linkage maps of intraspecific population of

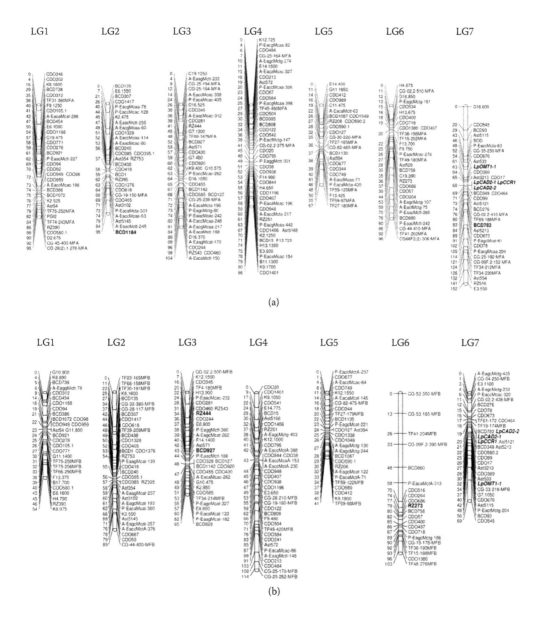

Figure 9.3 Two parent linkage maps (a: MFA parent; b: MFB parent) of the three-generation MFA × MFB population constructed separately using the AFLP, RAPD, RFLP, SSR, isozyme, and morphological markers. (From Sim, S. et al. 2007. *Phytopathology* 97:767–776.)

perennial ryegrass (Bert et al. 1999; Jones, Mahoney, et al. 2002; Jones, Dupal, et al. 2002; Armstead et al. 2002; Muylle et al. 2005; Gill et al. 2006) and of interspecific population between perennial ryegrass and Italian ryegrass (Hayward et al. 1998; Warnke et al. 2004) have been published using various DNA marker types such as isozyme, RAPD, AFLP, RFLP, STS, and SSR. A genetic map of Italian ryegrass has been developed using AFLP, RFLP, and telomeric repeat associated sequence (TAS) markers (Inoue, Gao, Hirata, et al. 2004). In addition, a functionally associated EST-SSR marker-based genetic map of ryegrass has been constructed (Faville et al. 2004). This map will be very useful in comparative genome analysis within and among taxa as well as in the detection of functionally associated markers linked to agronomically important complex traits.

9.5.2 Comparative Genome and QTL Mapping

Genetic linkage mapping and trait mapping using molecular markers have been studied in forage and turf grasses and provide the basis of marker-assisted selection (Hayward et al. 1998; Bert et al. 1999; Forster et al. 2001; Jones, Mahoney, et al. 2002; Curley et al. 2005; Sim et al. 2007) as well as comparative genome analysis with model cereals (Jones, Mahoney, et al. 2002; Sim et al. 2005). Recent studies of comparative genome mapping at the macrosyntenic level suggest that a high degree of genome conservation between ryegrass and other Poaceae species (oat, rice, and wheat) exists (Jones, Mahoney, et al. 2002; Sim et al. 2005) and that the ryegrass genome has a higher syntenic relationship with wheat than oat, despite a closer taxonomic relationship between oat and ryegrass. The conserved synteny observed among genomes of cereal crops and other grass species (Gale and Devos 1998; Devos and Gale 2000) makes it possible to transfer valuable genetic information to "orphan" crops like ryegrass from cereal crops that have more genetic resources available; this can facilitate the identification of candidate genes of agronomic importance in syntenic genome regions of ryegrass.

Effective utilization of important genetic information available in cereal crops will also lead to a better understanding of the genetic architecture of agronomical traits, like disease resistance, that are important targets for genetic manipulation and eventually crop improvement. Once DNA markers tightly linked to disease resistances or function attached markers such as EST-SSR or SNP (simple nucleotide polymorphisms) are found, those markers will be very effective for marker-assisted breeding, particularly in pyramiding multiple resistance genes from different sources into a single cultivar. For this reason, many trait-mapping studies have been conducted in ryegrass. Examples include QTL

resistance to gray leaf spot (*Magnaporthe grisea* (Hebert) Barr) (Curley et al. 2005; Miura et al. 2005) and crown rust (*Puccinia coronata* f. sp. *lolii*) (Dumsday et al. 2003; Fujimori et al. 2003; Hirata et al. 2003; Muylle et al. 2005; Sim et al. 2007; Studer et al. 2007; Schejbel et al. 2007);
forage quality (Humphreys, Turner, and Armstead 2003; Cogan et al. 2005; Xiong et al. 2006);
winter hardiness (Yamada et al. 2004; Xiong et al. 2006);
vernalization response (Jensen et al. 2005);
lodging (Inoue, Gao, and Cai 2004);
heading date (Armstead et al. 2004); and
morphological and developmental traits (Yamada et al. 2004; Brown et al. 2005).

Extensive synteny relationships among rice, oat, and ryegrass have enabled further genetic dissection of common traits, such as resistance to gray leaf spot (Curley and Jung 2004) and crown rust (Dumsday et al. 2003; Sim et al. 2005). For example, Sim et al. (2007) demonstrated that linkage group 7 QTL for crown rust (*P. coronata* f. sp. *Lolii*) in ryegrass is closely located in the syntenic genomic region where genes *Pcr* cluster, *Pcq2*, *Pc38*, and *Prq1b*-resistant crown rust (*P. coronata* f. sp. *avenae*) in oat (*Avena sativa* L.) were previously identified. This suggests that the ortholoci for resistance genes to different *formae speciales* of crown rust might be present between two distantly related grass species (Figure 9.4).

EST-RFLP markers have served as the tool for alignment of existing ryegrass maps with other grass maps; however, development of polymerase chain reaction (PCR)-based co-dominant EST-SSR or SNP markers will be needed in the future. These PCR-based markers are transferable within and beyond diverse taxa; they are likely to be more conserved and are highly polymorphic. The EST-derived markers will also offer the opportunity for gene discovery and increase the value of genetic markers by surveying variation in transcribed gene-rich regions of the genome. Because the rice genome has been sequenced, the ryegrass chromosomal regions syntenous with QTL for traits of interest can be searched for sequences matching their gene motifs in rice. Once identified, the candidate genes from rice can be used to design PCR-based function known markers for ryegrass

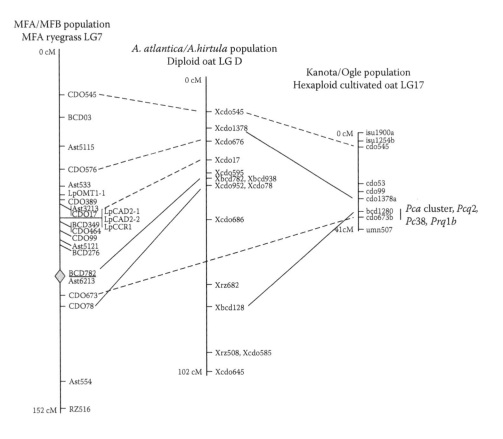

Figure 9.4 Comparative mapping of QTL and major genes for crown rust resistance in ryegrass and oat. The location of ryegrass crown rust QTL was indicated by a diamond shape on linkage group 7 of the MFA × MFB population, and the most significant marker associated with the QTL was underlined. Vertical line with gene names indicates approximate location of the previously reported oat crown rust major resistance genes. The dotted lines connected EST-RFLP markers generated by the same heterologous probes. (From Sim, S. et al. 2007. *Phytopathology* 97:767–776.)

so that individual genes underlying a QTL of interest can be utilized for marker-assisted selection in ryegrass breeding programs

The future development of a genetically anchored physical map of ryegrass chromosomes will be another essential area as more genes underlying agronomically important quantitative trait loci are identified. International efforts will be more desirable for building a relational comprehensive database to hold all of the physical and genetic map data and to display the physical and genetic integrated map with locations of genes/QTL for all genetically dissected traits.

9.6 CONCLUSION

Although the ryegrasses are an extremely important group of species, they are often overlooked when important grasses are considered. However, the ryegrasses are vitally important to both the forage and turfgrass industries, and therefore the genus may be underappreciated. The ryegrasses are complicated genetically because they are primarily outcrossing and have a highly heterozygous genome. However, the fact that ryegrasses are diploid has resulted in a considerable amount of genomics research being conducted with them in comparison to other species utilized as forage and turfgrasses. Molecular markers and genetic linkage mapping are beginning to play a big role in breeding efforts to improve ryegrasses for a number of important traits. Research in this area will undoubtedly continue, and markers will eventually play a big role in the development of future cultivars. The potential of association mapping approaches is just beginning to be explored, and it remains to be determined whether the relatively limited levels of linkage disequilibrium in ryegrass will permit its effective utilization (Skot et al. 2005).

REFERENCES

Armstead, I. P., L. B. Turner, M. Farell, L. Skot, P. Gomez, T. Montoya, et al. 2004. Synteny between a major heading-date QTL in perennial ryegrass (*Lolium perenne* L.) and the *Hd3* heading-date locus in rice. *Theoretical and Applied Genetics* 108:822–828.

Armstead, I. P., L. B. Turner, I. P. King, A. J. Cairns, and M. O. Humphreys. 2002. Comparison and integration of genetic maps generated from F_2 and BC_1-type mapping populations in perennial ryegrass. *Plant Breeding* 121:501–507.

Bert, P. F., G. Charmet, P. Sourdille, M. D. Hayward, and F. Balfourier. 1999. A high-density map for ryegrass (*Lolium perenne*) using AFLP markers. *Theoretical and Applied Genetics* 99:445–452.

Brown, R. N., R. E. Barker, S. E. Warnke, L. A. Brilman, M. A. Rouf Mian, G. Jung, and S. Sim. 2005. QTL analyses for morphological traits useful in distinguishing annual ryegrass and turf-type perennial ryegrass. *International Turfgrass Society Research Journal* 10:516–524.

Charmet, G., F. Balfourier, and V. Chatard. 1996. Taxonomic relationships and interspecific hybridization in the genus *Lolium* (grasses). *Genetic Resources and Crop Evolution* 43:319–327.

Cogan, N. O. I., K. F. Smith, T. Yamada, M. G. Francki, A. C. Vecchies, E. S. Jones, G. C. Spangenberg, and J. W. Forster. 2005. QTL analysis and comparative genomics of herbage quality traits in perennial ryegrass (*Lolium perenne* L.). *Theoretical and Applied Genetics* 110:364–380.

Curley, J., and G. Jung. 2004. RAPD-based genetic relationships in Kentucky bluegrass: Comparison of cultivars, interspecific hybrids, and plant introductions. *Crop Science* 44:1299–1306.

Curley, J., S. C. Sim, S. Warnke, S. Leong, R. Barker, and G. Jung. 2005. QTL mapping of resistance to gray leaf spot in ryegrass. *Theoretical and Applied Genetics* 111:1107–1117.

Darbyshire, S. J. 1993. Realignment of *Festuca* subgenus *Schedonorus* with the genus *Lolium* (Poaceae). *Novon* 3:239–243.

Devos, K. M., and M. D. Gale. 2000. Genome relationships: The grass model in current research. *Plant Cell* 12:637–646.

Dumsday, J. L., K. F. Smith, J. W. Forster, and E. S. Jones. 2003. SSR-based genetic linkage analysis of resistance to crown rust (*Puccinia coronata* f. sp. *lotti*) in perennial ryegrass (*Lolium perenne*). *Plant Pathology* 52:628–637.

Essad, S. 1954. Contribution á la sytèmatique du genre *Lolium*. Ministère de l'Agriculture. Annales de l'Institute Nationale Recherche Agronomie, Paris. *Amelioration des Plantes* 4:325–351.

Evans, G. 1976. Chromosome complements in grasses. *Nature* 118:841.

Faville M. J., A. C. Vecchies, M. Schreiber, M. C. Drayton, L. J. Hughes, E. S. Jones, K. M. Guthridge, et al. 2004. Functionally associated molecular genetic marker map construction in perennial ryegrass (*Lolium perenne* L.). *Theoretical and Applied Genetics* 110:12–32.

Ferguson, J. M., and D. F. Grabe. 1984. Separation of annual and perennial species of ryegrass by gel electrophoresis of seed proteins. *Journal of Seed Technology* 9:137–149.

Forster, J. W., E. S. Jones, R. Kölliker, M. C. Drayton, J. Dumsday, M. P. Dupal, K. M. Guthridge, N. L. Mahoney, E. van Zijll de Jong, and K. F. Smith. 2001. Development and implementation of molecular markers for forage crop improvement. In *Proceedings of 2nd International Symposium on Molecular Breeding Forage Crops,* ed. G. Spangenberg, 101–133. Dordrecht, the Netherlands: Kluwer Academic Publishers.

Fujimori, M., K. Hayashi, M. Hirata, K. Mizuno, T. Fujiwara, F. Akiyama, Y. Mano, T. Komatsu, and T. Takamizo. 2004. Linkage analysis of crown rust resistance gene in Italian ryegrass (*Lolium multiflorum* Lam.). In *Plant and Animal Genome XI Conference,* San Diego, CA, p. 203.

Gale, M. D., and K. M. Devos. 1998. Comparative genetics in the grasses. *Proceedings of National Academy of Science USA.* 95:1971–1974.

Gentner, G. 1929. Über die verwendbarkeit von ultra-violetten strahlen bei der samenprüfung. *Praktische Blätter für Pflanzenbau und Pflanzenschutz* 6:166–172.

Gill, G. P., P. L. Wilcox, D. J. Whittaker, R. A. Winz, P. Bickerstaff, C. E. Echt, J. Kent, M. O. Humphreys, K. M. Elborough, and R. C. Gardner. 2006. A framework linkage map of perennial ryegrass based on SSR markers. *Genome* 49:354–364.

Grattapaglia, D., and R. Sederoff. 1994. Genetic linkage maps of *Eucalyptus grandis* and *Eucalyptus urophylla* using a pseudotestcross: Mapping strategy and RAPD markers. *Genetics* 137:1121–1137.

Griffith, S. M., and G. M. Banowetz. 1992. Differentiation of *Lolium perenne* L. and *L. multiflorum* Lam. Seed by two esterase isoforms. *Seed Science Technology* 20:343–348.

Hayward, M. D., J. W. Forster, J. G. Jones, O. Dolstra, C. Evans, N. J. McAdams, K. G. Hossain, et al. 1998. Genetic analysis of *Lolium*. Identification of linkage groups and the establishment of a genetic map. *Plant Breeding* 117:451–455.

Hayward, M. D., N. J. McAdam, J. G. Jones, C. Evans, G. M. Evans, J. W. Forster, A. Ustin, et al. 1994. Genetic markers and the selection of quantitative traits in forage grasses. *Euphytica* 77:269–275.

Hirata, M., M. Fujimori, M. Inoue, Y. Miura, H. Cai, H. Satoh, Y. Mano, and T. Takamizo. 2003. Mapping of a new crown rust resistance gene, Pc2, in Italian ryegrass cultivar "Harukaze." In *Proceedings of Molecular Breeding of Forage and Turf 2003, 3rd International Symposium*, Dallas, TX, p. 15.

Holub, J. 1998a. Reclassification and new names in vascular plants. *Preslia* 70:112–113.

———. 1998b. Reclassification and new names in vascular plants. *Preslia* 70:111.

Humphreys M., L. Turner, and I. Armstead. 2003. QTL mapping in *Lolium perenne*. In *Plant and Animal Genome XI Conference*, San Diego, CA.

Inoue, M., Z. Gao, and H. Cai. 2004. QTL analysis of lodging resistance and related traits in Italian ryegrass (*Lolium multiflorum* Lam.). *Theoretical and Applied Genetics* 109:1576–1585.

Inoue, M., Z. Gao, M. Hirata, M. Fujimori, and H. Cai. 2004. Construction of a high-density linkage map of Italian ryegrass (*Lolium multiflorum* Lam.) using restriction fragment length polymorphism, amplified fragment length polymorphism, and telomeric repeat associated sequence markers. *Genome* 47:57–65.

Jenkin, T. J. 1933. Interspecific and intergeneric hybrids in herbage grasses. Initial crosses. *Journal of Genetics* 28:205–264.

———. 1935. Interspecific and intergeneric hybrids in herbage grasses. II. *Lolium perenne* × *L. temulentum*. *Journal of Genetics* 31:379–411.

———. 1953. Interspecific and intergeneric hybrids in herbage grasses. V. *Lolium rigidum* sens. Ampl with other *Lolium* species. *Journal of Genetics* 52:252–281.

———. 1954a. Interspecific and intergeneric hybrids in herbage grasses. VI. *Lolium italicum* Abr. Intercrossed with other *Lolium* types. *Journal of Genetics* 52:282–299.

———. 1954b. Interspecific and intergeneric hybrids in herbage grasses. VII. *Lolium perenne* with other *Lolium* species. *Journal of Genetics* 53:105–111.

———. 1955. Interspecific and intergeneric hybrids in herbage grasses. XVII. Further crosses with *Lolium perenne*. *Journal of Genetics* 53:442–466.

Jenkin, T. J., and P. T. Thomas. 1939. Interspecific and intergeneric hybrids in herbage grasses. III. *Lolium loliaceum* L. *rigidum*. *Journal of Genetics* 37:255–286.

Jensen, L. B., J. R. Andersen, U. Frei, Y. Xing, C. Taylor, P. B. Holm, and T. Lübberstedt. 2005. QTL mapping of vernalization response in perennial ryegrass (*Lolium perenne* L.) reveals co-location with an ortho-logue of wheat *VRN1*. *Theoretical and Applied Genetics* 110:527–536.

Jones, E. S., M. D. Dupal, J. L. Dumsday, L. J. Huges, and J. W. Forster. 2002. An SSR-based genetic linkage map for perennial ryegrass (*Lolium perenne* L.). *Theoretical and Applied Genetics* 105:577–584.

Jones, E. S., N. L. Mahoney, M. D. Hayward, I. P. Armstead, J. G. Jones, M. O. Humphreys, I. P. King, et al. 2002. An enhanced molecular marker-based genetic map of perennial ryegrass (*Lolium perenne*) reveals comparative relationships with other Poaceae genomes. *Genome* 45:282–295.

Jung, G. A., A. J. P. Van Wijk, W. F. Hunt, and C. E. Watson. 1996. Ryegrasses. In *Cool-season forage grasses*, ed. L. E. Moser, D. R. Buxton, and M. D. Casler, 605–641. Madison, WI: ASA, CSSA, and SSSA.

Larsen, A. L. 1966. A distinction between proteins of annual and perennial ryegrass seeds. *Proceedings of the Association of Official Seed Analysts* 56:47–51.

Miura, Y., C. Ding, R. Ozaki, M. Hirata, M. Fujimori, W. Takahashi, H. Cai, and K. Mizuno. 2005. Development of EST-derived CAPS and AFLP markers linked to a gene for resistance to ryegrass blast (*Pyricularia* sp.) in Italian ryegrass (*Lolium multiflorum* Lam.). *Theoretical and Applied Genetics* 111:811–818.

Muylle, H., J. Baert, E. Van Bockstaele, J. Pertijs, and I. Roldán-Ruiz. 2005. Four QTLs determine crown rust (*Puccinia coronata* f. sp. *lolii*) resistance in a perennial ryegrass (*Lolium perenne*) population. *Heredity* 95:348–357.

Nakamura, S. 1979. Separation of ryegrass species using electrophoretic patterns of seed protein and enzymes. *Seed Science Technology* 7:161–168.

Nyquist, W. E. 1963. Fluorescent perennial ryegrass. *Crop Science* 3:223–226.

Payne, R. C., J. A. Scott, and T. J. Koszykowski. 1980. An esterase isoenzyme difference in seed extracts of annual and perennial ryegrass. *Journal of Seed Technology* 5:14–20.

Ritter, E., C. Gebhardt, and F. Salamini. 1990. Estimation of recombination frequencies and construction of RFLP linkage maps in plants from crosses between heterozygous parents. *Genetics* 125:645–654.

Schejbel, B., L. B. Jensen, Y. Xing, and T. Lübberstedt. 2007. QTL analysis of crown rust resistance in perennial ryegrass under conditions of natural and artificial infection. *Plant Breeding* 126:347–352.

Sim, S., T. Chang, J. Curley, S. E. Warnke, R. E. Barker, and G. Jung. 2005. Chromosomal rearrangements differentiating ryegrass genome from the Triticeae, oat, and rice genomes using common heterologous RFLP probes. *Theoretical and Applied Genetics* 110:1011–1019.

Sim, S., K. Diesburg, M. Casler, and G. Jung. 2007. Mapping and comparative analysis of QTL for crown rust resistance in an Italian x perennial ryegrass population. *Phytopathology* 97:767–776.

Skøt, L, M. O. Humphreys, I. Armstead, S. Heywood, K. P. Skøt, R. Sanderson, I. D. Thomas, K. H. Chorlton, and N. R. Sackville Hamilton. 2005. An association mapping approach to identify flowering time genes in natural populations of *Lolium perenne* (L.). *Molecular Breeding* 15:233–245.

Soreng, R. J., and E. E. Terrell. 1997. Taxonomic notes on *Schedonorus*, a segregate genus from *Festuca* or *Lolium*, with a new nothogenus, x *Schedololium* and new combinations. *Phytologia* 83:85–88.

Soreng, R. J., E. E. Terrell, J. Wiersema, and S. J. Darbyshire. 2001. (1488) Proposal to conserve the name *Schedonorus arundinaceus* (Schreb.) Dumort. *Schedonorus arundinaceus* Roem. & Schult. (*Poaceae: Poeae*). *Taxon* 50:915–917.

Studer, B., B. Boller, E. Bauer, U. K. Posselt, F. Widmer, and R. Kölliker. 2007. Consistent detection of QTLs for crown rust resistance in Italian ryegrass (*Lolium multiflorum* Lam.) across environments and phenotyping methods. *Theoretical and Applied Genetics* 115:9–17.

Terrell, E. E. 1968. A taxonomic revision of the genus *Lolium*. Technical bulletin no. 1392. United States Department of Agriculture, Washington, D.C.

Thorogood, D. 2003. Perennial ryegrass (*Lolium perenne* L.). In *Turfgrass biology, genetics, and breeding*, ed. M. D. Casler, and R. R. Duncan, 75–105. New York: John Wiley & Sons, Inc.

Tyler, B. F., K. H. Chorlton, and J. D. Thomas. 1987. Collection and field-sampling techniques for forages. In *Collection, characterization and utilization of genetic resources of temperate forage grass and clover*, ed. B. F. Tyler, 3–10. IBPGR Training Courses: Lecture series 1. International Board for Plant Genetic Resources, Rome.

Warnke, S. E., R. E. Barker, L. A. Brilman, W. C. Young III, and R. L. Cook. 2002. Inheritance of superoxide dismutase (*Sod-1*) in a perennial x annual ryegrass cross and its allelic distribution among cultivars. *Theoretical and Applied Genetics* 105:1146–1150.

Warnke S. E., R. E. Barker, G. Jung, S. C. Sim, M. A. R. Mian, M. C. Saha, L. A. Brilman, M. P. Dupal, and J. W. Forster. 2004. Genetic linkage mapping of an annual x perennial ryegrass population. *Theoretical and Applied Genetics* 109:294–304.

Xiong Y., S. Fei, E. Charles Brummer, K. J. Moore, R. E. Barker, G. Jung, J. Curley, and S. E. Warnke. 2006. QTL analyses of fiber components and crude protein in an annual x perennial ryegrass interspecific hybrid population. *Molecular Breeding* 18:327–340.

Yamada, T., E. S. Jones, N. O. I. Cogan, A. C. Vecchies, T. Nomura, H. Hisano, Y. Shimamoto, K. F. Smith, M. D. Hayward, and J. W. Forster. 2004. QTL analysis of morphological-, developmental-, and winter hardiness-associated traits in perennial ryegrass. *Crop Science* 44:925–935.

Appendix

MAJOR CROPS WITH COMMON AND BOTANICAL NAMES GROWN WORLDWIDE

Common Name	Botanical Name
African bermudagrass	*Cynodon nlemfuensis* Vanderyst
Alfalfa	*Medicago sativa* L.
Alsike clover	*Trifolium hybridum* L.
Annual ryegrass	*Lolium multiflorum* Lam.
Arrowleaf clover	*Trifolium vesiculosum* Savi
Australian bluestem	*Bothriochloa bladhii* (Retz.) S. T. Blake
Bahiagrass	*Paspalum notatum* Flüggé
Barley	*Hordeum vulgare* L.
Bermudagrass	*Cynodon dactylon* (L.) Pers
Berseem clover	*Trifolium alexandrinum* L.
Big bluestem	*Andropogon gerardii* Vitman
Birdsfoot trefoil	*Lotus corniculatus* L.
Bitter panicgrass	*Panicum amarum* Elliott var. *amarulum* (Hitchc. & Chase) P. G. Palmer
Black medic	*Medicago lupulina* L
Blue grama	*Bouteloua gracilis* (Kunth) Lag. ex Griffiths
Bottlebrush grass	*Elymus hystrix* L.
Bromegrass	*Bromus inermis* Leyss.
Browntop millet	*Panicum ramosum* L.
Buckhorn plantain	*Plantago lanceolata* L.
Buckhorn plantain	*Plantago lanceolata* L.
Buffalograss	*Buchloe dactyloides* (Nutt.) Engelm.
Chicory	*Cichorium intybus* L.
Cicer milk-vetch	*Astragalus cicer* L.
Cowpea	*Vigna unguiculata* (L.) Walp.
Crab finger grass	*Digitaria sanguinalis* (L.) Scop.
Creeping foxtail	*Alopecurus arundinaceus* Poir.
Crested wheatgrass	*Agropyron desertorum* (Fisch. ex Link) Schult.
Crimson clover	*Trifolium incarnatum* L.
Crownvetch	*Coronilla varia* L
Dallis grass	*Paspalum dilatatum* Poir.
Eastern gamagrass	*Tripsacum dactyloides* (L.) L.
Fodder vetch	*Vicia villosa* Roth.
Foxtail millet	*Setaria italica* (L.) P. Beauv.
Giant fescue	*Festuca gigantea* (L.) Vill.
Grass-pea	*Lathyrus sativus* L.
Groundnut (Peanut)	*Arachis hypogaea* L.
Guineagrass	*Panicum maximum* Jacq.
Holy-clover	*Onobrychis viciifolia* Scop.

Common Name	Botanical Name
Hyacinth-bean	*Lablab purpureus* (L.) Sweet
Illinois bundleflower	*Desmanthus illinoensis* (Michx.) MacMill. ex B. L. Rob. & Fernald
Indian grass	*Sorghastrum nutans* (L.) Nash
Intermediate wheatgrass	*Elytrigia intermedia* (Host) Nevski
Italian ryegrass	*Lolium multiflorum* Lam.
Jerusalem-artichoke	*Helianthus tuberosus* L.
Johnsongrass	*Sorghum halepense* (L.) Pers.
Kenaf	*Hibiscus cannabinus* L.
Kentucky bluegrass	*Poa pratensis* L.
Kikuyu grass	*Pennisetum clandestinum* Hochst. ex Chiov.
King Ranch bluestem	*Bothriochloa ischaemum* (L.) Keng var. *songarica* (Rupr. ex Fisch. & C. A. Mey.) Celarier & J. R. Harlan
Kura clover	*Trifolium ambiguum* M. Bieb.
Ladino clover	*Trifolium repens* L.
Lentil	*Lens culinaris* Medik.
Limpo grass	*Hemarthria altissima* (Poir.) Stapf & C. E. Hubb.
Little bluestem	*Schizachyrium scoparium* (Michx.) Nash
Meadow fescue	*Festuca pratensis* Huds.
Nile grass	*Acroceras macrum* Stapf
Oat	*Avena sativa* L.
Orchard grass	*Dactylis glomerata* L.
Partridge-pea	*Chamaecrista fasciculata* (Michx.) Greene
Pea	*Pisum sativum* L.
Pearl millet	*Pennisetum glaucum* (L.) R. Br.
Perennial ryegrass	*Lolium perenne* L.
Phasey-bean	*Macroptilium lathyroides* (L.) Urb.
Pigeon-pea	*Cajanus cajan* L. Millsp.
Prairiegrass	*Bromus catharticus* Vahl
Quackgrass	*Elytrigia repens* (L.) Desv. ex Nevski
Rayado bundleflower	*Desmanthus virgatus* (L.) Willd.
Red clover	*Trifolium pratense* L.
Reed Canary grass	*Phalaris arundinacea* L.
Rhizoma perennial peanut	*Arachis glabrata* Benth.
Rice	*Oryza sativa* L.
Rose clover	*Trifolium hirtum* All.
Russian wildrye	*Psathyrostachys juncea* (Fisch.) Nevski
Rye	*Secale cereale* L.
Sand bluestem	*Andropogon hallii* Hack.
Scarlet runner bean	*Phaseolus coccineus* L.
Smooth bromegrass	*Bromus inermis* Leyss.
Sorghum	*Sorghum bicolor* (L.) Moench
Soybean	*Glycine max* (L.) Merr.
Strawberry clover	*Trifolium fragiferum* L.
Stylo	*Stylosanthes guianensis* (Aubl.) Sw.
Sulla sweet-vetch	*Hedysarum coronarium* L.

Common Name	Botanical Name
Switch grass	*Panicum virgatum* L.
Tall fescue	*Festuca arundinacea* Schreb.
Timothy	*Phleum pratense* L.
Triticale	x *Triticosecale* Wittmack
Velvet bean	*Mucuna pruriens* (L.) DC.
Western wheatgrass	*Elymus smithii* (Rydb.) Gould
Wheat	*Triticum aestivum* L
White clover	*Trifolium repens* L.
White sweet-clover	*Melilotus alba* Desr.
Woollypod vetch	*Vicia villosa* Roth subsp. *varia* (Host) Corb.
Yellow sweet-clover	*Melilotus officinalis* (L.) Lam.

Source: http://www.ars-grin.gov/cgi-bin/npgs/html/taxgenform.pl

Index

T

9 780367 386023